TOPICAL WORKSHOP ON LOW RADIOACTIVITY TECHNIQUES

To learn more about the AIP Conference Proceedings, including the Conference Proceedings Series, please visit the webpage
http://proceedings.aip.org/proceedings

TOPICAL WORKSHOP ON LOW RADIOACTIVITY TECHNIQUES

LRT 2004

Sudbury, Ontario, Canac   ember 2004

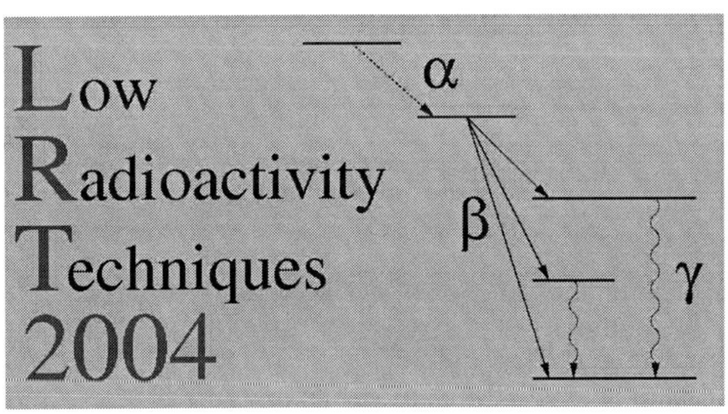

EDITORS

Bruce Cleveland
Richard Ford
Sudbury Neutrino Observatory Laboratory
Lively, Ontario

Mark Chen
Queen's University
Kingston, Ontario

SPONSORING ORGANIZATIONS

Sudbury Neutrino Observatory Laboratory - (SNOLAB)
Canada Foundation for Innovation - (CFI)
Canadian Institute of Particle Physics - (IPP)

AMERICAN INSTITUTE OF PHYSICS

Melville, New York, 2005

AIP CONFERENCE PROCEEDINGS ■ VOLUME 785

Editors:

Bruce Cleveland
Richard Ford

Sudbury Neutrino Observatory Laboratory (SNOLAB)
P.O. Box 159
Lively, Ontario P3Y 1M3
Canada

E-mail: bclevela@snolab.ca
ford@snolab.ca

Mark Chen
Queen's University
Department of Physics
Kingston, Ontario K7L 3N6
Canada

E-mail: mchen@post.queensu.ca

L.C. Catalog Card No. 2005931622
ISBN 0-7354-0274-4
ISSN 0094-243X

Printed in the United States of America

CONTENTS

V - REDUCED BACKGROUND AIR AND PURIFIED NOBLE GASES

VI - CLEANLINESS STUDIES AND CLEAN FABRICATION TECHNIQUES

VII - LIQUID SCINTILLATORS AND WATER PROPERTIES AND PURIFICATION

VIII - BACKGROUND STUDIES AND SIMULATIONS

POSTERS

Preface

The 2004 Workshop on Low Radioactivity Techniques (LRT2004) was held December 12th-14th at Laurentian University in Sudbury, Canada. The timing and venue for the workshop was to mark the beginning of excavation for the new international facility for deep underground particle astrophysics, SNOLAB, located in Sudbury, Canada. Funded through the Canada Foundation for Innovation (CFI), the facility, located 2070m below ground in INCO's Creighton Mine, will host multiple large-scale and small experiments in neutrino physics, dark matter, and other fundamental research.

Recent experimental proof of neutrino mass and flavour oscillations continues to enrich and expand the field of neutrino astrophysics. The next steps in neutrino research seek to fix the absolute mass scale, measure the mixing parameters and CP-violation, determine if neutrinos are Majorana or Dirac particles, and measure the flux of lower energy neutrinos from the primary reaction processes in the sun. The field also includes detectors for dark matter searches, and the detection of geo-neutrinos, reactor neutrinos and from possible beamed neutrino sources.

The next generation of proposed experiments will demand ultra-low detector backgrounds to reach the instrument sensitivities required. This will require novel techniques in the development, construction, operation and analysis of these experiments. Such issues include the low background radioactivity assay and purification of detector components and materials, and the development of high purity noble gases.

The goal of this workshop was to bring together experts in this field for presentations and discussion covering broadly the issues of low radioactivity techniques. The intention was to foster the collaboration and resource sharing required for the new generation of detectors being developed at underground facilities.

The workshop was a great success with representatives from North American, European, and Japanese facilities. Starting with overviews of existing low background facilities, the following sessions included low background techniques, materials screening, detectors and components, radon reduced air and noble gases, clean fabrication techniques, purification systems, and calculations of backgrounds.

The timing of this workshop also coincided with the retirement of Professor Gerd Heusser of the Max Planck Institute for Nuclear Physics in Heidelberg, Germany. Gerd attended the workshop as our invited guest of honour, and delivered the keynote lecture "Low Level Counting from Meteorites to Neutrinos". Gerd is recognized as a pioneer of low-level techniques, having worked in this field for almost 40 years and developed many of the techniques discussed at the workshop.

A lunch-time round-table discussion was held on ideas to organize resource sharing and to enhance the visibility of this emerging field. Issues including coordination of low-background screening facilities and the creation of materials or resource databases were discussed.

The intention is to make LRT a regular workshop to be held every two years. A preliminary organizing committee has been formed for the next workshop LRT2006 to be hosted by the Modane Underground Laboratory at the Centre Paul Langevin in Aussois, France in September 2006.

The workshop organizers would like to thank Eva Lacelle, Pat Brouse, and Shari Moss for invaluable assistance in administration and local arrangements. The workshop was funded in part by SNOLAB, the Canada Foundation for Innovation, and the Institute of Particle Physics.

Editors: Bruce Cleveland
Richard Ford
Mark Chen

I – OVERVIEW OF LOW RADIOACTIVITY MEASUREMENT FACILITIES

European underground laboratories: An overview.

Lino Miramonti

Physics Dept. of Milano University and National Institute of Nuclear Physics
via Celoria 16, 20133 Milano, Italy

Abstract. Underground laboratories are complementary to those where the research in fundamental physics is made using accelerators. This report focus on the logistic and on the background features of the most relevant laboratories in Europe, stressing also on the low background facilities available. In particular the report is focus on the laboratories involved in the new Europeean project ILIAS with the aim to support the European large infrastructures operating in the astroparticle physics sector.

Keywords: Deep underground laboratories, Low background techniques,
PACS: 01.52.+r

INTRODUCTION

Thanks to the underground facilities one reaches the low radioactivity background environment needed to search for rare events in order to study the fundamental laws of nature such as those governing the proton stability or the nature of the neutrino. In addition the study of astroparticle physics phenomena such as neutrino from astrophysical objects and dark matter searches are also possible. The underground laboratories are complementary to those where the research in fundamental physics is made using accelerators. In these last decades different underground laboratories have been excavated all over the world; they differ in size, in the support of technology available and in the way to access them. In Europe the most relevant laboratories (not in shallow site) are the Italian Laboratori Nazionali del Gran Sasso, the French Laboratoire Souterrain de Modane, the Spanish Laboratorio Subterraneo de Canfranc, the UK Boulby Underground Laboratory, the Centre for Underground Physics in Pyhäsalmi in Finland, the Ukranian Solotvina Underground Laboratory and the Russian Baksan Underground Laboratory. Table 1 summarizes the European underground laboratories. In the next chapters we will concentrate in particular on the first four laboratories, involved in the new Europeean project ILIAS with the aim to support the European large infrastructures operating in the astroparticle physics area.
The most relevant sources of background are the penetrating muons, the primordial radionuclides in the rock and in all material constituting the detector and/or its shielding and the radon gas.
In this report we are focusing on the logistic and on the background features of the labs, stressing also on the low background facilities available.

THE GRAN SASSO NATIONAL LABORATORY - LNGS

The Gran Sasso National Laboratory (Laboratori Nazionali del Gran Sasso - LNGS) is the world's largest underground research center; it is financed by the Italian National Institute for Nuclear Physics (INFN), which coordinates and finances research in nuclear, subnuclear and elementary particle physics. The excavation and construction of the underground halls began in mid eighties. The laboratory is located in the highway tunnel between Teramo and L'Aquila under the "monte Aquila" (Gran Sasso mountain) in the center of Italy. Three main experimental halls (20 m high, 18 m wide and 100 m long) host big scientific experiments. Including the connecting tunnels and emergency passageways, the total volume is 180000 m^3 and the area is 13500 m^2. Because of the great quantity of water inside the Gran Sasso mountain, the natural temperature is about 6-7 ^0C and the humidity is approximately 100%. In order to host scientific facilities, the experimental halls are waterproofed and heated; furthermore, a ventilation system provides about 35000 m^3 of air per hour from outside. The underground laboratory is connected to the external laboratory (situated about 8 km away) by means of fiber optical cables; the external structures, located in Assergi, include offices, laboratories, machine shop, seminar halls and a computer center for a total surface of about 12000 m^2.

CP 785, *Topical Workshop on Low Radioactivity Techniques,* edited by B. Cleveland, R. Ford, and M. Chen

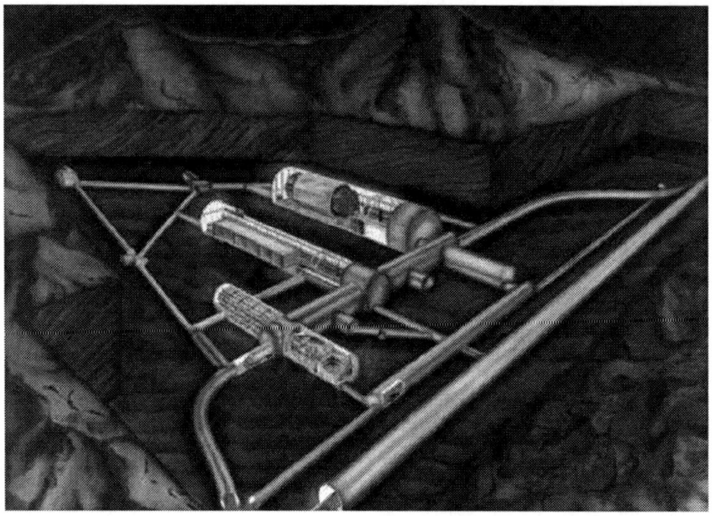

FIGURE 1. The Gran Sasso National Laboratory

The experimental activities presently ongoing at LNGS include all major research topics in the field of underground science: neutrino astrophysics (solar neutrinos (Borexino) and supernova neutrinos (LVD)), long baseline neutrino detection (CNGS project with the OPERA and ICARUS experiments), search for neutrino mass in neutrinoless $\beta\beta$ decay (Cuoricino/Cuore and Gerda), dark matter search (Dama, Libra, CRESST), nuclear astrophysics (LUNA), and several projects in the fields of geophysics, biology, and environmental sciences.

Background characteristics and facilities

The overburden rock is on the average about 1400 m, equivalent to about 3800 meter of water. The muon flux is reduced of about 6 orders of magnitude to a value of approximately 1.1 muons per square meter per hour, whereas the neutron flux is on the order of 10^{-6} neutrons per square centimeter per second depending on energy (see table 4). The rock of the Gran Sasso mountain has a density of $2.71 \pm 0.05\ g \cdot cm^{-3}$, and consists mainly of $CaCO_3$ and $MgCO_3$ [1]. Because of the presence of a particular type of rock (roccia nera marmosa), the content of uranium and thorium in hall A rock is about 10 to 30 times higher than in hall C. The primordial radionuclide content of the rock and of the concrete employed to cover the three big halls are reported in table 3. The low-level radioactivity topics carried out in the LNGS are: material selection and sample measurements (mainly with HPGe detectors); radon groundwater monitoring (Environmental and geophysical monitoring of the Gran Sasso aquifer); development and characterization of new detectors (for nuclear spectrometry of environmental radioactivity). At present, a 32 m^2 room is dedicated to low-level radioactivity measurements (for the future a new building with ca. 80 m^2 distributed on three floors is foreseen in hall A). This facility consists currently in nine high purity germanium detectors; in table 2 are reported the performances of the HPGe detectors at LNGS.

THE MODANE UNDERGROUND LABORATORY - LSM

The Modane Underground Laboratory (Laboratoire Souterrain de Modane - LSM) belongs to the French institutions of IN2P3-CNRS (Centre National de la Recherche Scientifique) and DSM-CEA (Commissariat à l'Energie Atomique). It is located in the highway tunnel connecting Italy to France at about 6.5 km from both entrances under the "pointe du Fréjus" at 1200 m over the sea level. This underground facility is also know as Fréjus underground laboratory. The underground laboratory consists of one main hall (11 m high, 10 m wide and 30 m long) and three secondary halls: a gamma spectroscopy hall of 70 m^2 and two little rooms of 21 m^2 and 18 m^2. The laboratory was created at the beginning of eighties in order to host a large calorimeter for proton decay searches. The natural temperature of the rock is of the order of 28 ^0C, but an air conditioning system, in addition to the ventilation system, reduces the temperature

TABLE 1. European underground laboratories. For the LNGS, LSM and LSC the overburden is the approximate mean depth, while for the IUS, CUPP and SUL the overburden is the vertical depth under flat surface.

Laboratory	Location	Overburden $[mwe]$	muon flux $[\mu\ m^{-2}\ h^{-1}]$
LNGS	Italy	3800	1.1
LSM	France	4800	0.17
LSC	Spain	2450	7.2
IUS	UK	2800	1.5
CUPP	Finland	4000	-
SUL	Ukraina	430	6.2

TABLE 2. Performances of the HPGe detectors at LNGS (Data @ 2002). Total and peak background count rate in $d^{-1}\ kg^{-1}_{Ge}$. [2]

	40-2700 keV	352 keV	583 keV	1461 keV
GeMi	611	5.6	2.1	5.2
GePV	482	2.8	2.1	3.2
GeOr	469	2.4	0.76	4.3
GePaolo	266	0.83	0.38	1.4
GeCris	87	<0.39	<0.29	1.0
GeMPI	30	<0.20	<0.15	0.36

at about 20 ^0C with a relative humidity under 50%. The external buildings are located in the Modane village (French side), and are composed of offices and workshops.
The present scientific programme of LSM include direct dark matter detection (Edelweiss), search for neutrinoless double beta decay (NEMO), and study of super-heavy nulei (SHIN). Several projects requiring ultra-low levels of background in the field of environmental and geophysical sciences are also ongoing.

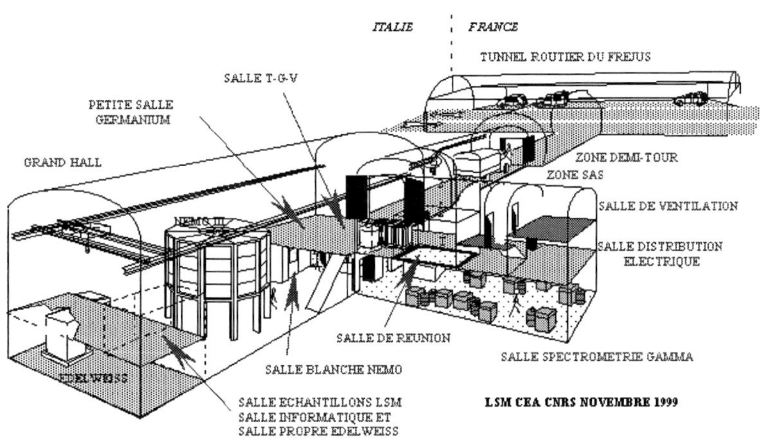

FIGURE 2. The Modane Underground Laboratory

TABLE 3. Primordial radionuclides content in rock and concrete in the European underground facilities.

	^{238}U	^{232}Th	K	Ref.
LNGS Hall A (rock)	6.80 ± 0.67 ppm	2.167 ± 0.074 ppm	160 ppm	[6]
LNGS Hall B (rock)	0.42 ± 0.10 ppm	0.062 ± 0.020 ppm	160 ppm	[6]
LNGS Hall C (rock)	0.66 ± 0.14 ppm	0.066 ± 0.025 ppm	160 ppm	[6]
LNGS all Halls (concrete)	1.05 ± 0.12 ppm	0.656 ± 0.028 ppm	-	[11]
LSM (rock)	0.84 ± 0.2 ppm	2.45 ± 0.2 ppm	$213 \pm 30\ Bq \cdot kg^{-1}$	[12]
LSM (concrete)	1.9 ± 0.2 ppm	1.4 ± 0.2 ppm	$77 \pm 13\ Bq \cdot kg^{-1}$	[12]
LSC	$2.2 \cdot 10^{-2}\ \gamma\ cm^{-2}\ s^{-1}$			[13]
IUS	70 ppb	125 ppb	1130 ppm	[9]
CUPP	27.8-44.5 $Bq \cdot m^{-3}$	4.0-18.7 $Bq \cdot m^{-3}$	267-625 $Bq \cdot m^{-3}$ (^{40}K)	[10]

Background characteristics and facilities

The rock overburden, essentially made of schist (with a density of about 2.73 $g \cdot cm^{-3}$) is about 1700 m, (corresponding to about 4800 mwe). The residual muon flux at the laboratory depth is of about 0.17 muons per square meter per hour. The neutron flux is $1.6 \cdot 10^{-6}\ n\ cm^{-2}\ s^{-1}$ for neutrons having energy up to 0.63 eV and $4 \cdot 10^{-6}\ n\ cm^{-2}\ s^{-1}$ for neutrons with energy ranging between 2 MeV and 6 MeV (see table 4). In table 3 are reported the primordial radionuclide content of the rock and of the concrete. Thirteen HPGe from 6 different laboratories of CNRS and CEA are available at LSM for material selection in fundamental physics searches and for samples measurement for environmental control, earth science, archaeology, biology and dating measurement.

THE CANFRANC UNDERGROUND LABORATORY - LSC

The Canfranc Underground Laboratory (Laboratorio Subterraneo de Canfranc - LSC) is located in the now closed to traffic railway tunnel of Somport (6.9 km long) connecting Spain to France in the Aragonese Pyrenees, at 1080 m above the sea level. The operating institution is the Zaragoza University. The underground facilities consist of one main hall called Lab3 (4.5 m high, 5 m wide and 20 m long) and a small laboratory called Lab1 formed by two small rooms of 18 m^2 each. A third laboratory, called Lab2 and now dismounted, consisting in a mobile pre-fabricated hut of about 20 m^2, is placed on the railway tracks and moved along the tunnel to operate at different overburden. The natural temperature of the laboratory is 7 ^0C and the relative humidity is about 90%. There is a constant flow wind coming from the French side which keeps the level of radon fairly low. The main laboratory is ventilated every four hours with fresh air at a rate of 1500 m^3 per hour. The air conditioning system provides average temperature of 21 ^0C and a relative humidity of about 60%. At the railway entrance (Spanish side) there is a building of the Spanish National Railway Company at disposal of the Zaragoza University for support of the activities. A pre-fabricated hut of about 20 m^2 is equipped for logistic link and monitoring.

The laboratory is presently being substantially enlarged both in size and logistic support (see below). Activities in the old lab included $\beta\beta$ decay search (IGEX-$\beta\beta$, Gedeon), and dark matter searches (Igex-DM, Rosebud); the scientific programme of the new laboratory is still under discussion.

Background characteristics and facilities

Lab3 is located the Tobazo's peak (1980 m high) under an overburden of about 900 m of rock (limestone, mainly calcium carbonate), which correspond to about 2450 mwe; (the overburden of Lab1 is about 200 m of rock which correspond to about 675 mwe). Lab1 is used now only to store detectors and other materials (which are so kept shielded from the cosmic radiation). The muon flux in the main hall is 7.2 muons per square meter per hour. The integrated neutrons flux from radioactivity is $3.82 \cdot 10^{-6}\ n\ cm^{-2}\ s^{-1}$, whereas the integrated neutron flux from muon-induced is $1.73 \cdot 10^{-9}\ n\ cm^{-2}\ s^{-1}$ (see table 4). The measured gamma flux inside the laboratory is about $2 \cdot 10^{-2}\ \gamma\ cm^{-2}\ s^{-1}$. An installation (AMBAR) for measuring low contents of radioactive contaminants in materials intended for low-background experiments is available.

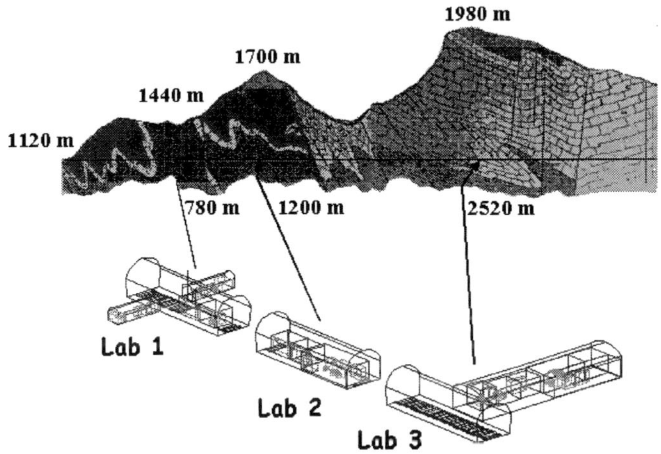

FIGURE 3. The Canfran Underground Laboratory

The new Canfranc Underground Laboratory

Civil works are underway (50 m from the old facility) for the construction of the new Canfranc Underground Laboratory. It will consist in two experimental halls: the main one, oriented towards CERN is 11 m high, 15 m wide and 40 m long for fundamental physics research, and a second one (8 m high, 10 m wide and 15 m long) as low background facility. An access corridor housing offices, workshops and clean rooms for a total surface of about 1000 m^2 completes the new laboratory.

THE BOULBY MINE LABORATORY - IUS

Unlike the previous laboratory, the IUS (Institute for Underground Physics) laboratory is not located into a tunnel but in a salt mine near Sheffield, in northern England. The Boulby mine laboratory started in 1988 for the search for non-baryonic dark matter candidates. It consists of different areas: a first room called *Stub 2*, excavated in 1988, of about 300 m^2, a second room called *Stub 2a*, excavated in 1995, with an area of about 150 m^2, an area called *H area*, excavated in 1998, with an area of about 900 m^2, and the new *JIF area*, excavated in 2003, with a total area of about 2500 m^2. The relative humidity is approximately 37% with an environment radon concentration of about 5 $Bq \cdot m^{-3}$. The temperature of the surrounding rock is about 35 ^0C, but in the experimental areas it is lowered to 28 ^0C thanks to an air conditioned system. Surface facilities, on the mine site, consisting in administration offices, computer facilities, conference room, workshop and storage space are available.
Research topics at Boulby are mostly concentrated on Dark Matter search with the Zeplin and Drift projects.

Background characteristics and facilities

The depth of the laboratory is about 1070 m, corresponding to approximately 2800 mwe; at this depth the muon flux is about 1.5 muons per square meter per hour. The calculated neutron flux is about $2 \cdot 10^{-6}$ $n\ cm^{-2}\ s^{-1}$ for neutrons having energy greater than 100 keV, and about $1 \cdot 10^{-6}$ $n\ cm^{-2}\ s^{-1}$ for neutrons with energy greater than 1 MeV (see table 4). Boulby mine laboratory has an intrinsically (cavern walls - halite) low background gamma ray activity from radionuclides contamination; the uranium and thorium content is of the order of 0.1 ppm, while the potassium concentrations is about 1130 ppm (see table 3).

In the new *JIF area*, there will be a low background facility (located in the so called *Stub A*) consisting in HPGe detectors and a NaI crystal from the NaIAD experiment for crude bulk tests of activity. Recently the crane, required to service the low background Ge detectors and ancillary equipment has been installed. Construction of the facility

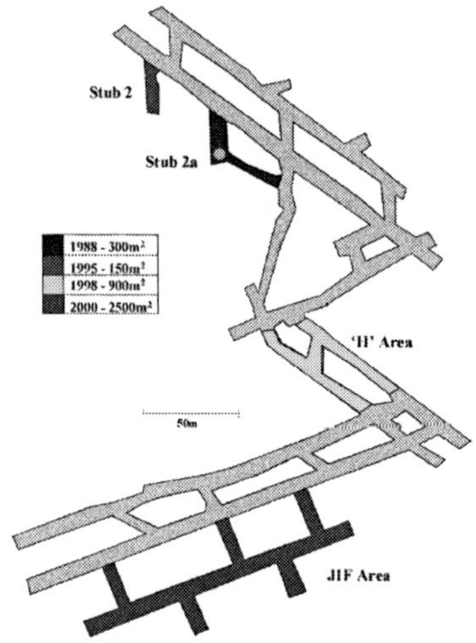

FIGURE 4. The Institute for Underground Science - Boulby Mine

TABLE 4. Neutron flux in the European underground facilities.

	Neutron flux $[10^{-6}\ n\ cm^{-2}\ s^{-1}]$	**Energy**	**Ref.**
LNGS	1.08 ± 0.02	0 - 0.05 eV	[6] [7]
	1.84 ± 0.20	0.05 eV - 1 keV	
	0.54 ± 0.01	1 keV - 2.5 MeV	
	0.27 ± 0.14	2.5 MeV - 5 MeV	
	0.05 ± 0.01	5 MeV - 10 MeV	
LSM	1.6 ± 0.1	0 - 0.63 eV	[12]
	4 ± 0.1	2 - 6 MeV	
LSC	3.82 ± 0.44	neutrons from radioactivity	[8]
	$(1.73 \pm 0.22 \pm 0.69)\ 10^{-3}$	muon-induced neutrons in rock	
IUS *	≈ 2	> 100 keV	[9]
	≈ 1	> 1 MeV	
SUL	< 2.7	not specified	[14]

* calulated neutron flux

clean room area recently been completed. The Boulby Mine facility hosts a 2 kg (400 cc) germanium detector, used for material radiopurity measurements (20 cm of lead as outer shielding and 8 cm of copper as the inner shielding plus radon steel box). The setup will be sensitive to contamination of samples at the level of 0.1- 0.2 ppb.

THE ILIAS PROJECT

Last year a new initiative supported by the European Union (within the 6th Framework Programme) has been started with the aim to support the European large infrastructures operating in the astroparticle physics area. This initiative, called ILIAS (Integrated Large Infrastructures for Astroparticle Science), involes 19 European institutions (see table 5) and is based on 3 groups of activities:

- Networking Activities
    - (N2) Deep underground science laboratories;
    - (N3) Direct dark matter detection;
    - (N4) Search for double beta decay;

TABLE 5. Europen institutions within the ILIAS project.

	Istitution
France	Commissariat à l'Energie Atomique
	Centre National de la Recherche Scientifique
Italy	Istituto Nazionale di Fisica Nucleare
	Istituto di Fotonica e Nanotecnologie Trento
	European Gravitational Observatory
Germany	Germany Max Planck Institut für Kernphysik
	Technische Universität München
	Max Planck Institut für Physik München
	Karls Universität Tubingen
Spain	Zaragoza University
UK	Sheffield University
	Glasgow University
	London University
Czech Rep	Czech Technical University in Prague
Denmark	University of Southern Denmark
Netherland	Leiden University
Finland	University of Jyväskylä
Slovakia	Comenius University Bratislavia
Greece	Aristot University of Thessaloniki

(N5) Gravitational wave research;

(N6) Theoretical astroparticle physics.

- Joint Research Activities (R&D Projects)
  (JRA1) Low background techniques for Deep Underground Science;
  (JRA2) Double beta decay European observatory;
  (JRA3) Study of thermal noise reduction in gravitational wave detectors.
- Transnational Access Activities
  (TA1) Access to the EU Deep Underground Laboratories.

The activities involving more extensively the deep underground laboratories are N2, JRA1 and TA1;

The N2 network aims to implement a mechanism for the joint structuring and coordination of the four European underground facilities, in order to provide a better service to users with more efficient use of resources.

Due to the strict correlation and complementarity of the research carried on in the different laboratories, Transnational Access project to the four European deep underground laboratories is coordinated with a unique management. Within this activity the laboratories should be seen as a single infrastructure with four different installations. The ILIAS TA1 activity aims to continue and improve the experience started by LNGS within the European 5th Framework Programme, extending the access possibilities to the other European deep underground laboratories (LSM, LSC and IUS).

The JRA1 activity is a vast R&D program on the improvement and implementation of innovative ultra-Low Background Techniques, and is carried out cooperatively in the four Deep European Underground Laboratories (LBT-DUSL). JRA1 should improve the quality of the services offered by underground laboratories to the scientific community in the sectors of astroparticle physics and rare-event physics, as well as improving the efficiency of the research carried on in the laboratories. Its principal objectives are the background identification and measurement (intrinsic, induced, environmental) and the study of the background rejection techniques (shielding, vetoes, discrimination). There are four different Working packages:

- Measurements of the backgrounds in the underground laboratories (WP1);
- Implementation of background MC simulation codes (WP2);
- Ultra-low background techniques and facilities (WP3);
- Radiopurity of materials and purification techniques (WP4).

Beside the four deep underground laboratories coordinated by the ILIAS project, other two deep facilities are presented in the next section: the Finland Centre for Underground Physics in Pyhäsalmi, and the Ukranian Solotvina

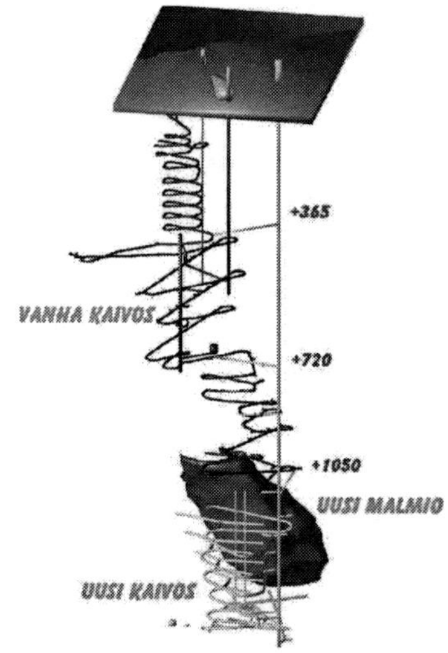

FIGURE 5. The Centre for Underground Physics in Pyhäsalmi

Underground Laboratory. In 2005 the Centre for Underground Physics in Pyhäsalmi will reach the ILIAS community.

CENTRE FOR UNDERGROUND PHYSICS IN PYHÄSALMI - CUPP

The Laboratory is located in the mine of Pyhäsalmi in the town of Pyhäjärvi in Finland. The project to host an underground laboratory in the mine was started in 1993 [3]. The facility consists of several halls at different depths. The natural temperature increases with depth: 6 ^0C at a depth of 100 m, 16 ^0C at a depth of 1000 m (increment of about 1 ^0C per 100 m). The relative humidity in non-vantilated halls is near 100%. The maximum depth is 1440 m that with an average density of 2.8 $g \cdot cm^{-3}$ corresponds to about 4000 mwe. The radon concentration in air is less than 30 $Bq \cdot m^{-3}$ with ventilation. The old part of the mine is no longer in use for mining, and will be plenty of free space to host and storage experiments. The new mine started to operate in July 2001, the largest cavern that can be easily constructed is 20 m high, 15 m wide and 100 m long. A preliminary study, including some background measurement and rock analysis has been made (see table 3) [4] [5]. The headquarters are in the Oulu University.

THE SOLOTVINA UNDERGROUND LABORATORY - SUL

The laboratory was constructed in 1984 by the Institute for Nuclear Research of the Ukrainian National Academy of Sciences. It is situated on the west of Ukraine, in a small town (Solotvina) in carpathian region just near the border with Romania. This facility is built in the operational salt (NaCl) mine 430 m underground (146 m below the sea level). It consist of one big hall (8 m high, 20 m wide and about 30 m long) and four small halls (3 m high, 3 m wide and about 3 m long) for a total area near to 1000 m^2. The natural temperature is 22-24 ^0C. The Laboratory is equipped with uninterruptible power supply in order to continue the measurements if mine power system fails. At a depth of 430 m (that correspond to about 1000 mwe) the cosmic ray flux is reduced by a factor of about 100, at a level of 62 muons per square meter per hour. The neutron flux is less than $2.7 \cdot 10^{-6}$ $n\ cm^{-2}\ s^{-1}$, and radon concentration in air is less than 33 $Bq \cdot m^{-3}$. Due to a low radioactive contamination of salt, the natural gamma-rays background in the SUL is 10-100 times lower than in other underground laboratories escavated in the rock.
The main scientific activities at the Solotvina laboratory are concentrated in the $\beta\beta$ decay search sector [14].

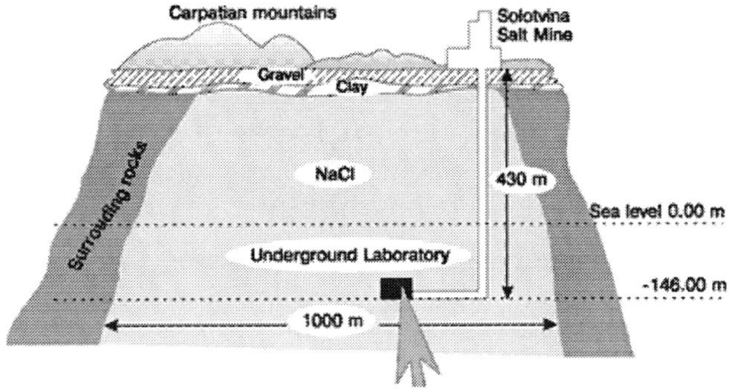

FIGURE 6. The Solotvina Underground Laboratory

ACKNOWLEDGMENTS

This work was supported by the Italian Institute of Nuclear Physics (INFN) and by the European Union (6th Framework Programme) in the context of the ILIAS project (contract number UE RII3-CT2004-506222). I would like to thank Nicola Ferrari (INFN-LNGS) for useful comments.

REFERENCES

1. P.G.Catalano et al., Men. Soc. Geol. It. 35 (1986) 647.
2. M. Laubenstein, private communication.
3. The preliminary plan for the deep laboratory, Kalliosuunnittelu Oy (Rockplan ltd) (2001).
4. N. Wang, The characteristics of uranium and radon migration underground, CUPP-07 (2000).
5. J.N. Abrurashitov, V.N. Gavrin, V.L. Matushko, and A.A. Shikhin, Measurement of the fast neutron background in the Pyhĺasalmi mine CUPP-11/2003.
6. H.Wulandari et al., Astroparticle Physics 22 313-322 (2004).
7. E.Belli et al., Il Nuovo Cimento 101A 959 (1989).
8. J.M.Carmona et al., Astroparticle Physics 21 523-533 (2004).
9. www.shef.ac.uk/physics/research/pppa/ilias/ILIASBoulbyJRA1N2-1104.ppt
10. N. Wang, The characteristics of uranium and radon migration underground, CUPP-07/2000; J.N. Abrurashitov, V.N. Gavrin, V.L. Matushko, and A.A. Shikhin, Measurement of the fast neutron background in the Pyhĺasalmi mine, CUPP-11/2003.
11. G.Bellini, Borexino Proposal to Gran Sasso Laboratory, 1991, p.184.
12. V.Chazal et al., Astroparticle Physics 9 193-172 (1998).
13. J. Morales, talk at IDM 2004, to be published by World Scientific.
14. Yu. G. Zdesenko et al., Underground Physics'87: Proc. of the 2 Int. Symp., Baksan Valley, USSR, August 17-19, 1987. – Moscow, Nauka, (1988), p. 291.

Japanese and Eastern Facilities

M. Nakahata

Kamioka Observatory, ICRR, University of Tokyo, Higashi-Mozumi, Kamioka, Gifu 506-1205, Japan

Abstract. The underground facilities in Japan and Korea are reviewed. Those facilities are Kamioka Observatory, Oto Cosmo Observatory, Ogoya Underground Laboratory, and Kashiwa Underground Laboratory in Japan and YangYang Underground Laboratory in Korea. Features of those facilities and radon reduction systems at Kamioka Observatory are presented.

Keywords: underground facility, low background
PACS: 95.85.Ry

INTRODUCTION

Detectors for astroparticle physics aiming at detection of rare phenomena are placed at underground. As demonstrated by the observation of supernova neutrinos by Kamiokande and IMB in 1987 and the discovery of atmospheric neutrino oscillations by Super-Kamiokande(SK) in 1998 and solar neutrino oscillations by SK and SNO in 2001, underground experiments are able to perform neutrino astrophysics and elementary particle physics beyond the standard model. Searchs for dark matter and neutrinoless double beta decay are also important subjects and several underground facilities in Japan and Korea are prepared for those topics. Low background environment is required for the observation of low energy phenomena such as solar neutrinos, dark matter and double beta decay. In this report, low background features at Japanese and Korean facilities and radon reduction systems at Kamioka Observatory are presented.

JAPANESE AND KOREAN FACILITIES

Figure 1 shows the location of Japanese underground facilities. They are Kashiwa Underground Laboratory, Ogoya Underground Laboratory, Oto Cosmo Observatory, and Kamioka Observatory.

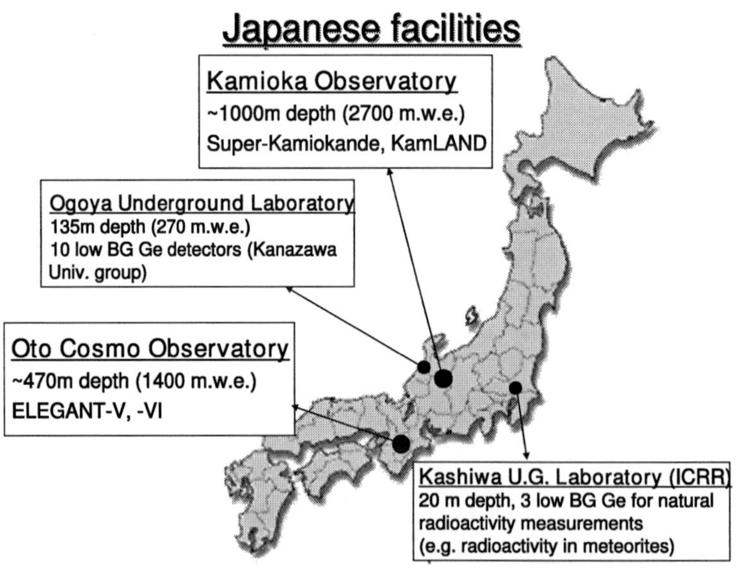

FIGURE 1. Location of Japanese underground facilities

CP 785, *Topical Workshop on Low Radioactivity Techniques,* edited by B. Cleveland, R. Ford, and M. Chen

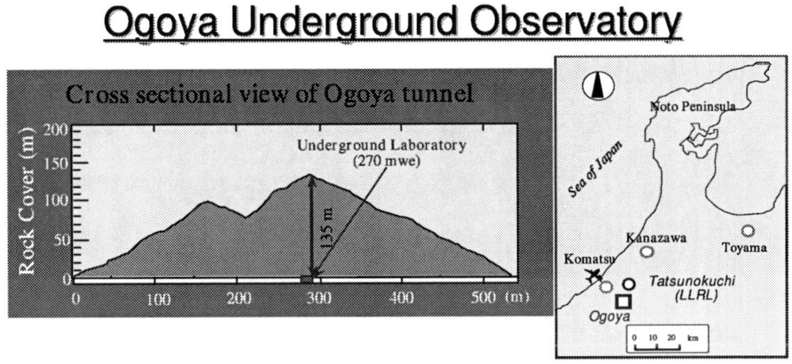

FIGURE 2. Ogoya Underground Laboratory.

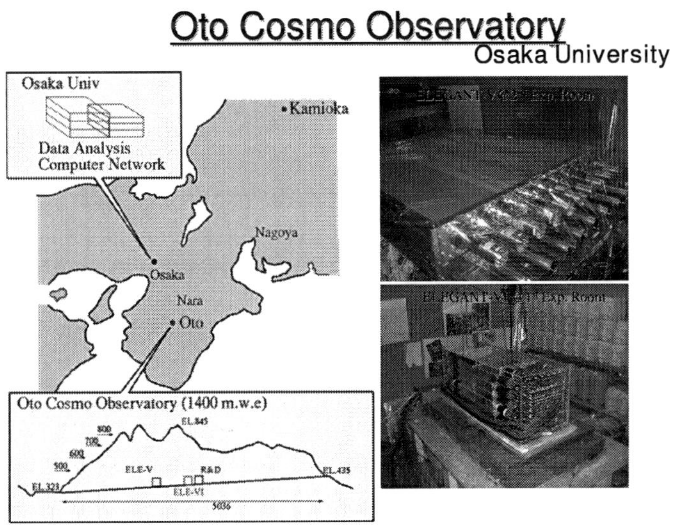

FIGURE 3. Oto Cosmo Observatory.

Kashiwa Underground Laboratory is placed at the Kashiwa campus of Institute of Cosmic Ray Research(ICRR), the University of Tokyo. It is placed at 20m underground and three low background germanium detectors serve for natural radioactivity measurements such as the radioactivity measurement in meteorites.

Ogoya Underground Laboratory[1] is run by Kanazawa university. It is located at 135 m underground (270 m.w.e.) from the top of the mountain as shown in Fig.2. Muon (neutron) intensity at the laboratory is 1/200 (1/300) of the intensity at surface. Ten low background germanium detectors are placed in the laboratory and environmental radioactivity measurements are being performed. Some of the achievements at the laboratory are ^{198}Au measurement after JCO criticality accident, the discovery of natural ^{152}Eu, and the ^{60}Co and ^{152}Eu measurements at Hiroshima.

Oto Cosmo Observatory[2] is in Osaka area and run by Osaka university(Fig.3). It is placed at 467m underground in a tunnel which was made for railway (but never been used for railway). An advantage of Oto Cosmo Observatory is that natural continuous air flow through the tunnel makes the radon concentration to the level of $\sim$10 Bq/m^3 all the year through. The cosmic ray muon flux is $8\times10^{-7}/cm^2/sec$ at the observatory. The neutrino flux is 4×10^{-5} $/cm^2/sec$ for thermal and $\sim$ 10^{-5} $/cm^2/sec$ for fast neutrons. The ELEGANT V detector was placed in the observatory and double beta decay of ^{100}Mo and dark matter search using NaI(Tl) were performed. The MOON-I detector is now being constructed at the former ELEGANT V area. Another experiment in the observatory is ELEGANT VI which uses CaF_2 crystals for double beta decay of ^{48}Ca and for dark matter search.

YangYang Underground Laboratory (Y2L)[3] is located in mountain area of east Korea (Fig.4). It is run by Dark

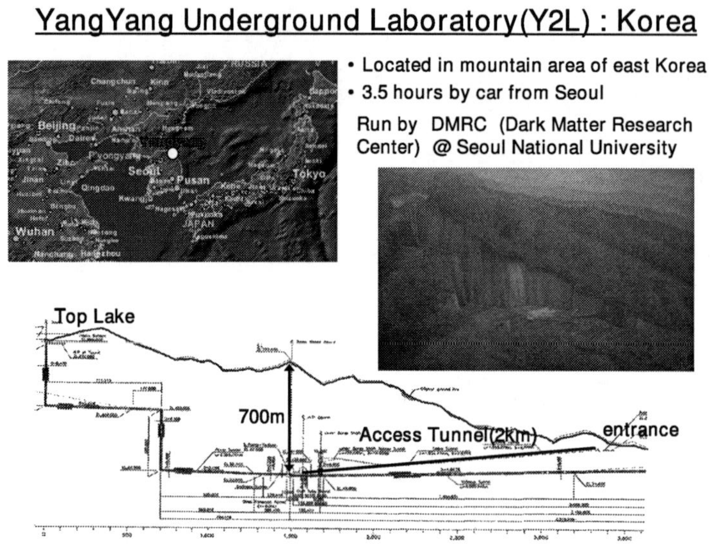

FIGURE 4. YangYang Underground Laboratory.

Matter Research Center(DMRC) at Seoul National University. The laboratory is placed in a tunnel for hydroelectric power station. Water is pumped up to a lake at the top of the mountain during nighttime and electricity is generated during daytime. The vertical depth of Y2L is 700 m and the length of the access tunnel is 2 km. The temperature in the laboratory is 20$\sim$25 centigrade and humidity is 35$\sim$60 %. Radioactivity concentration in the rock around the laboratory is ^{238}U $<$ 0.5ppm, ^{232}Th = 5.6$\pm$2.6 ppm and K_2O=4.1%. Radon concentration in the laboratory is 40$\sim$150 Bq/m^3. The cosmic ray muon flux is $4.4\times10^{-7}/cm^2/sec$ and The neutrino flux is 8×10^{-7} $/cm^2/sec$ at Y2L. A dark matter detector using CsI(Tl) crystal (KIMS) is placed in the laboratory and a new dark matter detector using HP germanium with very low threshold ($\sim$0.1keV) is being prepared. A R&D work for neutrinoless double beta decay using $CaMoO_4$ crystal is also going on.

Kamioka Observatory[4] is placed in Mt.Ikenoyama. The top of the mountain is 1369 m above see level. The access to the observatory is through a horizontal tunnel at about 350 m above see level. Figure 5 shows cross section of Mt.Ikenoyama. The experimental area is about 1000 m underground from the top of the mountain and the minimum distance from the Super-Kamiokande detector to the surface of the mountain is about 800 m. Figure 6 shows detectors in Kamioka Observatory. The Super-Kamiokande detector, the KamLAND detector, the XMASS prototype detector, a gravitational interferometer with 100 m baseline and an interferometer for geophysics measurement are placed in the observatory. The rock around the observatory is mainly gneiss and radioactivities in the rock are ^{238}U =$\sim$1.0 ppm, ^{232}Th = $\sim$3.2 ppm and K=1.6%. Figure 7 shows a measurement of the environmental gamma-ray flux in Kamioka Observatory. The cosmic ray muon flux is $2\times10^{-7}/cm^2/sec$ and the neutrino flux is $\sim 8.3\times10^{-6}$ $/cm^2/sec$ for thermal and $\sim 1.2\times10^{-5}$ $/cm^2/sec$ for fast neutrons, respectively.

A new experimental site is planned to be excavated at Kamioka Observatory. Figure 8 shows a preliminary design of the new experimental site. Several future dark matter and double beta decay experiments will be placed in the site.

RADON REDUCTION IN KAMIOKA OBSERVATORY

One problem in Kamioka Observatory is the strong time variation of radon concentration in the mine air. Because there are many tunnels at various altitudes in Mt.Ikenoyama, the direction of the air flow in the mine changes as shown in Fig.9. Radon concentration in the mine is as high as $\sim$3000 Bq/m^3 in summer time, while the concentration in winter time is a few tens of Bq/m^3. In order to reduce radon level at experimental sites, fresh air is always supplied to the sites using a high power air pumping system as shown in Fig.10. The fresh air supply to SK is with a flow rate of 50 m^3/min using a 22kW blower. The air duct to SK is 30 cm diameter and 1700 m length. The fresh air supply to KamLAND area is through a pipe of 20 cm diameter and 2145 m length and the flow rate is 15 m^3/min using a 15

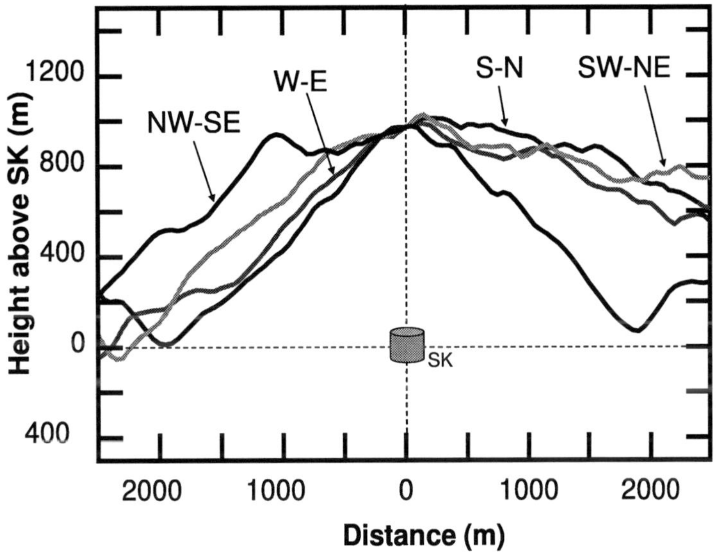

FIGURE 5. Cross section of Mt.Ikenoyama where Kamioka Observatory is placed.

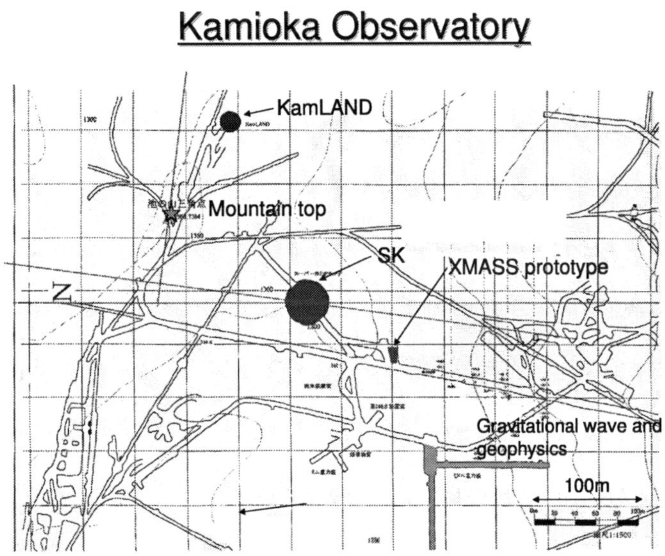

FIGURE 6. Experiments at Kamioka Observatory.

kW blower. Figure 11 shows the radon concentration in the supplied air and in the dome of the Super-Kamiokande detector. Although there is slight time variation of radon concentration at the dome, the radon level is less than 80 Bq/m^3.

In the Super-Kamiokande experiment, we need air with much less radon concentration for the detector. The air is used to fill cover air in the SK tank and some buffer tanks in the water purification system of SK. The air is also used for removing radon from water as described later. Figure 12 shows a schematic diagram of the system to make radon-less-air[5]. Key components in the system are four charcoal columns of 2 m^3 tanks at normal temperature and a 63L charcoal column cooled down to -40 centigrade. The production rate of radon-less-air is 18 m^3/hour. Figure 13 shows radon concentration at various position in the SK site. The radon concentration in the radon-less-air is $(2 \sim 3) \times 10^{-3}$ Bq/m^3 and that in the SK tank is $(10 \sim 20) \times 10^{-3}$ Bq/m^3. No time variation of radon concentration is observed as shown in the figure.

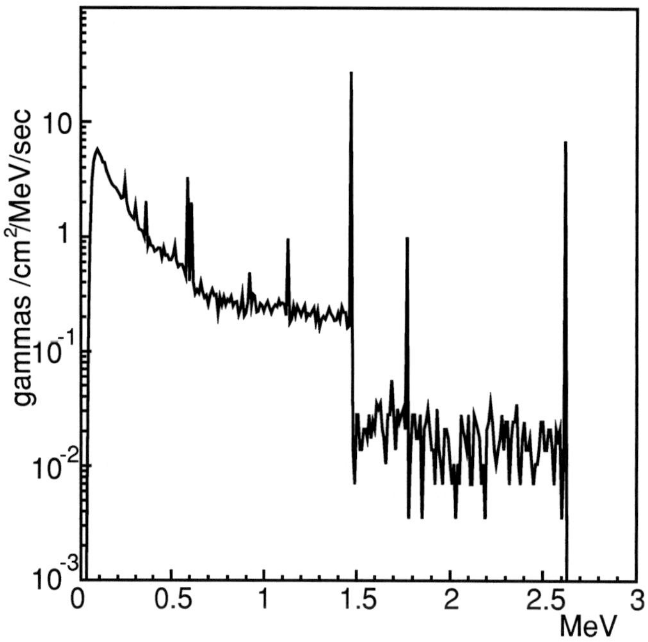

FIGURE 7. Environmental gamma-ray flux in Kamioka Observatory.

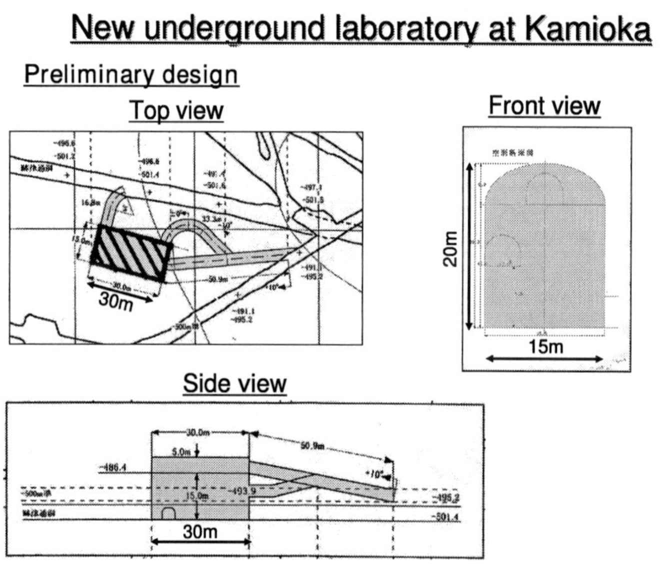

FIGURE 8. Planned new experimental site at Kamioka Observatory.

Figure 14 shows a schematic diagram of the water purification system of SK[5]. Radon in water is removed by the vacuum degasifier and the membrane degasifier. A tank to artificially dissolve radon-less-air is placed before the vacuum degasifier. It is to dilute radon concentration in order to improve radon reduction efficiency of the vacuum degasifier. The membrane degasifier is a system of 30 columns filled with hollow fiber membrane. Water flows outside the fibers in the columns and inside of the fibers are evacuated with a radon-less-air flow in the fiber. The radon

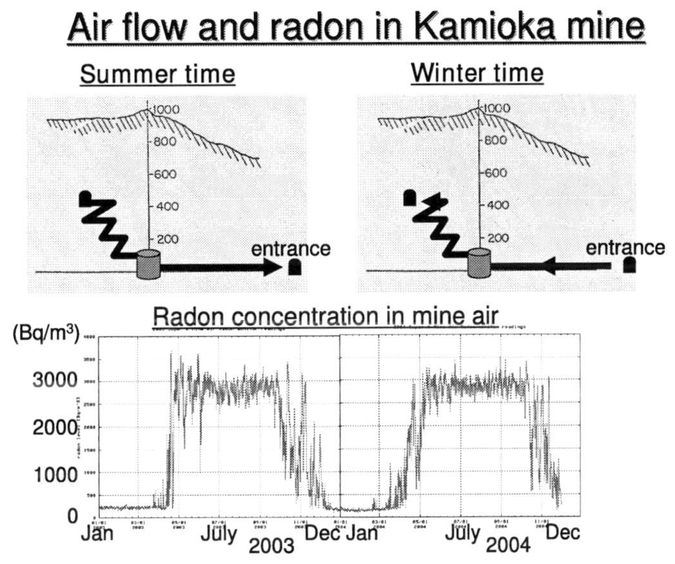

FIGURE 9. Radon flow in Kamioka Observatory and time variation of radon concentration in the mine.

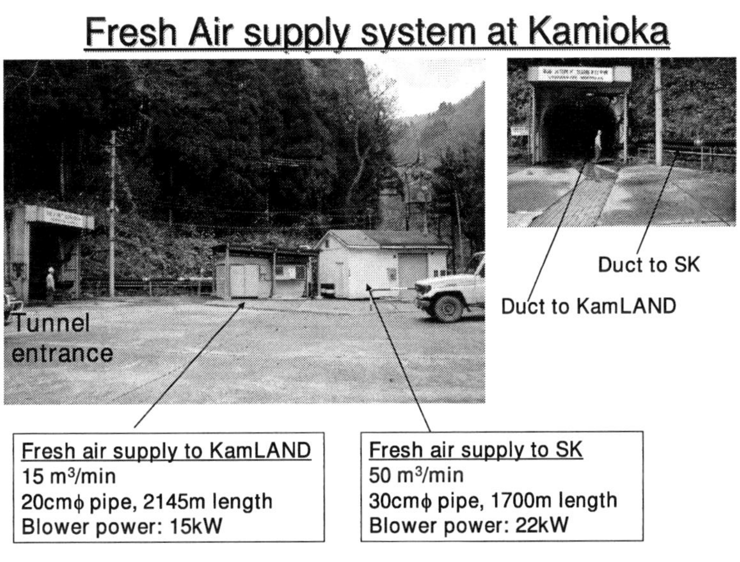

FIGURE 10. Rresh air supply system for Super-Kamiokande and KamLAND.

concentration in the supply water to the SK tank is less than 0.7×10^{-3} Bq/m^3. Radon could be supplied to the SK tank by radium in water. Ionized radium is removed by the ion-exchangers and radium in particles are removed by the Ultra-Filters(UF) and the Reverse-Osmosis(RO) units.

SUMMARY

Japanese and Korean underground facilities are reviewed. The depth of those facilities, the muon and neutron fluxes, the radon level, and the experiments in those facilities are summarized in Table 1. The radon reduction systems at Kamioka Observatory are also reported.

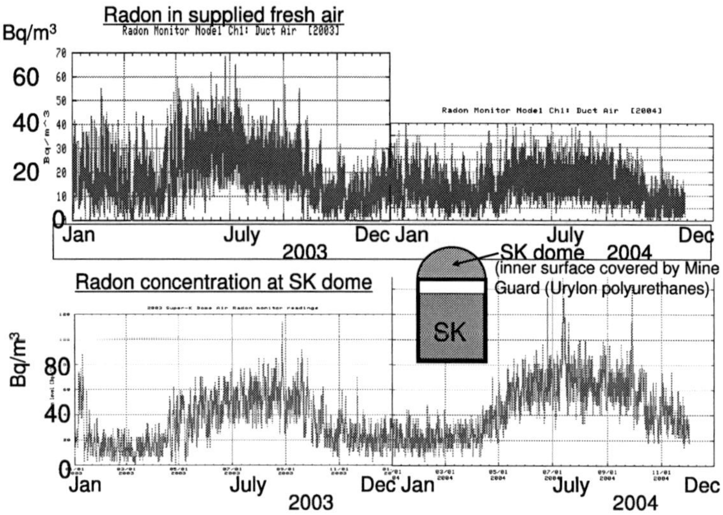

FIGURE 11. Radon concentration at SK dome.

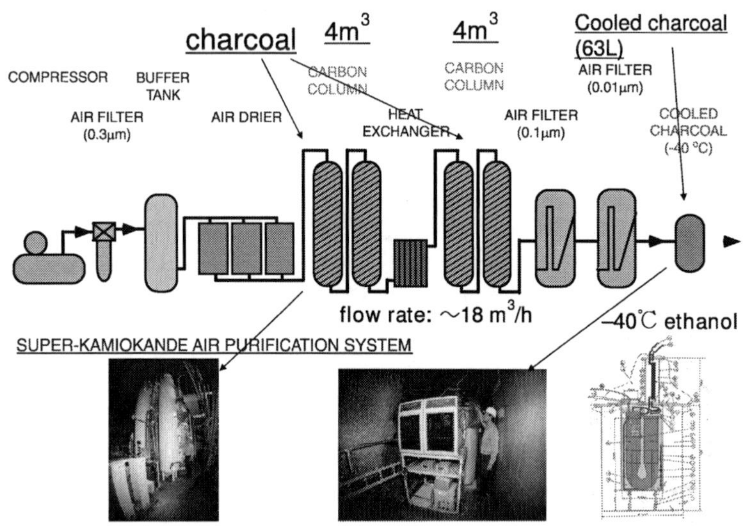

FIGURE 12. System to produce radon-less-air at SK.

TABLE 1. Summary of Japanese and Korean facilities.

facility	depth	muon flux ($/cm^2/sec$)	radon level (Bq/m^3)	neutron ($/cm^2/sec$)	experiment
Kamioka (Japan)	1000 m	2×10^{-7}	20∼60 (exp. area)	8.3×10^{-6}	SK KamLAND
YangYang (Korea)	700 m	4.4×10^{-7}	40∼150	8×10^{-7}	KIMS
Oto Cosmo (Japan)	467 m	8×10^{-7}	∼10	4×10^{-5}	ELEGANT CANDLES
Ogoya (Japan)	135 m	5×10^{-5}	∼25	$\sim 2\times10^{-5}$	Ge

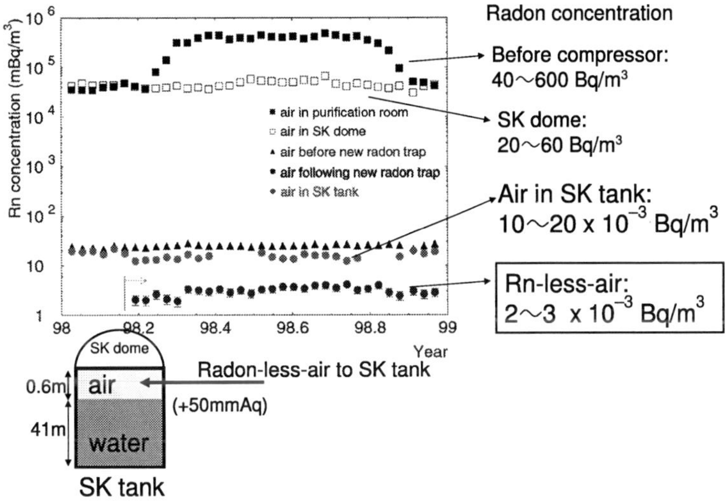

FIGURE 13. Radon concentration at various places in Super-Kamiokande.

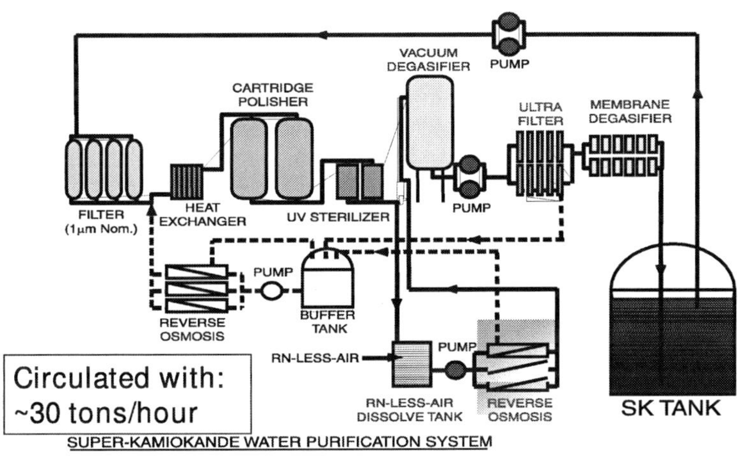

FIGURE 14. Water purification system of Super-Kamiokande.

REFERENCES

1. K. Komura and Y.Hamajima, Applied Radioation and Isotope 61 (2004) 185-189.
2. H. Ohsumi et al., "The New Underground Laboratory 'Oto Cosmo Observatory' and ELEGANT", proceedings of 15th RCNP Osaka International Symposium on Nuclear Physics Frontiers with Electroweak Probes, p274-279, March 1996.
3. J.W. Kwak et al., "Status of KIMS experiment searching for WIMP with CsI(Tl) crystals", proceedings of IDM2004 held at Edinburgh.
4. Y. Fukuda et al., N.I.M. A501(2003)418-462.
5. Y. Takeuchi et al., Phys. Lett. B452(1999)418-424.

Status of Low Background Counting Facilities in the USA

Priscilla Cushman

School of Physics and Astronomy
University of Minnesota

Abstract. The present status of several low background counting facilities located in the US is surveyed.

Keywords: underground laboratories, low background techniques.
PACS: 01.52+r

1. INTRODUCTION

Low level counting and materials purity are vital to the success of most underground experiments, yet the infrastructure to support these needs is in its infancy. Part of the problem is the funding structure, which only supports the scientific goals of individual experiments. Each experiment by itself may require a small, but important screening program, forcing each experiment to invest in its own instrumentation. It would be far more cost effective to share these resources in a properly managed user facility.

In Europe and Japan, the longer history of underground experiments at a few sites has resulted in a de facto collection of screening devices (mostly high purity germanium detectors) locally available to existing and future experiments, as well as detectors which have outlived their original purpose and can be redirected toward screening. In an effort to organize and share resources, the European Union has instituted ILIAS (Integrated Large Infrastructures for Astroparticle Physics). Cooperation with this entity and the creation of North American integrative structures is overdue. The new NSF DUSEL initiative may lead in this direction. Hopefully the process itself can be used to form useful collaborations and move toward the creation of low background counting centers on this continent.

Often a series of increasingly sensitive steps are required for any individual application, starting with component pre-screening to identify materials that themselves might contaminate the ultra-sensitive detectors. Thus, an underground site becomes much more useful if it includes detectors capable of less sensitive applications with quick turn-around. As a function of depth and increasing sensitivity, the situation in the US today is a decentralized set of detectors dedicated to specific research projects at surface sites, a few shallow depth sites which have evolved into user facilities, and a few deep sites which are struggling to provide the ultra-low counting needs for the next generation of double beta decay, solar neutrino, and dark matter experiments. Since many of these collaborations have an international component, there is the option of meeting these needs in Europe and Japan, however, not every collaboration is in this situation. There has not, to date, been much dialogue with other fields that might also benefit from the ultra-low counting sensitivities available at deep sites. This may be changing with the current DUSEL initiative and would greatly enhance the user base and motivation for a few centralized underground low background counting facilities in the US.

CP 785, *Topical Workshop on Low Radioactivity Techniques,* edited by B. Cleveland, R. Ford, and M. Chen

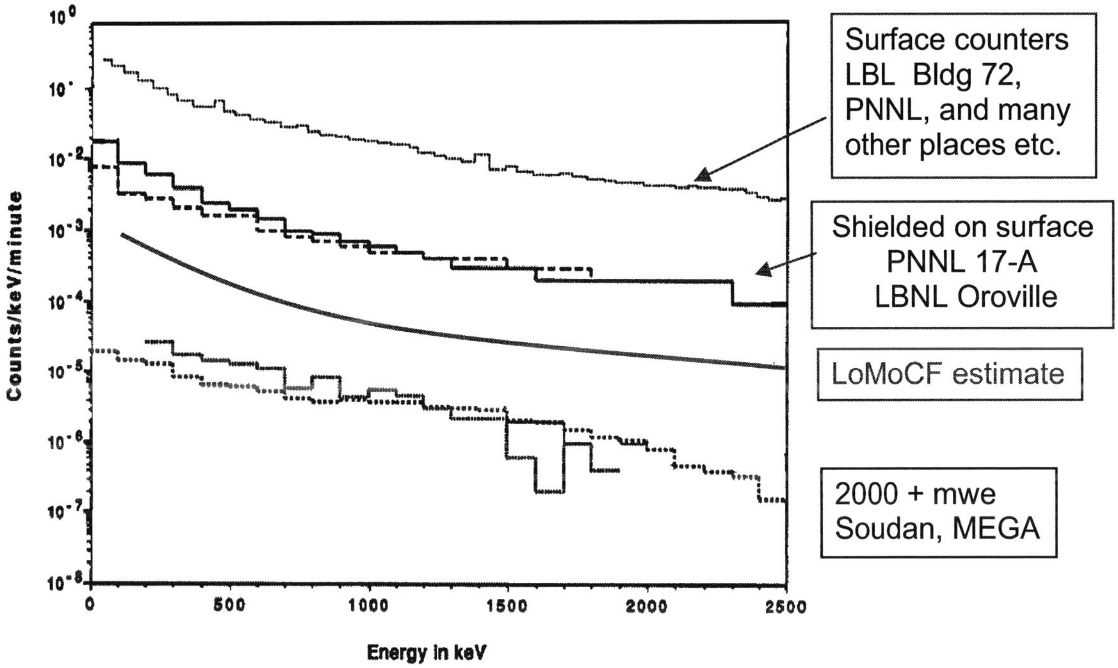

FIGURE 1. Estimated sensitivity that can be attained before becoming background-limited. (Courtesy of H. Miley). LoMoCF stands for The Lower Monumental Counting Facility (LoMoCF).

2. ABOVE-GROUND TECHNIQUES

For many less sensitive applications, low level counting is not necessarily the best technique for screening materials. Radioisotope identification can be done using surface techniques and mass spectrometry. Most research universities and many companies offer services for a fee. University fees are often very cheap for faculty, and companies offer quicker turn-around for a higher charge. A quick summary of the more common techniques available:

1. Surface Analysis

A. Ion Beam Analysis (IBA) utilizes high-energy ion beams to probe elemental composition non-destructively as a function of depth to several microns with a typical depth resolution of 100-200 angstroms. The energy distribution of backscattering ions (RBS = Rutherford Backscattering Spectroscopy) quantifies the depth distribution for a given element. Distinctive characteristic X-rays emitted from the different target elements (PIXE = Particle-induced X-Ray Emission) upon the beam bombardment ensure the accurate identification of similar mass elements. They can also have additional simultaneous measurements using forward-recoil spectrometry (FReS) and a NaI scintillator detector for Nuclear Reaction Analysis (NRA). These techniques can identify elements to 1 ppm, but not in the bulk.

B. Electron beams for Auger Electron Spectroscopy are less sensitive; the best can find trace contaminants at the 0.1% level within 1 nm of the surface. It is possible to obtain depth profiles of 100 nm or more with a sputter gun. Spot size can sometimes be small enough (5 microns) to do x-y probe of contaminants on the surface.

2. Mass spectroscopy

A. Commercial machines that use plasma (ICPMS) or thermal ionization (TIMS) to release ions into the mass spectrometer channel can reach ppb levels for many elements. The sample (or surface) must be dissolved in an acid and this is what is analyzed.

B. Secondary Ion Mass Spectroscopy (SIMS) releases ions from the surface by ion beam bombardment and then they pass through a mass spectrometer for analysis. This technique can be as good as ppb of atoms on the surface, probed to a depth of 10 nm, but it is inherently a destructive technique.

C. For small samples, Accelerator Mass Spectroscopy (AMS) is the most sensitive, but it is also expensive and not always available locally.

All of these techniques have drawbacks. For example, ^{40}K is a common surface contaminant which is hard to distinguish from argon in RBS and has large background in mass spec techniques which use argon plasmas. Poor ionization yield from electron recoils off such surface betas can mimic a nuclear recoil in the CDMS dark matter experiment, so these betas are very dangerous. For this reason novel beta counters are being developed. Counting of decay particles will thus always find a place in the suite of techniques, since this provides a direct measure of the effect in question. The signal from many trace isotopes can be enhanced using neutron activation analysis (NAA), which then observes counts from short-lived isotopes produced by epithermal neutron capture. This requires sending the samples to a reactor and the use of a radiochemistry lab to process the material, but the samples still end up being counted in a high purity germanium detector (HPGe).

3. SHALLOW SITES

Low level counting that can be done in shielded environments on the surface is usually performed by an HPGe or NaI crystals equipped with a PMT. These are commercial devices that are usually dedicated to a specific project. All the national labs have such detectors and many will negotiate a "use for others" contract with outside users, which involves training on their machines and time spent on site performing the tests. Many university groups also have dedicated machines. The great need here is for coordination and integration of these facilities. The two which have been used by nuclear astrophysics projects are in California and Washington.

The Institute for Nuclear and Particle Astrophysics at Berkeley manages a user facility with a shielded building at Lawrence Berkeley Lab (Bldg 72) and site with an 80 mwe overburden at the Oroville Dam, 140 miles away. Bldg 72 is surrounded by low activity concrete and currently houses NaI counters and a 130% HPGe detector. The Oroville site maintains an 80% p-type HPGe. Both have a well-developed infrastructure with clean room preparation and sample handling procedures, as well as dedicated on-site technical help. This facility was responsible for most of the materials screening for the SNO experiment (screened virgin and machined acrylic, plastic resins, PMT envelopes, internal components, stainless steel, etc.). There have been more detectors employed in the past and there is room for more in the future if funding permits. The detectors currently in use will not benefit from a deeper site as they are limited by their own background radiation.

Direct Counting Sensitivities	Berkeley Site	Oroville Site
Uranium and Daughters	0.5 ppb	50 ppt
Thorium and Daughters	2.0 ppb	200 ppt
K	1.0 ppm	100 ppb
^{60}Co	0.04 pCi/Kg	0.004 pCi/Kg

FIGURE 2. Sensitivities which can be achieved at the Berkeley facilities.

Pacific Northwest National Lab operates a shielded facility on site in a set of buildings which include 17-A ULB. These detectors serve mostly PNNL projects, including national security components. A "use for others" contract could be arranged project by project. The University of Washington and PNNL manage a 4' x 8' space accessible via a 4 person elevator located under the Lower Monumental Dam, which is operated by the US Army Corp of Engineers and is about an hour's drive from PNNL.

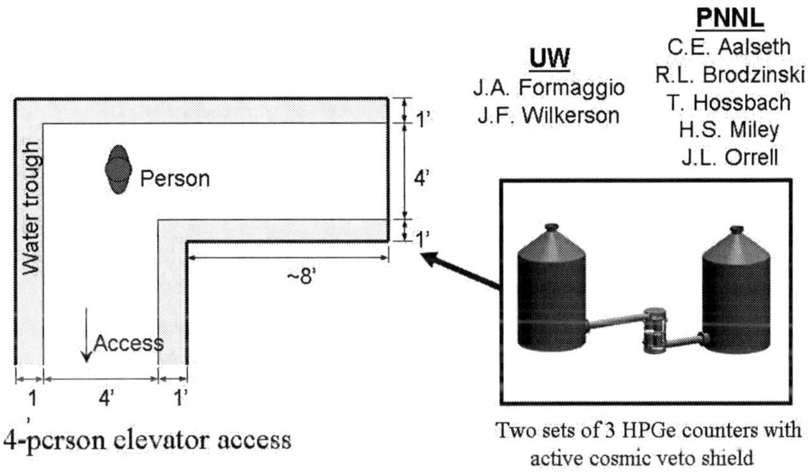

FIGURE 3. Schematic of the Lower Monumental Dam Facility and the screening HPGE to be installed at that location. At 37 m overburden, it was used for cosmogenic isolation storage of copper stock and MEGA parts, but has recently been instrumented with two sets of triple-pack HPGe counters with active cosmic veto shield in the coincident configuration shown. The site will be mostly used by the Majorana collaboration.

4. DEEP SITES

Three deep sites have made infrastructure investments leading to potential low background counting centers. The Waste Isolation Pilot Plant in Carlsbad, NM is at a depth of 1600 mwe in a salt dome. WIPP is a DOE facility with a fully developed infrastructure and a heavy Los Alamos National Lab presence with an Earth and Environmental Science office located at Carlsbad. Although fundamental science is only an add-on to its primary mission, LANL and WIPP established a clean-room to house the SEGA/MEGA experiment (Segmented Enriched Germanium Array and Multi Element Germanium Assay). Another neutrinoless double beta decay experiment, EXO (Enriched Xenon Observatory), is being tested at Stanford and will move to WIPP in several years.

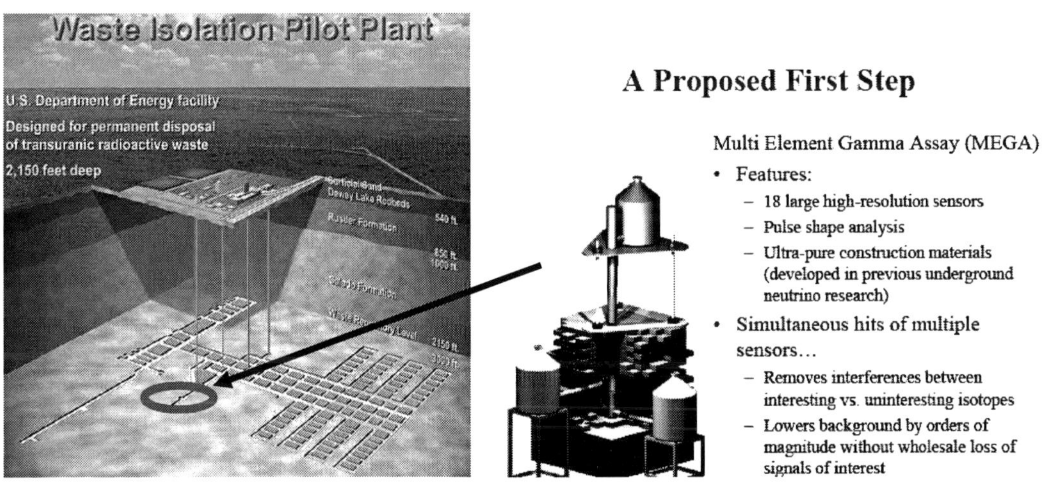

FIGURE 4. MEGA detector being currently installed at WIPP along with SEGA and additional infrastructure (clean rooms and DAQ) which may be generally useful for low background counting applications.

MEGA is itself the ultimate ultra-low level counting facility, but it is an effort directed by the needs of the Majorana group, not a user-oriented facility. Nevertheless, its presence, along with the EXO experiment, creates the nucleus of a user group that will require local screening services. One of the main components in the WIPP DUSEL proposal is a low background counting facility.

A second site is the Kimballton mine http://www.phys.vt.edu/~kimballton, an active limestone mine operated by the Chemical Lime Company in Giles County, Virginia. Virginia Tech is the lead institution developing this site for science. The solar neutrino experiment LENS (Low Energy Neutrino Spectrometer), which can observe the low energy neutrinos from p-p fusion, is slated to be installed in a location 7000 ft under Butt Mountain. Kimballton's proximity to the Naval Research Labs make it a promising site for secure applications involving homeland security and treaty verification and plans are already underway to make this a reality.

The third deep site is the Soudan mine in northern Minnesota. This iron mine is at 2090 mwe and is now an historical state park run by the Department of Natural Resources, rather than a working mine. The Soudan lab houses the MINOS neutrino experiment and the CDMS dark matter search and is managed by a team which includes the University of Minnesota and Fermilab. The old proton decay calorimeter has been removed, but the proportional tubes which line the 40' x 35' x 100' cavern remain behind and have been repaired. A new integrated data acquisition system combines epoch information and pps from GPS with coincidences through the shield to provide any user with muon information tagged by geographical location and time stamp. This was designed from the beginning as a user facility and currently has space available for other experiments. An existing surface building provides remote computer access to the underground experiments and additional space for offices, storage and preparation for users.

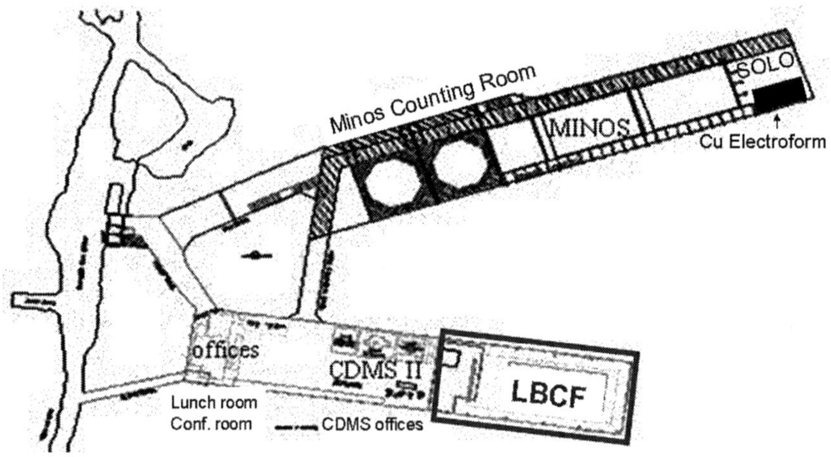

FIGURE 5. Layout of the Soudan Underground Lab, showing where the new LBCF is located, as well as the SOLO HPGe setup and the new Cu electroforming clean room..

Potential users are directed to http://www.hep.umn.edu/LBCF for details and scheduling. New occupants include COUPP, a bubble chamber sensitive to weakly interacting dark matter particles (see http://cfcp.uchicago.edu/research/highlights/highlight_2004-08-20.html), copper electroforming, and a Medtronics soft memory error experiment. The electroforming facility was funded using an SBIR and includes a special screening detector of the sensitivity required to test the purity of the electroformed copper. In Phase II, it is expected to move toward a self-sustaining commercial enterprise that can produce ultrapure copper for the shields and cryostats. The detector can also be put to use screening other user-supplied materials.

There already exists a set of high purity germanium detectors (HPGe) which have been operating for several years as a screening facility named SOLO. It is operated by Brown University (http://particleastro.brown.edu/SOLO) and PNNL at the Soudan mine, primarily for the Majorana and XENON collaborations, though CDMS has also screened poly and copper in it. Details on SOLO can be found in Rick Gaitskell's talk from this conference. A new 100% efficient HPGe is being added this spring to increase throughput and sensitivity. Eventually, its existing cryostat may be replaced by the electroformed copper. Since CDMS has need of low energy beta screening, collaborators are developing two technologies directed toward solving this problem: a neon gas MWPC and a cloud chamber with digital readout and pattern recognition. These devices are scheduled for transport into the mine by the end of 2006 and would become available to outside users.

This is a staged approach to a final low background counting facility that could serve the needs of development, screening, and testing of proposed DUSEL experiments before DUSEL is actually online. Since all experiments require additional passive shielding for neutrons and gammas generated by natural radioactivity and by muon interactions in the cavern rock, it would be most efficient to share this common shielding. A large cost-effective shield is a water shield built with commercial water tank technology. In cooperation with an engineering firm and several tank manufacturers, we have designed a large purified water shield into which detectors are lowered from the top. The background will be limited only by water purity, and thus will provide a lower background environment than standard Pb and plastic shielding. This will provide as much as 4 meters of shielding in the center of the tank, reducing the high energy neutron flux by two orders of magnitude. It will eliminate the need for expensive lead (including ancient lead) shielding around the instruments. A clean room will be assembled on top of the tank with an adjoining mezzanine for assembly and staging. There will be room for cubic meter size experiments to be lowered into the tank and dedicated screeners installed in other ports.

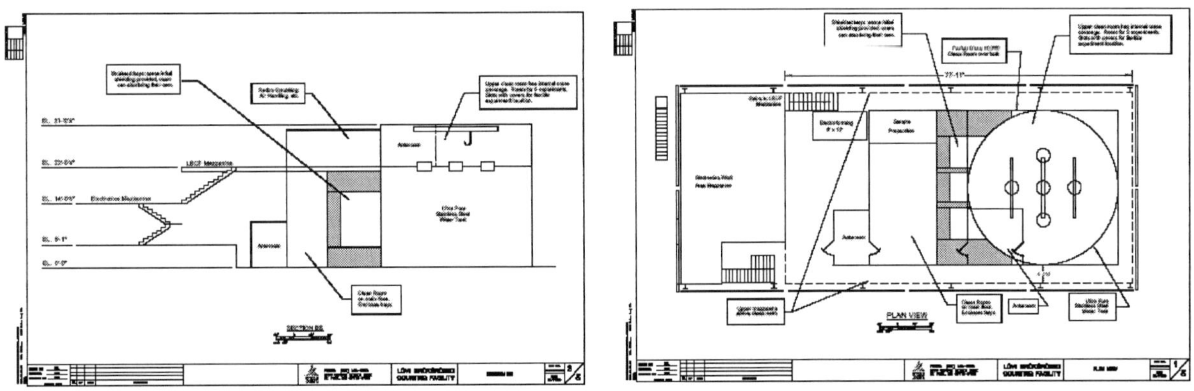

FIGURE 6. Elevation and plan views of the final Soudan Low Background Counting Facility.

A clean room will also be built adjacent to the tank on the main floor. Since the water tank already provides excellent shielding from the south wall, a set of shielded bays will be located at its base, inside the clean room coverage. This room will serve as an experimental area for prototypes which require deep underground environments, reducing or eliminating their need for expensive shielding. A copy of the system developed by the Princeton Borexino group will be used to provide breathable low-Rn air to the two clean rooms via two large (250 Kg) charcoal columns using the vacuum swing" technique to regenerate alternate columns. An overall Rn reduction factor of 10^4 is expected.

5. FUTURE DIRECTIONS

The final low background counting facility envisioned for DUSEL has been clearly described in the Homestake Reference Report and accompanying Science Book and is shown below. Its functionality must exist in some form during the DUSEL era. There are several ways in which this could happen. The counting/screening capabilities at existing sites could be augmented to serve complementary functions. If there is coordination between sites, this may be an option which preserves local communities while serving the whole. Economy of scale would prefer a unified one-site model with a complete range of services. This would happen naturally if one of these deep sites is chosen to be the final DUSEL. However, if an undeveloped site is chosen for DUSEL, the additional expense of creating a dedicated on-site LBCF would have to be weighed against a coordinated augmentation of existing underground facilities.

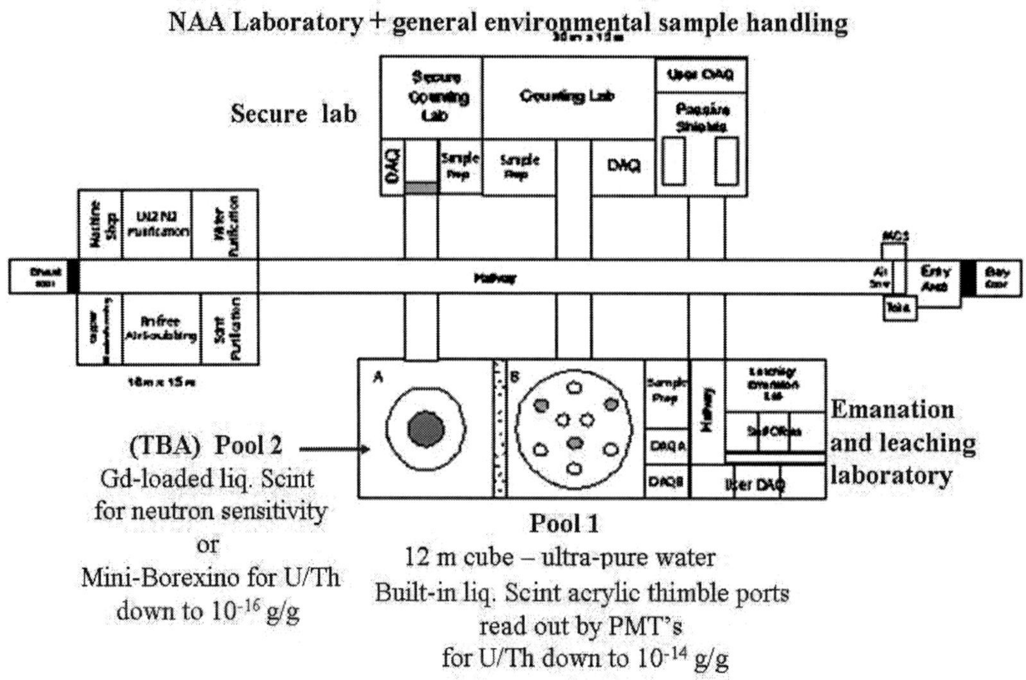

FIGURE 7. The type of low background counting facility envisioned for DUSEL. Figure E-16 from the Homestake Reference Design..

The situation in the US could be dramatically improved by coordinating the background screening efforts of nuclear and particle astrophysics experiments in order to schedule screening and to maximize throughput. Collaborations could jointly propose centralized counting facilities in the several deep sites that exist. These efforts at funding would be strengthened by broadening our base to include disciplines, such as geology, hydrology, microbiology that also require radiometric measurements and have a stake in underground science. The existence of a user base such as this, along with an integrated website, would allow such counting facilities to support themselves with user fees and take advantage of the economy of scale in shielding and infrastructure. Integration efforts have begun with this conference, with the DUSEL working groups, and in a Synergies in Low-Background, Ultra-sensitive Radiation Techniques to take place this summer July 25-26 http://www.hep.umn.edu/lbcf/workshop.

SNOLAB
An International Facility for Underground Science

A. Hime

Physics Division, MS H803, Los Alamos National Laboratory
Los Alamos, NM 87545, USA

Abstract. SNOLAB, an international facility for underground science, is presently under construction at a depth of 6000 meters of water equivalent (m.w.e.) at Inco's Creighton mine near Sudbury, Ontario, Canada. Building on the success of the Sudbury Neutrino Observatory (SNO) experiment, the creation of SNOLAB will provide the deep-site infrastructure required of next generation particle-astrophysics experiments in pursuit of low-energy solar neutrinos, neutrinoless double beta decay, and cosmological dark matter. Following an enthusiastic response from the scientific community to a call for Letters of Interest (LOI's) in staging experiments at SNOLAB, an initial evaluation process is now complete and an exercise is underway to define an initial suite of experiments and the longer term scientific program for this new facility.

Keywords: Sudbury, Underground Laboratory, Neutrino, Neutrino Mass, Solar Neutrinos, Double Beta Decay, Dark Matter, Cosmic Ray Muons, Fast Neutrons
PACS: 01.52.+r, 12.15.-y, 23.40.-s, 26.65.+t, 95.35.+d

INTRODUCTION

SNOLAB[1] is being developed at the site of the Sudbury Neutrino Observatory (SNO) at the Creighton Mine near Sudbury in Canada. Sudbury is located about 400 km north of Toronto and is accessible by regular flights from Toronto, Ottawa, or Sault Ste. Marie, or by major highways from any of these locations. The mine is about 25 km west of the city. The city of Sudbury is a major infrastructure center for northern Ontario and is home to Laurentian University, English and French language colleges, and regional hospitals in addition to the industrial support base. The Creighton Mine is an active nickel and copper mine owned and operated by Inco. At the start of the SNO project the Creighton Mine boasted the deepest continuous shaft in the western world. The currently active shaft runs from surface to a depth of 7000 feet and mining is progressing below this level. The SNOLAB facility is located on the 6800 foot level, a depth of 2000 m or 6000 m.w.e.

The Sudbury basin is thought to be the product of a major meteor strike that occurred 1.8 billion years ago. The geological pattern exhibits a series of concentric layers resulting from this impact. There are very rich mineral deposits in the crater rim that are mined for nickel, copper, cobalt, precious metals, and many of the rare earth elements. Mineral production has been valued at ~$\$10^{11}$ and known reserves are of a similar value. The area has about 4000 km of underground mine drifts at depths to almost 8000 feet. The Creighton mine is located on the south rim of the crater. The ore body lies between the footwall rock that is granite and the hanging wall that is Norite. The SNOLAB is located in the Norite hanging wall.

Because of the great mineral wealth, the geology of this area has been studied extensively. A collection of review articles on aspects of the geology and ore deposits of the Sudbury Structure have been published by the Ontario Geological Survey[2] wherein E.G. Pye notes that "the Sudbury Structure is one of the most scientifically and economically interesting geological structures on our planet." At the time of this volume, the meteor origin for the structure was not universally accepted. A more modern collection of papers is found in ref.[3].

CP 785, *Topical Workshop on Low Radioactivity Techniques*, edited by B. Cleveland, R. Ford, and M. Chen

The SNO detector[4] was built to solve the *Solar Neutrino Problem*. The collaboration sought to determine if the deficit of neutrinos from the Sun observed in previous experiments was due to defects in our understanding of the Sun or to new neutrino properties. In 1984, Herb Chen of UC Irvine identified heavy water as an ideal medium for resolving this problem as it was a very good target for detecting electron neutrinos through a charged current interaction and it could detect all active neutrino types through a neutral current process[5]. To realize this, a detector would have to contain about 1000 tonnes of heavy water, be deep underground to escape the cosmic ray flux, be well shielded from local radioactivity, and be made of materials that reached unprecedented levels of radioactive purity. A collaboration of US, Canadian, and UK scientists was formed to do the necessary R&D to accomplish these tasks and the result was the SNO detector. The detector is located 2000 m below ground in a 22-m-diameter, 35-m-high cavern. This is the world's largest cavern at this depth designed for human occupancy. The laboratory housing this cavern and the required infrastructure is established as a deep clean room so that radioactivity in dust would not contaminate the results. The construction of the detector was completed in 1999 and the results published to date demonstrate that the solar neutrino problem is resolved due to active flavor transformation of electron neutrinos born in the Sun into muon and tau neutrinos detected at Earth[6,7]. Because of the uniqueness of the experiment, the measurement is being carried out in three systematically different ways. In the first phase the detector was filled with pure heavy water[6]. In the second phase 2000 kg of pure NaCl was added to the water to enhance the detection of the neutral current process[7]. In the third and final phase, the NaCl has been removed and an array of ^{3}He filled proportional counters has been added to give a completely separate detection mode for the neutral current reaction. Operations of the SNO detector will cease at the end of 2006 at which point the heavy water will be returned and the infrastructure will be available for another experiment.

In 2001, the Canada Foundation for Innovation (CFI) announced a program to fund facilities in Canada that would attract scientists from around the world to carry out research projects in Canada. Buoyed by the success of SNO, a Letter of Intent to establish an International Facility for Underground Science was submitted. This proposal was accepted in 2002 and funding was released in spring of 2003. The objective was to turn SNO from a single project into a long-term facility in which several experiments, presumed to be in different phases of R&D, construction or data taking, could share a common infrastructure and provide a stimulating research environment. The scientific driving force for this facility was to build on the discoveries of, and the technologies developed for SNO. The obvious areas of interest were to push to lower energy for solar neutrino studies, to search for double beta decay, and to search for cosmological dark matter. Each of these areas of research needs the very deep site and clean environment that was created for SNO. With funding from the Federal government for a new underground laboratory, the Province of Ontario was asked to provide funds for a surface laboratory to support the underground activities. This too has been provided. The surface building is expected to be ready in June of 2005 while the full underground laboratory is expected to be complete in 2007.

While the scientific development for SNOLAB has focused upon pursuits in particle-astrophysics, the laboratory is open to proposals for any form of research that can benefit from the infrastructure being established. Investigations have begun into research areas outside of particle-astrophysics and how SNOLAB would be complementary to other underground facilities around the globe, including the DUSEL project under development in the United States. Some particular types of projects are probably not suited to the SNOLAB site because of interference with the active mining. It would be challenging to create a 10^6 m^3 space at SNOLAB. Novel techniques would have to be developed to support such caverns at such depth. Some additional method for disposing of the rock would have to be found – either by changing the mine to a cemented rock fill operation or creating a new shaft dedicated to hoisting the rock. There are probably sites better suited to such projects. There are a number of research projects at the mine related to mining in stressed rock and at depth. However, some of the larger geotechnical projects might interfere with the active mine. Such possibilities will be investigated as the project definitions become clearer.

The SNOLAB project is managed by the SNO Institute (SNOI). SNOI is an Institute of Queen's University having a Board of Trustees representing the universities of Queen's, British Columbia, Carleton, Guelph, Laurentian, and Montreal, as well as Inco. The assets and liabilities of the Institute are shared by the university members of the Institute. The Institute Board appoints a Director for the Laboratory (currently David Sinclair) with responsibility to manage the construction and operation of the facility. The Board has also established a Scientific and Technical Review Committee to advise the Board on matters relating to the facility. The Board has also appointed an Associate Director for the facility, Andrew Hime. The Director is advised on science policy by an Experiments Advisory Committee (EAC). Barry Barish serves as Chair of this committee and Hime serves as the Secretary. There is a Scientific Executive Committee (SEC) that will evolve into a Users Committee, which is presently chaired by Hime.

SCIENTIFIC DEVELOPMENT

Development of the infrastructure requirements and scientific program for SNOLAB has evolved through a series of workshops organized by the SEC and by invitation to the international scientific community. The first workshop was established in Sudbury on August 20, 2002 and served largely as an informational meeting to inform the community on the basic opportunities at SNOLAB and to solicit interest in mounting experiments at this new facility. A second workshop was organized in Ottawa from November 21-22, 2002 that invited proponents of perspective experiments and initiated a program to define the infrastructure needs for SNOLAB in support of these experiments. An official call for LOI's in staging experiments at SNOLAB was issued on January 26, 2004[1]. This call was met with great enthusiasm with a total of 18 potential experiments presently under consideration (see Table 1) and was followed by a third and very successful workshop held in Sudbury from May 12-14, 2004 which served to refine the infrastructure needs and conceptual layout of the laboratory, and to begin the experiment evaluation process. We are now planning our fourth workshop to again be hosted in Sudbury in the new SNOLAB surface building from August 14-18, 2005. This workshop will focus upon a finalization of the infrastructure needs for the laboratory and to firm up our visions of a scientific roadmap for SNOLAB.

TABLE 1. Experimental Interest in SNOLAB

Experiment	Collaboration	Scientific Thrust
Majorana	US, Canada	Double Beta Decay
GerDA	Germany	Double Beta Decay
EXO	US, Canada, Switz., Russia	Double beta Decay
COBRA	UK, Germany, Italy	Double Beta Decay
Lithium Detector	Russia	Solar Neutrinos
SNO++	Canada, US, UK	Solar and Geo-Neutrinos
Noble Liquid Tracking	US	Solar neutrinos
LENA	Germany	Solar Neutrinos and Proton Decay
CLEAN	US	Solar Neutrinos and Dark Matter
CDMS	US	Dark Matter
PICASSO	Canada	Dark Matter
ZEPLIN	UK, US, Russia, Portugal	Dark Matter
XENON	US	Dark Matter
DEAP	US	Dark Matter
DRIFT	UK, US	Dark Matter
HALO	Canada, US	Supernova Neutrinos
NASTOS	France, Germany, Italy	Neutrino Oscillations
TRIGA	US	Neutron-Antineutron Oscillations

Following this series of workshops and the receipt of LOI's, the conceptual layout for the underground laboratory spaces (see Figure 1) evolved from a single, monolith-type hall to three distinct "classes" of space referred to as, the Rectangular Hall, the Ladder Laboratories, and the Cryopit. The SNO cavern and existing experimental facility will become available for new endeavors once decommissioning of the SNO experiment is complete. In addition to the existing SNO cavern and experimental area, SNOLAB will provide an additional 36,000 square feet of new excavations for a total of 45,000 square feet of clean underground space. A medium to large scale experiment will be accommodated in the Rectangular Hall. A set of Ladder Laboratories can accommodate 2-3 experiments of smaller footprint (e.g. double beta decay and smaller scale DM detectors) where the overhead height provided by the larger cavities is not required. Finally, the Cryopit would provide a cylindrical space for a large-scale cryogenic experiment that can be physically isolated safely from personnel and other ongoing and parallel experiments. The new surface building presently under construction will serve in support of the underground program with office space, computing facilities, and meeting rooms. Both high bay and standard laboratory space will be available together with infrastructure for clean assembly and transport of equipment underground. A white paper is presently under draft by the SEC that will detail the infrastructure of both underground and surface facilities, outline means for operations at SNOLAB, and eventually evolve to a user's manual for this new laboratory. A preliminary description of the surface and underground facilities for SNOLAB accompanies the original call for LOI's and can be located in ref.[1].

LOI's and eventual proposals to stage experiments at SNOLAB are reviewed by the EAC following the seven guidelines provided in the original request for LOI's:

1.Scientific Merit,
2.Infrastructure Requirements,
3.Progress on R&D,
4.Technical Feasibility,
5.Safety,
6.Funding & Schedule, and
7.Participation & Management.

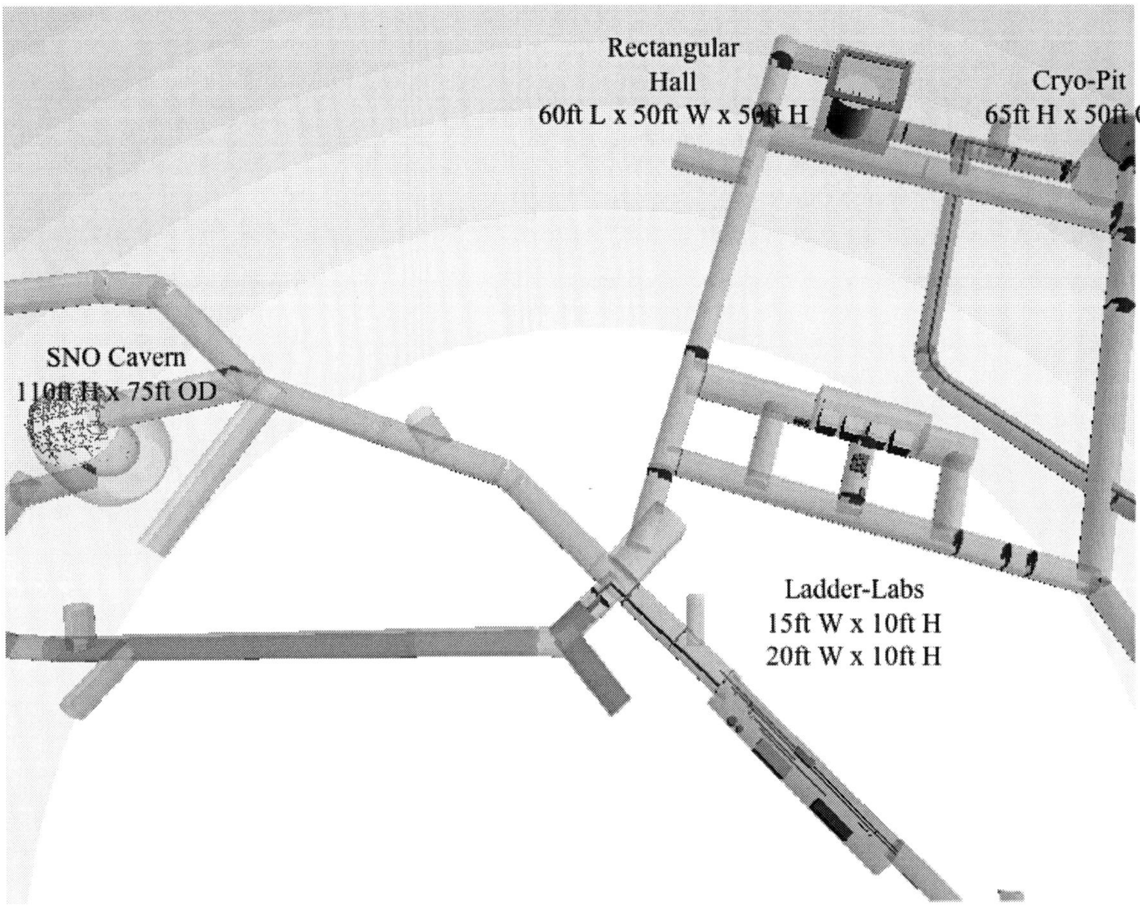

FIGURE 1: Conceptual Layout of the Underground Laboratory.

We want to develop a scientific program that exploits the unique opportunity of SNOLAB depth and infrastructure, and that recognizes the long-term scientific goals for major thrust areas in particle-astrophysics, including solar neutrinos, neutrinoless double beta decay, and cosmological dark matter. We want to develop a program with an initial suite of experiments capable of making the next steps in the field while fostering new detector technologies and collaborations that can lead to future generation experiments with reach well beyond the present state of the art. Below we summarize the present vision to establish such a program at SNOLAB. It must be emphasized, however, that this process is underway and prone to change given uncertainties in funding for experiments presently under review by the funding agencies and the need for research and development for projects attempting to demonstrate technical feasibility.

Dark Matter

The direct search for the Dark Matter (DM) in the Universe occupies one of the most important quests in modern particle-astrophysics. While a definitive WIMP (Weakly Interacting Massive Particle) signal eludes the most sensitive of modern experiments, the limits achieved are now just reaching into the regions predicted by viable models of Super-Symmetry (SUSY). A positive WIMP signal would have profound implications in elementary particle physics, astrophysics, and cosmology. The absence of a signal in experiments of ever increasing sensitivity would continue to constrain physics beyond the Standard Model and our basic understanding of the DM in the Universe.

The CDMS-II experiment has produced the world's best limits in a direct search for spin-independent WIMPs at the Soudan Mine. The current limits from CDMS-II (~4 x 10^{-43} cm^2 for a 100 GeV WIMP-nucleon cross-section following ~20 kg days of Ge exposure) set the standard for these experiments and benefit from essentially zero background after effective signal processing and pulse-shape discrimination cuts[8]. We note that these limits are in contradiction with the positive claim from the DAMA experiment for spin-independent WIMPs assuming a standard set of galactic-halo parameters. There remains a small region of parameter space for spin-dependent WIMP-nucleon interactions corresponding to the DAMA signal that have not yet been tested by other direct-search experiments. Indirect searches for WIMP annihilation in the Sun from the Super-Kamiokande experiment rule out a significant part of the DAMA allowed region, however, it is desirable to make further direct tests through WIMP-nucleus scattering experiments.

The challenge for future DM searches lies in the ability to achieve sensitivities some three orders of magnitude or better beyond the present state of the art. Such sensitivity would reach deep into the space predicted by SUSY models and be highly complementary with accelerator searches planned for the Large Hadron Collider (LHC) at CERN. The CDMS collaboration will attack this challenge by relying upon their well-defined technology using cryogenic Ge and Si detectors and the ability to discriminate ionizing radiation events from nuclear recoil events characteristic of a WIMP signature. Improved sensitivity relies upon the ability to scale up target mass significantly and to mitigate backgrounds associated with "surface" beta-activity and cosmic ray induced fast-neutrons. At this workshop, Bauer announced significant improvements in understanding the origin of this surface activity and in analysis cuts to remove this contamination from the signal region. Estimates indicate that, at 1000 kg days of exposure (some 50 times that achieved to date and the equivalent of running the existing experiment at full capacity for ~ 1 year) cosmic ray-induced neutron background will limit the sensitivity of the experiment at the Soudan overburden of ~2090 m.w.e. Consequently, future experiments searching for WIMP DM will benefit enormously from the 6000 m.w.e. overburden present at SNOLAB.

The PICASSO experiment aims to exploit a novel approach to spin-dependent WIMP searches using supersaturated fluorinated halocarbons in water-based gels. PICASSO is unique from other proposed DM searches due to its high sensitivity to spin-dependent interactions on ^{19}F. With a prototype detector presently operating in the SNO underground facility, the PICASSO collaboration proposes a 3-phase program that would definitively test the DAMA result and improve the sensitivity for spin-dependent WIMP interactions significantly, provided radioactive background can be controlled at the desired levels.

Generally speaking, next generation experiments reaching well into the SUSY predictions for the WIMP-nucleon cross-section will need to be sited deep underground if backgrounds associated with cosmic ray induced fast neutrons are to be adequately suppressed (see Figure 2). With sufficient overburden and local shielding, significant experimental effort focuses on techniques to separate nuclear-recoil events from ubiquitous gamma ray and electron background that can enter the sensitive region of a detector and masquerade as a false DM signal. We note that there are several efforts world wide to develop highly sensitive WIMP detectors using a variety of novel techniques, with particular emphasis on liquid-cryogen targets of Xe, Ar, and Ne (see Figure 3). It is prudent to consider a scientific road map for DM that will optimize the community's ability to achieve the scientific goals for the field and to understand how elements of the program are best suited for SNOLAB. The "Holy Grail" for the field is the positive observation of a WIMP signal, in which case detector technologies such as that under development by the DRIFT collaboration will be required to identify the signature with the "WIMP-Wind" in our galaxy and thus begin the field of "WIMP-Astronomy". Such visions represent a significant technological challenge to develop very large target mass, ultra-low energy threshold tracking detectors capable to reconstruct the direction of the WIMP-recoiling nucleus candidate events.

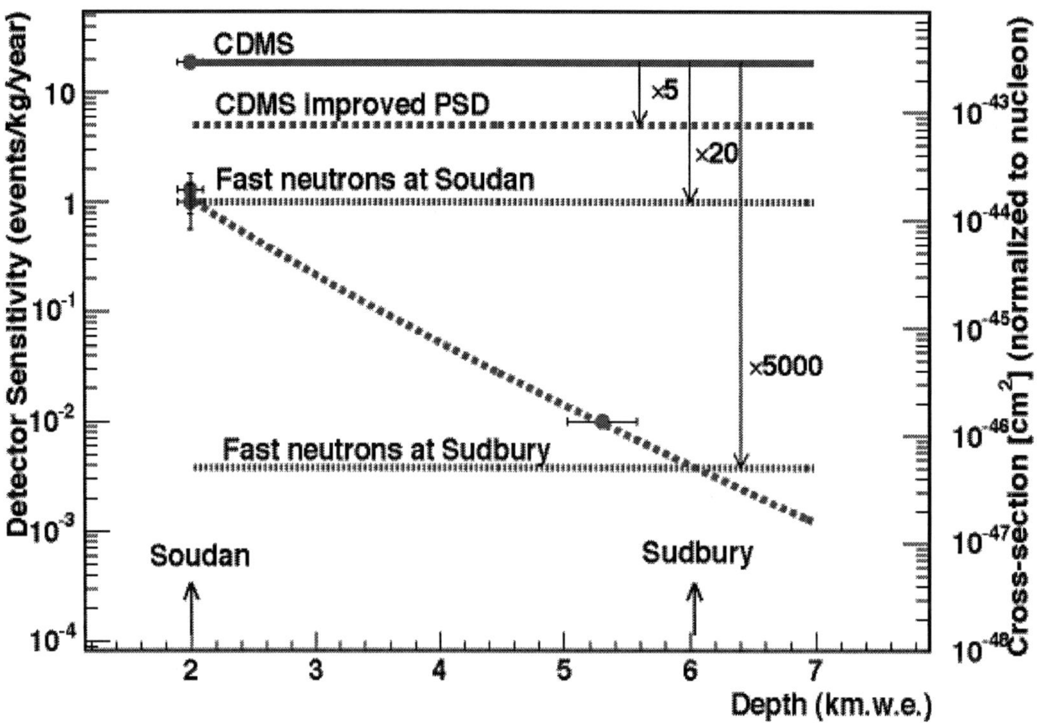

FIGURE 2: Depth sensitivity relationship for WIMP dark matter searches (ref.[9]) showing that, with radioactive contamination under control, experiments will need to operate deep underground in order to mitigate background associated with cosmic ray induced high energy (fast) neutrons.

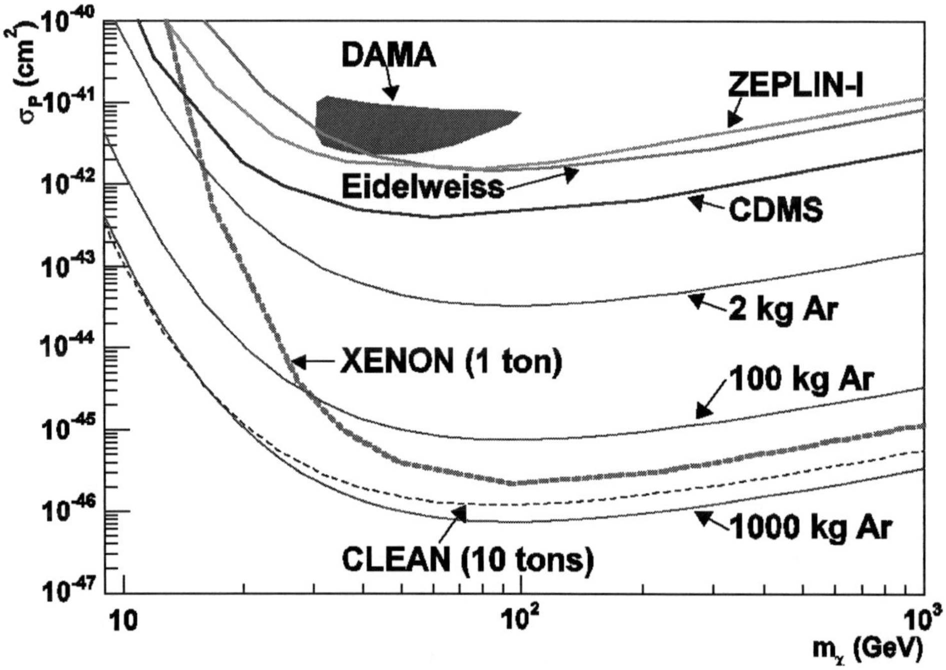

FIGURE 3: Sensitivity curves for existing (CDMS, ZEPLIN, Eidelweiss, DAMA) and planned WIMP dark matter experiments using liquid cryogen targets of Ar, Xe, and Ne (ref. [10]).

Neutrinoless Double Beta Decay

The pursuit of Neutrinoless Double Beta Decay ($\beta\beta_{0\nu}$) is one of the most important quests of neutrino physics today. It is the only known practical way in which to elucidate the fundamental character of the neutrino, that is whether or not the neutrino is its own anti-particle. Intimately connected with this question is the absolute mass scale for the neutrinos. If the mass scale is small relative to ~500 meV, $\beta\beta_{0\nu}$ may be the only means to measure this scale. With the discovery of neutrino oscillations in solar and atmospheric neutrino experiments, we now know that neutrinos have non-zero masses and that nontrivial mixing occurs in the lepton sector. This provides a unique opportunity for $\beta\beta_{0\nu}$ searches with rather specific scientific goals in mind. Unfortunately, nature has not yet been so kind to define this scale uniquely. The existing experimental data define three general scales for the effective Majorana neutrino mass (Figure 4):

1. A degenerate mass regime where the three active flavors of neutrinos have almost equal mass around ~100 to 500 meV. $\beta\beta_{0\nu}$ half-lives are expected to be of order 10^{25} years for ^{76}Ge and it is this regime where a positive signal is claimed based on a subset of the Heidelberg-Moscow collaboration (herein referred to as KDHK) using enriched ^{76}Ge[11].
2. An inverted mass hierarchy with an effective Majorana mass scale of ~20 to 50 meV as defined by the atmospheric neutrino mass splitting. Typical half-lives are expected to be of order 10^{27} years for ^{76}Ge and it is this level of sensitivity driving the next generation of experiments.
3. A regular mass hierarchy at a scale of ~1 to 5 meV. Typical half-lives are expected to be of order 10^{29} years for ^{76}Ge and beyond the reach of any envisioned experiment. To put this into perspective, an experiment exploiting enriched ^{76}Ge with ~100 tons of material and essentially zero background would be required to reach this level of sensitivity. This is likely an impractical step based upon the existing technology, indicating the need to demonstrate new detector technologies exploiting different candidate isotopes.

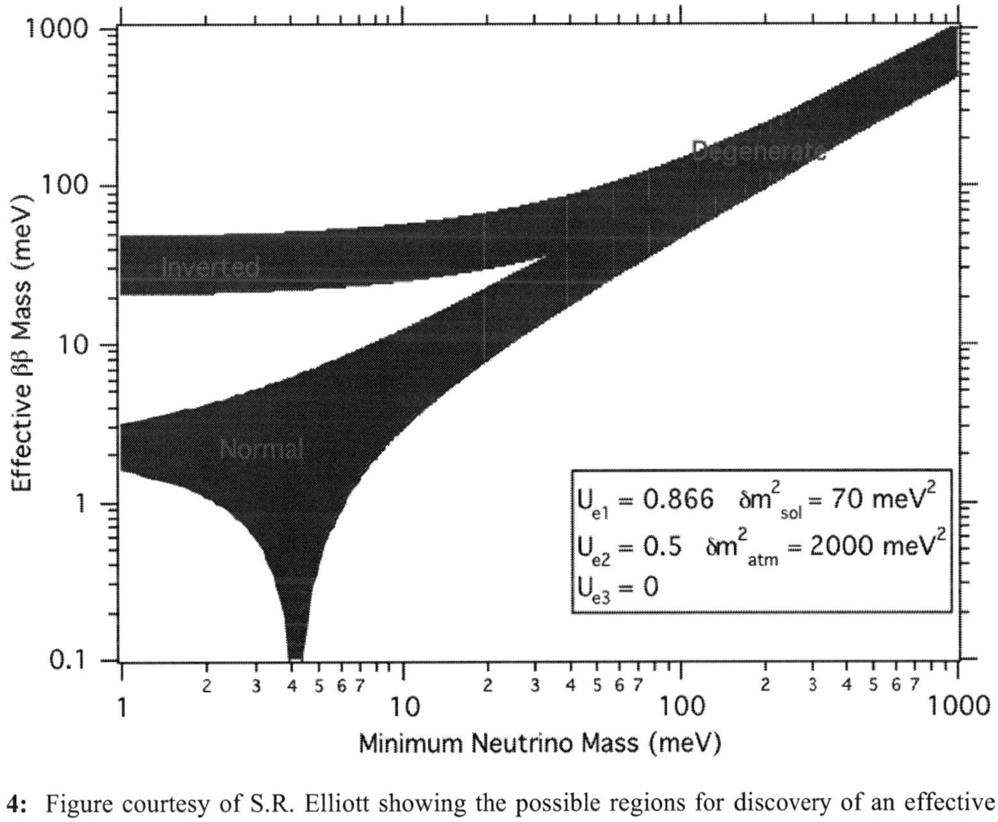

FIGURE 4: Figure courtesy of S.R. Elliott showing the possible regions for discovery of an effective Majorana neutrino mass using existing constraints on the neutrino mixing matrix deduced from solar and atmospheric neutrino experiments.

The field of $\beta\beta_{0\nu}$ is mature and a number of competing but complementary techniques are under development around the globe with the aim to achieve unprecedented sensitivity. SNOLAB has received several LOI's for dedicated $\beta\beta_{0\nu}$ experiments and it is prudent to consider a scientific road map for this important endeavor that will nurture the development of novel techniques while optimizing the scientific success of the program. An ideal program would involve an experiment (or experiments) taking a phased approach that could provide an uncontroversial test of the KDHK claim and have the capability to be scaled in target mass with sensitivity to the inverted mass hierarchy. Reaching into the regular mass hierarchy regime is likely not possible without first taking these intermediate steps.

The Majorana collaboration proposes a phased approach starting at the 120-180kg kg mass scale which would serve to test the KDHK claim and make a precision measurement in the degerate mass hierarchy regime. The experiment would be designed in a modular fashion to allow scaling to the 0.5 to 1 tonne scale of enriched ^{76}Ge and capable of precision measurements in the regime of the inverted mass hierarchy. In parallel, the GERDA collaboration is pursuing the possibility to submerge "naked" Ge crystals in a large volume of liquid nitrogen or argon to serve as external shielding and cooling while simultaneously removing background associated with conventional cryostats.

The EXO collaboration proposes a 200- liter liquid-Xe detector enriched in ^{136}Xe. Unique to the EXO approach is the possibility of tagging the daughter Ba-ions leading to a "background-free" search for $\beta\beta_{0\nu}$. If the Ba-tagging can be successfully demonstrated, this prototype experiment would test the KDHK claim and begin to have sensitivity to the inverted mass hierarchy. Scaling to the ton scale would definitively cover the phase space allowed at present for an inverted mass hierarchy and potentially lead to a viable approach to the regular mass hierarchy. Should we be so fortunate to observe $\beta\beta_{0\nu}$, the potential for tracking and vertex reconstruction in EXO would allow the study of $\beta\beta_{0\nu}$ physics such as angular correlations that would be important in elucidating the mechanism for $\beta\beta_{0\nu}$.

The COBRA experiment offers a unique opportunity to exploit a large array of Cd-Zn-Te crystals as a "solid state TPC" with the potential for excellent energy resolution, segmentation, and pulse-shape discrimination to identify $\beta\beta_{0\nu}$ events and discriminate background. The additional sensitivity to EC and β^+ modes is particularly interesting from the point of view of elucidating the nature of $\beta\beta_{0\nu}$ if observed.

Solar Neutrinos

We now know that the deficit of electron neutrinos observed in pre-SNO solar neutrino experiments is due to active flavor transformation induced by matter-enhanced neutrino oscillations. The experimental measurement of the total active flux of ^{8}B neutrinos is now some two times more precise than the predictions of the Standard Solar Model (SSM). It remains a remarkable achievement that our understanding of stellar evolution accurately predicts a flux varying so sensitively with the details of the Sun and comprising only 0.01% of the total neutrino flux emitted by the Sun. Experimental constraints on the low energy pp and ^{7}Be fluxes, which comprise some 98% of the solar neutrino flux, come only from integral measurements made in the radiochemical gallium and chlorine experiments and at present are not as precise as the quoted predictions of the SSM.

The real time detection of low-energy (pp, pep, CNO, and ^{7}Be) solar neutrinos is necessary to improve our fundamental knowledge of both neutrino properties and the basic tenets of stellar evolution. The measurement of pep and pp solar neutrinos, together with the 1.5% predicted accuracy in flux, would precisely constrain the solar mixing angle θ_{12}, accurately compare the solar luminosity measured with neutrinos to that measured in light, and accurately test the existing paradigm of solar neutrino oscillations as being dominated by matter effects at high energy *versus* vacuum transitions at low energy. The direct and precise measurement of the pp, pep, and ^{7}Be fluxes are critical to this latter exploit and perhaps the only means to directly demonstrate the effect of matter enhanced neutrino oscillations (see Figure 5).

The possible detection of pep and CNO neutrinos is unique and likely impossible without the depth of SNOLAB in a detector using organic scintillator such as that proposed for SNO$^+$. The real time detection of low-energy and dominant pp-fusion solar neutrinos represents the "holy grail" of a long and productive history of solar neutrino physics. The CLEAN experiment aims to exploit a large and ultra-pure fiducial volume of liquid neon together with novel pulse-shape discrimination capability in a dual-purpose detector of low-energy solar neutrinos and WIMP dark matter. The CLEAN detector would be served well by SNOLAB due to the requirement of at least 5000 m.w.e. over-burden to mitigate cosmogenic activation of the neon and to eliminate background associated with cosmic ray induced fast neutrons.

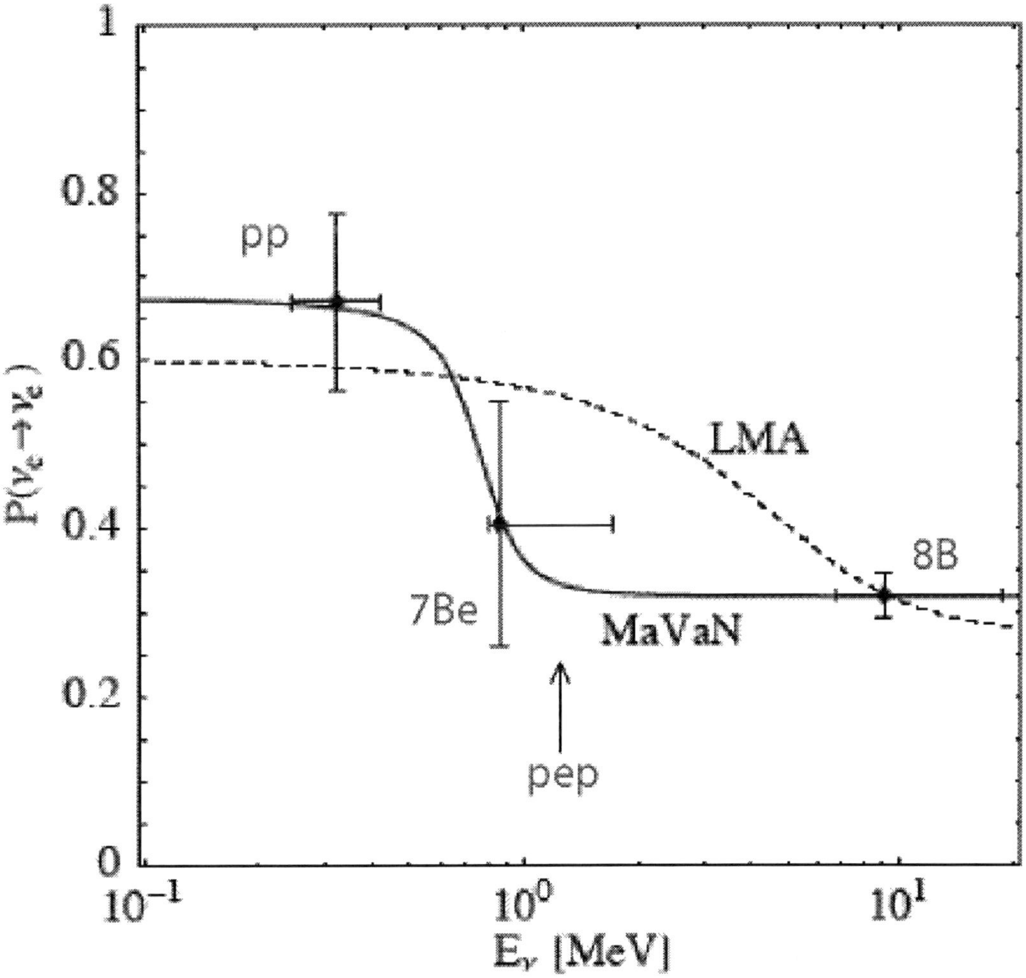

FIGURE 5: Figure adapted from ref.[12] showing the energy dependence of the solar neutrino survival probability for the now standard Large-Mixing-Angle solution (LMA) compared with the possibility of Mass-Varying-Neutrinos (MaVaN). Precision in real time measurements of the solar neutrino spectrum including the low-energy pp, 7Be, and pep components will test the basic tenants of stellar evolution, the standard paradigm of LMA neutrino oscillations, and potentially search for exotic phenomena beyond the standard model such as sterile neutrinos, new FCNC interactions, and MaVaN.

SUMMARY

The challenges ahead in the fields of low-energy solar neutrinos, neutrinoless double beta decay, and cosmological dark matter are daunting. Detectors are under development with the aim of making leaps in sensitivity by several orders of magnitude beyond the present state of the art. Meeting this challenge will require unprecedented improvements in the radiopurity of construction materials and detectors operating under extremely clean conditions deep underground. With SNOLAB under construction and this first workshop in Sudbury on Low Radioactivity Techniques, it is perhaps fitting to end with the motto “Get Clean and Go Deep”.

ACKNOWLEDGMENT

Many thanks must go the organizers and participants of this very informative and productive workshop that is likely the first in a fine series. This participant is grateful for the support of the US Department of Energy and for Los Alamos Directed Research & Development funds.

REFERENCES

1. www.snolab.ca
2. Ontario Geological Survey Special Volume 1, 1984, ISBM 0-7743-9423-4.
3. Ontario Geological Survey Special Volume 5, 1992, ISBM 0-7778-2105-2.
4. SNO Collaboration, NIM A449, 172 (2000).
5. H.H. Chen, PRL 55, 1534 (1985).
6. SNO Collaboration, PRL 87, 071301 (2001); PRL 89, 011301 (2002); PRL 89, 011302 (2002).
7. SNO Collaboration, PRL 92, 181301 (2004); nucl-ex/0502021.
8. CDMS Collaboration, PRL 93, 211301 (2004).
9. D-M. Mei and A. Hime, work in progress.
10. M.G. Boulay and A. Hime, astro-ph/0411358; M.G. Boulay, A. Hime, and J. Lidgard, nucl-ex/0410025.
11. H.V. Klapdor-Kleingrothaus *et al*, Phys. Lett. B586, 198 (2004).
12. V. Barger, P. Huber, and D. Marfatia, hep-ph/0502196.

II – LOW BACKGROUND COUNTING TECHNIQUES

Low level counting from meteorites to neutrinos

Gerd Heusser

Max-Planck-Institute for Nuclear Physics, POB 103980, D-69029 Heidelberg Germany

Abstract. The development in low level counting at Heidelberg with NaI(Tl) crystals, proportional counters and Germanium detectors is reviewed throughout the course of almost 40 years of experience. Research subjects changed from cosmogenic radionuclides in meteorites to solar neutrinos and double beta decay. Driven by screening measurements for these rare event experiments, the sensitivity in single gamma counting has gained almost 3 orders of magnitude. With Ge spectrometry the μBq/kg range is now accessible. It is discussed how further improvements can be realized. There is potential to reach a sensitivity at the level of 10 to 100 nBq/kg for cryogenic liquid type Ge-spectroscopy, a technique which the next generation ^{76}Ge double beta decay experiment GERDA is based on.

Keywords: Low Level counting techniques, gamma spectroscopy, NaI(Tl) coincidence spectrometry, Ge spectrometry, proportional counting
PACS: 06.30.-k, 07.85.Nc, 23.40.-s, 26.65.+t, 29.30.Kv, 29.40.Cs

INTRODUCTION

The development in low level counting during the last 3 to 4 decades is described from a more personal viewpoint. It is focused on the development of detectors and methods in which the author was engaged, partly as contributions within larger collaborations, such as the solar neutrino experiments GALLEX/GNO [1, 2], BOREXINO [3] and the double beta decay experiments HEIDELBERG-MOSCOW [4] and GERDA. [5]. Consequently, the results presented - hopefully to be excused – contain a relatively large fraction of my own citations. The main emphasis is put on developments which have not been fully published yet, however not exclusively, in order to transmit the intended message and for the sake of readability. More detailed information can be found in the two review articles [6, 7]. The reader is begged to forgive the picture-book like retrospective, which was chosen to condense the information, and enhance the sometimes poor quality of the pictures. Multi-coincidence NaI(Tl) arrays, proportional counters, Ge detectors and rare gas mass spectrometry are the instruments used in the reviewed low radioactivity background techniques.

Similar to other colleagues, the initial interest in low level counting was connected with investigations of cosmogenic isotopes in meteorites. They allow revealing the exposure geometry and the exposure history of the meteorite from recovered fragments. The state of the art 30 to 40 years ago for gamma counting was based on NaI(Tl) detectors, and since the isotopes of main interest such as ^{26}Al, ^{22}Na and ^{60}Co decay by positron emission or cascading gammas, coincidence spectrometers have been used. In the coincidence mode the background is much more strongly suppressed than the counting efficiency for these isotopes, so that they can be more easily detected and separated from isotopes decaying by single gamma emission.

NaI(Tl) – SYSTEMS AND PROPORTIONAL COUNTERS

Fig. 1A depicts a coincidence spectrometer built around 1970. Its special feature is the inner mercury shield contained in a double wall cylinder made from Plexiglas®. Mercury is known to be highly radiopure since it can be well purified by distillation, and the cosmogenic production rate of the only long living isotope ^{194}Hg (T1/2 = 520 y)

CP 785, *Topical Workshop on Low Radioactivity Techniques,* edited by B. Cleveland, R. Ford, and M. Chen

is rather small. However in shallow depth laboratories background contributions from neutron induced ^{200*}Hg has to be considered. Materials screening of the construction materials was at that time limited to a small NaI(Tl) spectrometer and a shielded beta detector. A light guide of Na(pure) separates the photomultiplier tube (PMT) from the active crystal to shield the crystal against the 1.46 MeV γ of ^{40}K in the glass of the PMT. The selection on low K PMT's was rather limited at that time. MgO of less than 5 ppm K is used as a reflecting material and the stainless steel envelope of the crystal was selected to be low in ^{60}Co. In Fig. 1B the background spectra of a single NaI(Tl) under different shielding conditions is shown. Despite the light guide the background is dominated by ^{40}K of the PMT glass envelope. The addition of the mercury reduces only the lower energy region, presumably due to some ^{226}Ra contamination in the proportional counter array. If the veto is active the annihilation peak is lowered and the continuum is reduced. The total count rate is similar as given in Table 1 for the same detector but installed within another shield (Fig. 2). Peak counting efficiencies of a few % could be reached for the coincidence mode operation, e.g. 1.45 % for ^{60}Co (1173 keV x 1332 keV) in case of a typical 100 g sample at a background rate of 30 counts per day. This translates into a sensitivity of some mBq/kg. Today such values are easily reachable with a medium size low level Ge-spectrometer in the single mode detection.

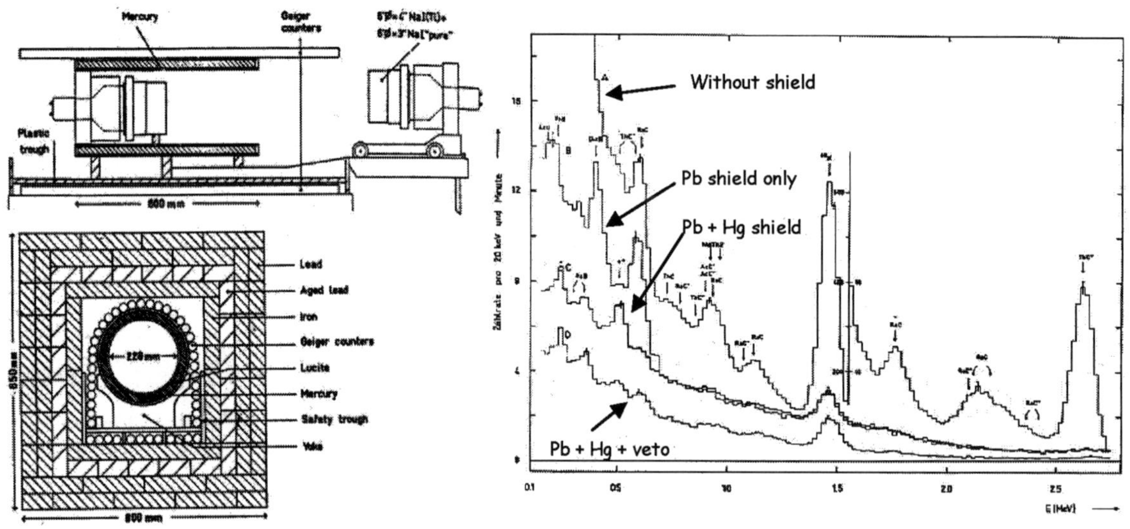

FIGURE 1A. (LEFT): NaI(Tl) coincidence spectrometer with an inner mercury shield and a proportional counter veto
FIGURE 1B. (RIGHT): Background spectra of a single NaI(Tl) detector under different conditions

The necessity to discriminate low ^{22}Na concentrations against much higher ^{60}Co and ^{26}Al activities required to construct a triple coincidence spectrometer (Fig. 2 A). Its central unit is a large pair well type detector facing to one of the old 6"Øx4" NaI(Tl) detectors. Plastic scintillators encase completely the detector arrangement as a cosmic veto system. The outer shield is similar to the other spectrometer except that the supporting structure for the roof layers is formed by the steel frame of the plastic scintillator plates instead of old marine iron. Material screening for low radioactive concentrations of the construction materials was much more extended, especially for the crystal gladding materials (reflector and metal casing) and the phototubes. OFHC copper was chosen as casing material and analytical grade MgO as reflector. A Low level Ge(Li) spectrometer was available for this purpose [8]. In Table 1 the total background count rate for various coincidence modes is given for the unvetoed and the vetoed case. For both cases the triple coincidence results in a reduction factor close to 140 compared to the two fold coincidence mode. As compared to the former spectrometer the background for the characteristic peaks of ^{26}Al and ^{22}Na (in triple coincidence mode) could be reduced by a factor 3.3 at even slightly higher efficiencies. Interference of ^{60}C0 with the ^{22}Na peak (1.27x0.51x0.51 MeV) is lowered by 17.5 times. ^{40}K from the PMT glass still dominates the single background spectrum, although at a lower level. More details of the triple coincidence spectrometer and its performance are given in [9]. The coincidence modes are much more sensitive to neutrons than the single mode. This was discovered by accident when a 50 MBq ^{241}Am source was used in the same laboratory at variable distance to the spectrometer. The background variation in the triple coincidence was about twice higher than on the two fold one. Each alpha source is a potential neutron source due to (α,n) reactions on light element impurities in the source.

The same spectrometer without the smaller NaI(Tl) detector is shown in Fig. 2A with miniature gas proportional counters in the well. This is the configuration for which it originally was designed in order to serve as a beta-gamma

TABLE 1. Background count rate of the NaI(Tl) triple coincidence spectrometer.

detector/coincidence condition	total count rate > 100 keV [cpm]	
	veto off	veto on
half of 10" Ø x 9"	537	167
6" Ø x 4"	201	67
2 fold coincidence 1a + 1b	214	22
3 fold coincidence 1a + 1b + 2	1.6	0.16

coincidence spectrometer in the GALLEX project [1]. In the coincidence mode ^{68}Ge and ^{69}Ge can be distinguished from ^{71}Ge since they decay, in addition to electron capture, partly also by β^+ and γ-emission (decay of the daughter ^{68}Ga in case of ^{68}Ge), whereas ^{71}Ge decays purely by electron capture. ^{68}Ge and ^{69}Ge are indicators for side reactions to the solar neutrino capture reaction ^{71}Ga (ν_e, e-) ^{71}Ge. The spectrometer was also used to calibrate the counting efficiency of the proportional counter for ^{71}Ge via the beta-gamma-gamma coincidence method applied to ^{69}Ge [10]. The signature of ^{69}Ge-electron capture inside the counter is identical to that of a ^{71}Ge decay. First, the absolute activity A of ^{69}Ge (admixed to the proportional counter gas in form of GeH_4) is determined by the count rates c_x of the different single or coincidence modes based on:

$$A = \frac{c_{\beta} \bullet c_{\gamma\gamma}}{c_{\beta\gamma\gamma}} = \frac{\varepsilon_{\beta} \bullet \varepsilon_{(a)511} \bullet \varepsilon_{(b)511} \bullet A^2}{\varepsilon_{\beta} \bullet \varepsilon_{(a)511} \bullet \varepsilon_{(b)511} \bullet A}$$

Next the individual efficiencies ε_x (a and b stands for the two half's of the split NaI(Tl)), can be deduced from that number and from ε_{x511} measured with another ß$^+$ source. The ^{69}Ge method was finally chosen as the main calibration procedure for GALLEX [11] and GNO [12].

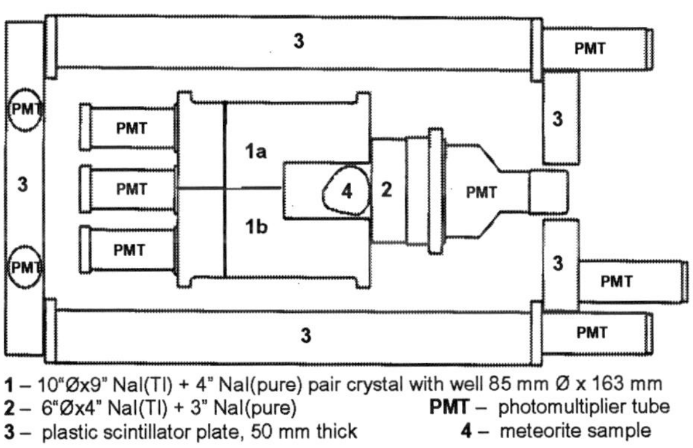

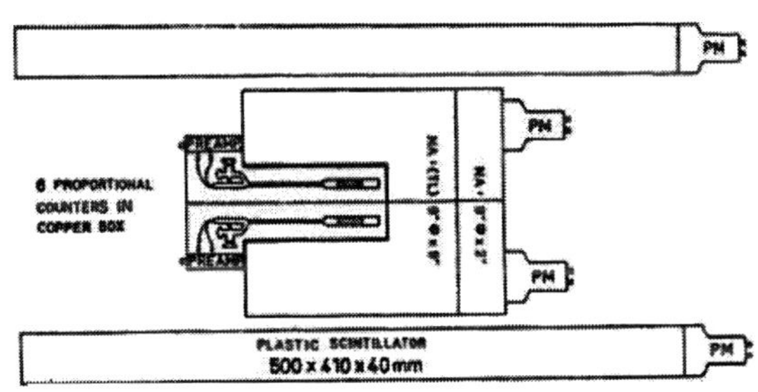

FIGURE 2A. (LEFT): cross section of the triple coincidence NaI(Tl) spectrometer with plastic scintillator veto shield.
FIGURE 2B. (RIGHT): same spectrometer with miniature proportional counter as beta gamma coincidence spectrometer.

A second very similar[a] NaI(Tl) pair well type detector was constructed for the GALLEX spectrometer at the underground laboratory LNGS [13]. In Fig. 3 it is shown inside the GALLEX shielding tank embedded in Cu plates as an inner shield. One box with the proportional counter is shown in the well of the NaI(Tl). An exploded view of the counter and how it is placed in the box is shown on the left side. The tube on top is slid over the two lead beds. On the opposite side of the NaI(Tl) is a passive shield made of the same pure Cu (K/Th/U < 2/1/1 mBq/kg) as that around the NaI(Tl). Due to the higher radio-purity and thus lower background, most GALLEX/GNO runs were counted in the positions of the passive side.

Only recently a miniature x-ray source (picture and lower drawing on the left side in Fig. 3) was mounted between the passive and the active side for counter calibration [14]. The generator and the x-ray tube combination

[a] Ø 254 x 229 mm + Ø 254 x 102 mm NaI "pure", well Ø 85 x 163 mm

(in a brass shielding cylinder) are seen in the middle of the picture and as the horizontal cylinder in the lower drawing. The beam exiting from the rim of the cylinder in the axial direction can be positioned by two stepping motors (smaller cylinders in lower drawing) to each counter position on the passive side (symbolized by the holes in the right plate of the drawing). A cerium target, excited under 45° by the x-rays of the generator, is the beam source. The induced spectrum in the proportional counter gas (mixture of GeH_4 and Xe) features three peaks at 1.03, 5.09 and 9.75 keV. They well match the energy region of the two ^{71}Ge peaks at 1.2 and 10.4 keV.

All parts of the motion system, mainly made from brass, had been carefully checked for radiopurity by Ge-spectrometry (see below). Before the installation of the x-ray source, each counter had to be calibrated in front of the Cu shield with a ^{153}Gd source exciting a cerium target. For that purpose the shield had to be opened to remove the counter from its shielded position and to insert the Gd-source through an air lock system, mounted on the folding plastic sleeve (Fig. 3). For Rn suppression the inner tank was pressurized with nitrogen gas. For more details of the GALLEX spectrometer see [13]. Background performance of the proportional counters can be found in [12, 13, 15].

These Davis type proportional counters are also used for highly sensitive Rn detection in BOREXINO [16-18]

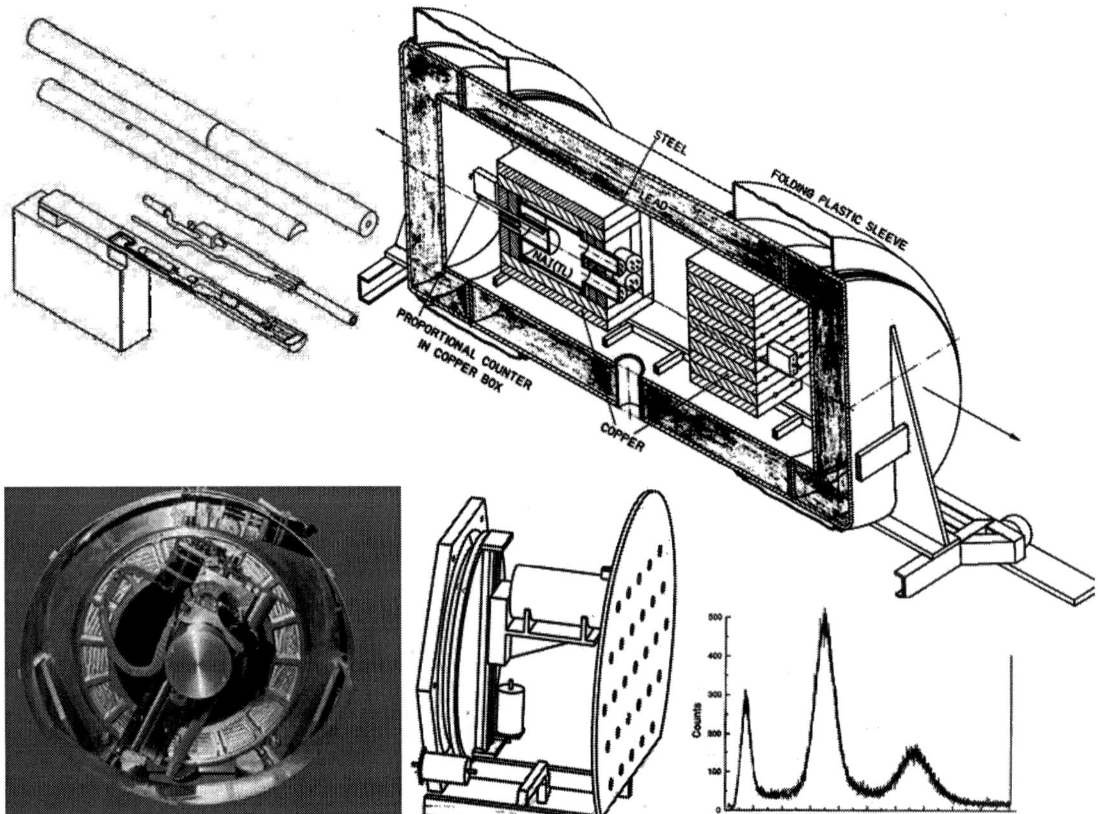

FIGURE 3. GALLEX shielding tank with an active side (NaI(Tl) and a passive side (Cu plates), on the upper left – an exploded view of a proportional counter in its box, lower row - movable x-ray calibration source and a calibration spectrum of a proportional counter measured with this source.

Ge - SPECTROMETRY

General background studies and the growing demand on measurements for the low activity projects [1-4] resulted in several generations of low level Ge-spectrometers, of which 3 are shown in Fig. 4. The two on the left side are installed at the shallow underground laboratory of the MPI Heidelberg. The overburden of about 15 m water equivalent (m w. e.) reduces strongly the cosmic ray neutron flux, the main background source for Ge-detectors at

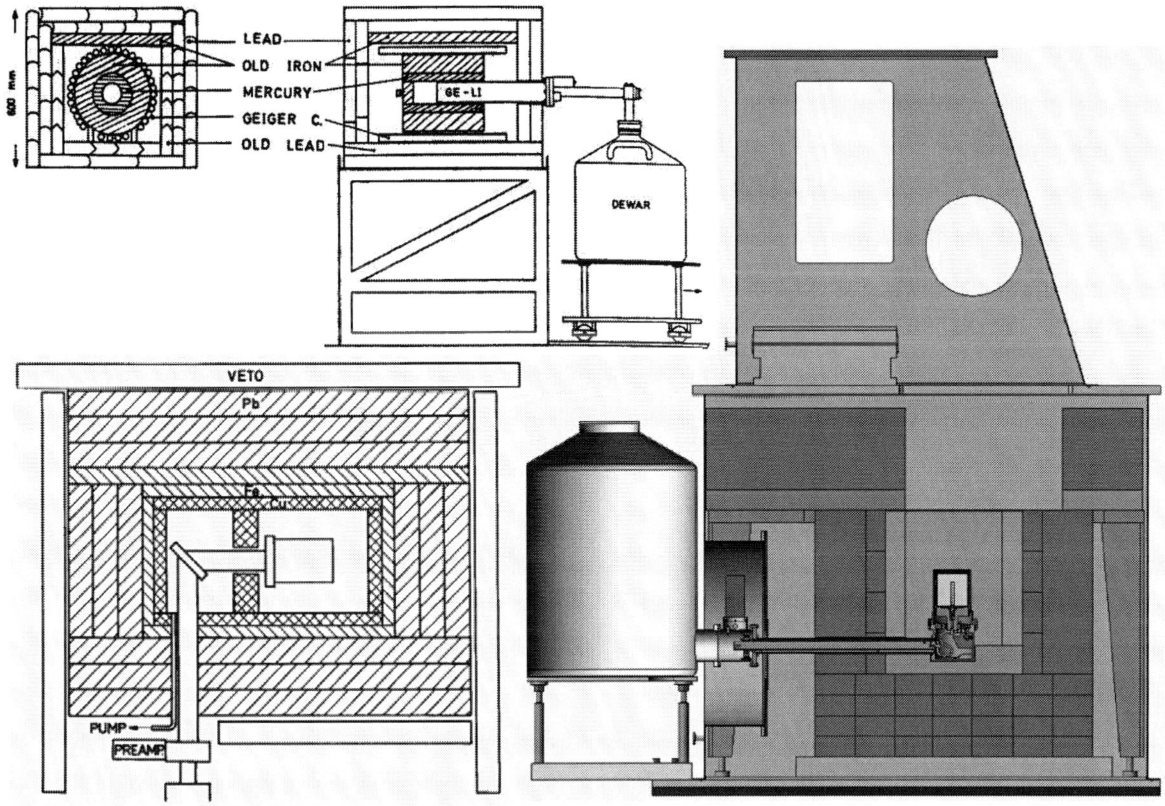

FIGURE 4. Three generations of low level Ge-spectrometer with the latest, "GeMPI" at the right side (see text).

sea level. The muon flux is reduced only by about a factor of 3. Since the suppression factor of a cosmic veto device is much higher for muon induced background events than for neutron induced events its effect is more pronounced at the reduced neutron flux of the low level laboratory. In addition it also suppresses events caused by neutrons from muon interactions with the lead shield, especially if placed outside of the lead shield (Fig. 4, lower spectrometer on the left) [19]. The suppression factor of the veto detector is limited by the achieved radiopurity level in the detector system, but finally by in-situ cosmic ray induced activities with half-lives exceeding the tolerable veto blocking time. In case of Ge detectors these are mainly the neutron induced isomeric states of ^{75m}Ge, ^{77m}Ge and ^{71m}Ge. Improvements in background have been achieved in iterative steps by selecting the construction materials of the next spectrometer with the experience of the former one.

After several hundreds of measurements one can state that there are only a very few materials which are reliably (batch independent) free of contaminations at the mBq/kg level. As will be discussed below, copper seems to be one of the exceptions.

Experience collected with the earlier Ge-spectrometer systems and with the enriched detectors of the HEIDELBEG MOSCOW (HDM) experiment [4] resulted in the latest iterative step - the spectrometer called GeMPI [20, 21]. GeMPI is operated at the LNGS underground laboratory at 10^6 times reduced muon flux. The spectrometer is shown in Fig. 4 on the right side. Most parts of the cryostat system have been fabricated at the MPI from NOSV grade copper[1], which was stored underground (minimum 15 m.w.e.) shortly after electrolysis and between machining steps. Also for the crystal, the cosmic ray exposure at sea level was minimized to a few weeks after crystal pulling. The Rn protection system consists of a vacuum tight steel case, and an air lock system to insert samples without air (Rn) intake (both with pure nitrogen atmosphere). The achieved background level for GeMPI in combination with a large sample capacity of up to 15 l around the 2.2 kg Ge-crystal allows measuring concentrations of the gamma active chain members down to 10^{-12} g/g of U/Th equivalent (12.3 μBq/kg ^{238}U, 4.06 μBq/kg ^{238}Th)

[1] copper of 99.9975 % purity, produced by Norddeutsche Affinerie, Hamburg, Germany

and 10^{-9} g/g of K (31 μBq/kg ^{40}K). Compared to the HDM detectors it was possible to reduce the ^{226}Ra contamination by at least a factor 5 and the total specific background count rate by 2, whereas ^{228}Th and ^{40}K are only slightly improved [21].

Table 2 lists results of some selected measurements with GeMPI. Copper, lead and iron are very commonly used materials to shield or construct detectors in neutrino, double beta, and dark matter experiments. Therefore information on their radiopurity at the sensitivity level of GeMPI is of general interest.

The tolerable level of contamination at locations beyond the range of beta's (bremsstrahlung) and alphas from the detector is defined by γ-activity and therefore Ge-spectroscopy is the ideal screening tool. In case of the natural decay chains interest is focused on the most gamma active progenies ^{214}Pb/^{214}Bi (^{238}U-chain) and ^{228}Ac, ^{212}Pb/^{212}Bi, ^{208}Tl (^{232}Th-chain). The high resolution spectroscopy with its spectral information gives Ge-spectroscopy an almost visualizing ability and is thus superior to other screening methods which measure only atomic concentrations as e.g. in mass spectrometry, atomic adsorption, x-ray excitation or neutron activation. Concerning the U/Th decay chains, these other techniques are only applicable to the parent nuclides, which are rarely in equilibrium with their gamma active progenies in chemically processed construction materials. Moreover, Ge-spectroscopy is able, within certain limits, to observe deviations from secular equilibrium in the U/Th series. This is important, e.g. in the case of the ^{228}Ra/^{228}Th sub-chain, where the time for noticeable changes in activity ratios can occur within the lifetime of an experiment. In addition the measurements can mostly be performed in a non-destructive way without labor intensive sample pre-treatment.

Details of the measurements and of the samples presented in Table 2 are given in [21]. The low concentration of ^{226}Ra and ^{228}Th in lead is astonishing and makes lead suitable also for closer locations to detectors. ^{40}K seems to be lower in ancient roman lead than in modern lead. Also for the modern LC2 lead (about 0.4 mBq/kg ^{210}Pb) used in the HDM experiment a ^{40}K concentration of about (0.31 ± 0.3) mBq/kg was found by neutron activation [23]. ^{210}Pb is not discussed here, but the bremstrahlung from it's daughter ^{210}Bi can be easily shielded [19]. The concentrations of ^{226}Ra and ^{228}Th in the stainless steel sample and the old ship iron are rather low compared to other steel and stainless steel samples, which can reach several mBq/kg (see e.g. [17]). The very high radiopurity of the Cu sample makes Cu a unique material for extreme low level experiments provided that the NOSV quality (1-4 ppm O_2) is representative for electrolytic copper in general. From the detection limit of 10 μBq/kg for ^{60}Co (Table 2) and the cosmogenic production rate of 2.1 mBq/kg at sea level [21] it can be inferred that the copper was exposed for less than about 2 weeks after electrolyses. In the lower part of Table 2 the primordial concentration in the copper of the individual cryostats of the HDM detectors, as obtained by Monte Carlo simulations, is compared with the measured one of the same NOSV Cu quality. This Monte Carlo method is exploiting the different degrees of attenuation for γ-ray emissions of different energies from different materials and different locations in the experimental set-up. The result of this analysis is that the most probable location of the primordial activities is the copper of the detector cryostats [22]. The comparison with the bulk measurement of the same copper quality (Table 2) results (with one exception – ^{228}Th in ANG 2) much higher concentrations. The most plausible explanation of this difference and of the large spread among the HDM cryostats is that the activity is surface correlated and not homogeneously distributed in the Cu as assumed in the Monte Carlo method.

TABLE 2 Radionuclide concentrations in typical shielding materials measured with GeMPI and evaluated in HDM detectors

Sample	Weight	Time	Specific Activity [μBQ/KG]			
	[kg]	[d]	^{226}Ra (U)	^{228}Th (Th)	^{40}K	Various
lead (DowRun)	144.6	101.7	< 29	< 22	440 ± 140	98 ± 24 ^{207}Bi; 180 ± 20 ^{60}Co; (2.7 ± 0.4)x10^{7} ^{210}Pb
lead (Boliden)	144.3	75.0	< 46	< 31	460 ± 170	< 13 ^{207}Bi; < 11 ^{60}Co; (2.3 ± 0.4)x10^{7} ^{210}Pb
lead (Roman)[2]	22.1	37.2	< 45	< 72	< 270	< 19 ^{207}Bi; < 25 ^{60}Co; < 1.3 10^{6} ^{210}Pb
stainless steel[3]	38.1	80.1	600 ± 200	200 ± 100	1800 ± 600	18000 ± 1000 ^{60}Co
old ship iron	47.29	42.4	150 ± 40	460 ± 140	1000 ± 400	230 ± 40 ^{54}Mn; < 18 ^{60}Co; < 30 ^{137}Cs; < 20 ^{58}Co
copper NOSV	125.0	100.7	< 16	< 19	< 88	< 10 ^{60}Co
Activity localized in the NOSV Cu of the cryostat of the individual HDM detectors according to Monte Carlo simulations [22]						
ANG 1			168 ± 8	84 ± 7	236 ± 61	
ANG 2			91 ± 4	10 ± 3	78 ± 22	
ANG 3			105 ± 5	84 ± 5	927 ± 46	
ANG 4			115 ± 3	87 ± 4	199 ± 4	
ANG 5			100 ± 4	26 ± 4	1632 ± 49	

[2] a sample of this lead has been measured for the experiment CUORE [16]

[3] roll of foil (0.125 mm thick)

Future developments

From the above finding it is obvious that a strong reduction of surface and mass of the several layers of cladding material around the crystal, as in the HDM detectors and in GeMPI, may open the possibility for further background reduction. One option is to immerse the almost bare crystal in liquid nitrogen [7]. The highly radio-pure cryogenic liquid serves as the innermost shield against external radiation and as coolant. Very little material, up to four orders of magnitude less in mass compared to the HDM detectors, is needed to hold and contact the crystals. This idea is now being realized in the ^{76}Ge double beta decay experiment GERDA [5], a continuation of a former proposal [25, 26]. GERDA aims for, together with other suppression methods, a background reduction of two to three orders of magnitude in comparison to HDM, presently the most sensitive double-beta decay experiment.

The high statistics (about 70 kg y) spectrum of the HDM ßß experiment is shown in Fig. 5 (lower spectrum) together with the unshielded spectrum of one of the enriched detectors measured at LNGS [27]. Despite the five orders of magnitute different background level, the line composition is very similar in both spectra. Both are dominated by lines of the primordial radionuclide chains U/Th and of K; only ^{137}Cs (from the Chernobyl accident) and ^{60}Co from cosmogenic activation of the Cu in the cryostats are more prominent in the HDM spectrum. Consequently a further reduction of primordial concentrations is rather realistic by lowering the mass and the surface of the cladding material around the Ge-crystals. Even with material savings of factors of 10^3-10^4, radio-purity in the μBq/kg range will be required to reach the envisioned low background levels. Since material screening at that level with GeMPI is very time consuming, two additional spectrometers of this quality are being built [28]. A method to control surface radiopurity down to 0.5μBq/m^2 is ^{222}Rn assay of emanation. Details of this method are given in [16, 29]. There are also % contributions from cosmic activation of the Ge crystals to the continuum of the HDM spectrum. The reduction of this activation will be a further challenge for GERDA.

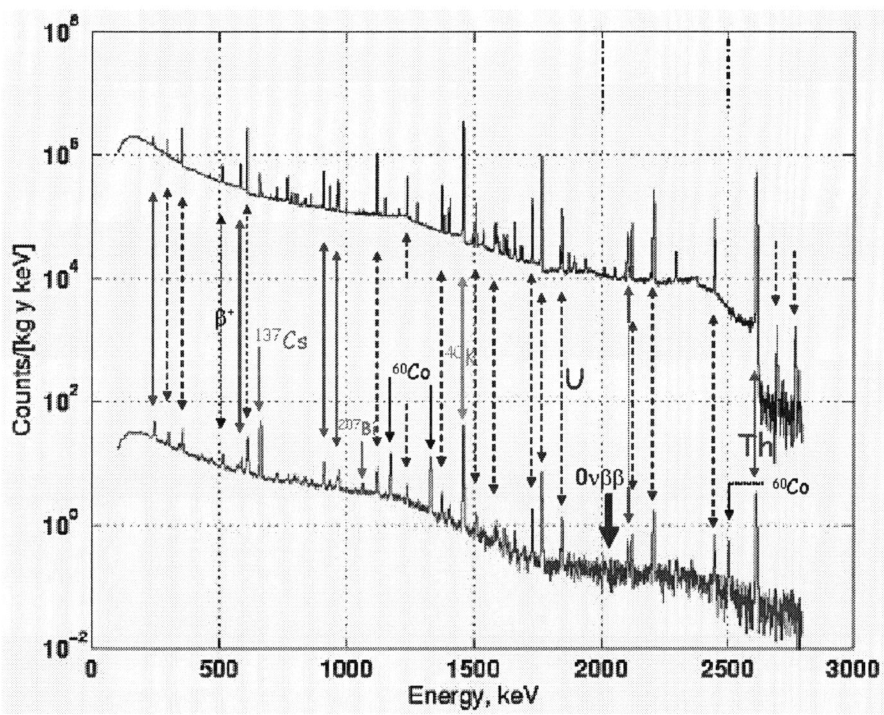

FIGURE 5. Spectra comparison: upper spectrum – ANG 2 unshielded at LNGS;
lower spectrum – full data set of all HDM detectors

Information on the radiopurity of (liquid) nitrogen is available from investigations for BOREXINO [17]. Table 3 summarizes concentrations of rare gas radionuclides in various nitrogen qualities measured recently at Heidelberg. The activity of ^{39}Ar and ^{85}Kr is obtained via the measurement of Ar and Kr concentrations in nitrogen by rare gas mass spectrometry, which is then converted by the known specific activities in the atmosphere (1 ppm Ar = 1.19 μBq/kg; 1 ppt Kr = 0.86 μBq/kg). For details see [16]. Radon is counted in miniature proportional Davis type counters (see above) after being concentrated from large N_2 volumes on charcoal at -196° C [16, 18]. Higher qualities of nitrogen contain uncritical low concentrations of ^{39}Ar and ^{85}Kr (three lower lines in Table 3) and ^{222}Rn can be further purified, if required, by charcoal column adsorption [18] as demonstrated in the reduction between the first and second line of Table 3. RPN2 stands for regular purity nitrogen and HPN2 for nitrogen after passing through the charcoal column. Information on rare gas purification in this context can be found in [30, 31]. In all these rare gas measurements the control of the procedure blank is decisive for the detection sensitivity. This requires keeping atmospheric contaminations insignificantly low and in case of Rn, additionally reducing emanation from ^{226}Ra impurities. This was best achieved by assaying Ar, Kr and Rn from larger liquid nitrogen reservoirs. Likewise the success of BOREXINO and GERDA will, among other challenges, depend on the ability to minimize atmospheric contamination and surface emanation.

If once cryogenic liquid type Ge-spectroscopy is established, a further step in screening sensitivity would be reachable, may be even down to the nBq/kg range. Limitations will arise from the increased measuring periods needed to arrive at useful counting statistics from such low activity concentrations, if not compensated by larger sample masses and multi-Ge detector arrays.

TABLE 3. Specific activity in nitrogen [μBq/kg]

Nitrogen sample	^{39}***Ar*** [a)]	^{85}***Kr*** [a)]	^{222}***Rn*** [b)]
RPN_2 Borexino	12	34	40
HPN_2 Borexino	12	26	0.4
HPN_2 Borexino liq.extr.			< 0.3
Linde Worms (7.0)	0.017	0.07	1
SOL Mantua (6.0 -7.0)	0.006	0.04	
Westfalen Hörstel (6.0)	0.0006	0.06	
Air	***~1.x10^4***	***~1x10^6***	***~1x10^7***

ACKNOWLEDGMENTS

All colleagues of the collaborations GALLEX/GNO, BOREXINO, HEIDELBERG –MOSCO and GERDA are thanked for the help and support along these developments. The author is especially indebted to W. Hampel, T. Kirsten, M. Laubenstein, H. Neder, W. Rau, S. Schoenert, H. Simgen, M. Wojcik, and G. Zuzel. The skillful work of the technical staff of the MPI workshops is greatly acknowledged.

REFERENCES

1. GALLEX Collaboration, Hampel, W., et al., Phys. Lett. B447, 127-133 (1999).
2. GNO Collaboration, Altmann, M., et al., Phys. Lett. B490, 16 (2000).
3. BOREXINO Collaboration, Alimonti, G., et al., Astrop. Phys. 16, 205 (2001).
4. HEIDELBERG-Moscow Collaboration, Guenther, M., et al., Phys. Rev. D 553, 54 (1997).
5. GERDA Collaboration, Abt, I., et al., arXiv: hep-ex/0404039 (2004).
6. Heusser, G., Proc. 3rd Intern. Summer School, Low-Level Measurements of Radioactivity in the Environment, edited by M. Garcia-Leon, R. and R. Garcia-Tenorio, Huelva, 1993, World Scientific, Singapore, 1994, pp. 69-112.
7. Heusser, G., Low-radioactivity background techniques, Annu. Rev. Nucl. Part. Sci., 543-590 (1995).
8. Heusser, G., et al., Proc. 2nd Int. Conf. on low-level counting, High Tatras (1980), Phys. Appl. 8 (1982) 127.
9. Heusser, G., Nucl. Instrum. Methods Phys. Res. B17, pp. 423-426 (1986).
10. Heusser, G., Hübner, M., Schlotz, R., American German Gallium Collaboration Internal Report 1980.
11. GALLEX Collaboration, Hampel, W., et al., Phys. Lett. B 338, 384-396 (1996)
12. GNO collaboration, Altmann, M., et al., to appear in Physics Letters B (2005), arXiv: hep-ex/0504037.
13. Heusser, G., Trends in Astroparticle Physics, ed. P. Ch. Bosetti, Teubner, Leibzig, 1994, pp. 33-49.
14. Heusser, G. and Lackner, R., to be published.
15. Wink, R., et al., Nucl. Instr. Meth. A329, 541-550 (1993).
16. Zuzel, G., this volume.

17. BOREXINO Collaboration, Arpesella, C., et al., Astrop. Phys. 18, 1-25 (2002).
18. Heusser, G., et al., Appl. Rad. Isot. 52, 691-695 (2000).
19. Heusser, G., Nucl. Instrum. Meth. Pys. Res. B17, 418-422 (1986); B58, 79-84 (1991) ; B83, 223-228 (1993).
20. Neder, H., et al. Appl. Rad. Isot. 53, pp. 191-195 (2000).
21. Heusser, G. and Laubenstein, M., to appear in Radionuclides in the Aquatic Environment, ed. P.P. Povinec, Elsevier Amsterdam, 2005
22. Dörr, C., Klapdor-Kleingrothaus, H. V., Nucl. Instr. Meth. A513, 596-621(2003)
23. Penicka, E., private communication
24. Ardito, R., et al., CUORE: A Cryogenic Underground Observatory for Rare Events, arXiv: hep-ex/0501010 (2005)
25. Klapdor-Kleingrothaus, H. V., et al., MPI-Report MPI-H-V 26-(1999)
26. Baudis, L., et al., Nucl. Instr. Meth. A426, pp. 425-435 (1999)
27. Chkworets, O., private communication
28. Rugel, G., et al., Nuclear Physics B (Proc. Suppl.) 143, 564 (2005).
29. Rau, W., and Heusser, G., Appl. Rad. Isot. 53, 371-375 (2000)
30. Simgen, H. et al., Appl. Rad. Isot. 61, 213-217 (2004)
31. Simgen, H., this volume

Low-level ^{14}C measurements and Accelerator Mass Spectrometry

A. E. Litherland[a], H. E. Gove[a,b], R. P. Beukens[a], X-L. Zhao[a] and W. E. Kieser[a]

a IsoTrace Laboratory, University of Toronto, Toronto, ON, Canada M5S 1A7
b Department of Physics and Astronomy, University of Rochester, Rochester, NY, 14627-0171, USA

Abstract

Accelerator Mass Spectrometry (AMS) and isotope enrichment were used in 1991 to estimate that the ^{14}C content of methane in natural gas was $\leq 1.6\times10^{-18}$ of the total carbon. The low content of ^{14}C in underground hydrocarbons was verified later in the remarkable results from the Borexino test scintillation counter for solar neutrino studies. Since then studies of the ^{14}C background problem have demonstrated that much of the background originally observed in the AMS measurements can, in principle, be eliminated. However, many difficulties and other backgrounds are to be faced as the limit for AMS is pushed still further towards possibly a ratio of $< 10^{-21}$. These will be discussed.

PACS: 07.75.+h
Keywords: Accelerator mass spectrometry; Radiocarbon; neutrino detectors

1. Introduction

It is now 55 years since Libby et al [1] reported the first detection of cosmic ray generated radiocarbon, or ^{14}C, by beta ray counting. At about 0.25Bq/g the radioactivity of a contemporary carbon sample was then just measurable. That activity combined with a mean-life of 8033 years gives an atom ratio for $^{14}C/C$ of about 1.2×10^{-12}. It is worth noting that prior to the first publication J. Zacharias (unpublished) had already suggested to Libby that mass spectrometry was a more suitable technique than beta ray counting for a long-lived radioisotope like ^{14}C. Arnold [2] reports that the suggestion was considered but the measurement by mass spectrometry of isotope ratios in the range 10^{-12} to 10^{-13} was then far beyond the state of the art and counting methods worked. About 50 years later the group using the Borexino test scintillation detector made the most remarkable beta ray counting measurement, at the Gran Sasso underground laboratory in Italy, using some $4.8m^3$ of high purity liquid scintillation counter [3]. It was reported that a ratio of $^{14}C/C$ equal to $(1.94 \pm 0.09)\times10^{-18}$ had been measured, a far lower ^{14}C abundance than had been measured by beta ray counting previously. However, by 1990 a similar low ratio for $^{14}C/C$ of $< 1.6\times10^{-16}$ had already been measured [4,5] by the method first suggested by Zacharias and this measurement was done as a precursor to the construction of the Borexino test detector. This was done by a variant of mass spectrometry known as accelerator mass spectrometry (AMS), which was developed as a result of the widespread use of accelerators for nuclear physics purposes [6] and their obvious mass spectrometric properties.

In this paper the further development of AMS reaching towards the more distant goal of measurements near the $^{14}C/C \sim 10^{-21}$ level will be discussed. This goal is chosen because estimates [3] of the level of radiocarbon deep underground indicate that it may

CP 785, *Topical Workshop on Low Radioactivity Techniques,* edited by B. Cleveland, R. Ford, and M. Chen

be near the lower limit in rocks with low levels of uranium and thorium. The actual level of the ^{14}C in material that has been isolated for say >190,000 years from that generated by cosmic rays in the atmosphere is, of course, unknown but could in theory have a $^{14}C/C$ ratio $< 10^{-21}$. The actual ratio will depend upon the depth of the material below the earth's surface, if the background is generated by cosmic ray muons, and the actinide content if alpha particles or neutrons generate the ^{14}C. Finally contamination by carbon from the biosphere would then be significant at the ppb level.

Three topics will be discussed or summarised briefly in this paper. In section 4 the background counting rate, introduced in the present AMS itself will be discussed. In section 5 the continuing background study of the ^{14}C introduced by contamination of ion source materials is outlined. Finally in section 6 the problem of the introduction of ^{14}C in the source material during sample handling is mentioned.

2. Accelerator Mass Spectrometry (AMS)

The state of the art of the mass spectrometry of ^{14}C has advanced in parallel with that of beta ray counting since 1950. Beta ray counting has reached an accuracy, for recent samples, approaching ±0.2% but with an age limit (with great difficulty) of near 57,000 years or 1.2×10^{-15} for the $^{14}C/C$ ratio. Two noteworthy developments in the mass spectrometry of radiocarbon were first, the realisation that the negative ion of nitrogen could not be detected [7] during acceleration of negative ions in the first tandem accelerator at Chalk River. It had been confidently predicted theoretically that N^-, along with the molecular nitrogen negative ion N_2^-, would be unstable and an effort was soon launched [8] to utilise this apparent fact. Unfortunately the negative ion chosen for the effort in low energy mass spectrometry was $^{14}C^{15}N^-$, which is relatively easy to make. The method could not be made to work because of the presence of persistent backgrounds. It was later discovered [9,10] that the ion N_2^- has a metastable excited state, which along with other molecular ions interfered with the detection of $^{14}C^{15}N^-$. The second essential step was the development of the high-current solid-carbon caesium sputter ion source [11,12]. At Chalk River in 1960 it was noted that, although the N^- ion might be unstable, ion sources gave good C^- ion beams even when no CO_2 was fed into a radio-frequency ion source. This was clearly due to a huge memory of the use of previous gases in the ion source, which was a situation that could make rare isotope measurements with a tandem accelerator, using such ion sources, very difficult indeed [13]. In 1972 the development of the caesium sputter ion source technology, for bombarding solid carbon samples to produce many μA of C^- ions, turned out to be the key to making radiocarbon measurements utilising existing tandems. Measurements of ^{14}C soon followed [14], by what came to known as AMS, after it had been noted that the accompanying copious molecular isobars were also destroyed during tandem acceleration. A general description of subsequent events in AMS is given in a comprehensive book [6] on the subject.

The essentials of the AMS of radiocarbon and other long-lived radioactive isotopes have been described many times [6,15] but they can be summarised briefly as a method in three essential steps. The first is the complete removal of the ^{14}N isobar atoms by the use of the negative ions of the carbon atom. The second is that the intense molecular isobars, such as ^{13}CH and $^{12}CH_2$ (both form negative ions readily), are destroyed during the charge changing involved in tandem acceleration. Finally the unique ^{14}C positive ions are detected at ion energies of several MeV, which besides providing the possibility of ion identification in an ionisation spectrometer also ensures a very low dark current compared with low energy mass spectrometry. Dark currents of less than 1 count per day are possible. In AMS there must also be adequate mass and charge spectrometry, which will be discussed in section 4.

The very first measurements by AMS of the ^{14}C in reactor grade graphite [14,15] demonstrated a surprisingly low value of the ratio for $^{14}C/C$ of near $\sim 10^{-16}$. This low level is still not understood well and it is undoubtedly a mixture of graphite contamination, ion source memory of previously analysed samples and artefacts generated by the mass spectrometry. The study of such low levels of radiocarbon was not a priority in the early years of AMS.

3. The First Measurement of ^{14}C in Natural Gas

In 1990 R. S. Raghavan, in preparation for the construction of the Borexino solar neutrino scintillation detector, wanted to know just how much radiocarbon there might be in the source materials for making such a detector. It was hoped that the source materials would have a ratio $^{14}C/C < 10^{-18}$, because of the long temporal separation of the underground hydrocarbons from the biosphere. This would then make the detection of some of the lower energy solar neutrinos possible. However, such measurements had not previously been attempted using AMS. Here it is appropriate to introduce the tenth life of radiocarbon instead of the half-life. With a half-life of 5730 years the tenth life becomes close to 19,000 years. Consequently, if a hydrocarbon has been separated from new ^{14}C for about 114,000 years the decay would have lowered the $^{14}C/C$ ratio from 1.2×10^{-12} to about 1.2×10^{-18}. Such a ratio was considered to be a good starting point for a scintillation counter for detecting the solar pep, CNO neutrinos and those following the electron capture by ^{7}Be. The problem was that the lowest ratios for $^{14}C/C$, obtained from AMS measurements at that time, were about 5×10^{-16}. The origin of this limit was unknown at that time. It was known to partly due to the influence of residual ^{13}C and ^{12}C ions in the final counter. In addition it was suspected to be partly due to a memory effect, the contamination of the source material, contamination introduced during the preparation of the solid graphite-like carbon used in the ion source or contamination in the materials of construction of the ion source or target supports.

Raghavan then conceived the daring idea that isotope separation might be useful. Here it must be remembered that the enrichment of carbon isotopes is now carried out on an industrial scale. The enrichment is carried out at the Isotec Company [17], which produces, for example, CO gas enriched in ^{13}C with the ratio of $^{13}C/^{12}C \sim 100$. The level of ^{14}C in this material was unknown but Isotec estimated that the ratio of $^{14}C/^{12}C$ would be enhanced by a factor of near 200. The CO gas was produced from underground methane and liquefied prior to transportation from Louisiana to Ohio, where it was used in the isotope separation by liquid distillation. As one of the reactions creating ^{14}C by secondary cosmic ray neutrons is the $^{17}O(n,\alpha)^{14}C$ reaction, some ^{14}C could be created during the time spent as liquid CO. The dilute gas of neutrons created at the surface of the earth by cosmic rays and nearby actinides together with the residual primary cosmic rays themselves could therefore generate ^{14}C in the liquid CO during and after enrichment. The actual level of ^{14}C in the CO is expected to be low and is still unknown. A rough estimate is a ratio of $\sim 10^{-12}$ but that depends upon many details.

CO gas was submitted to the IsoTrace laboratory and converted into a graphite-like material for analysis in the AMS equipment. The result reported by Beukens to Raghavan in 1991 was that there was an upper limit for the $^{14}C/C$ ratio in the CO gas of $\leq 1.6\times10^{-18}$. The limit was determined mainly by the need to subtract the contributions from the ^{13}C ions in the final detector. This result was described briefly in 1993 [5] and the implications discussed [18] in 2000. Subsequently it was shown [3] that the level of ^{14}C at the Borexino test facility was consistent with this value although a very different starting material was used. A counter containing some 5Mg of carbon was analysed.

4. The present AMS Background at IsoTrace

A huge number of molecular ions, such as $^{13}CH^-$ and $^{12}CH_2^-$, accompany the rare ^{14}C ions into the accelerator of the mass spectrometer. They are each about 3×10^{-4} of the $^{12}C^-$ atomic ions [19]. These molecular ions are completely destroyed during conversion to positive atomic ions in the tandem accelerator but the molecular fragments must be removed by the subsequent mass spectrometry. Fortunately the high-energy mass spectrometry can accomplish this, as has been discussed in some detail recently [20,21]. A problem for the majority of AMS laboratories with tandem accelerators with terminal voltage less than 3MV is that the final ^{12}C, ^{13}C and ^{14}C positive ions have very similar values of dE/dx. As a result they cannot easily be distinguished in the final ionisation detector, if they have the same energy. This is discussed further below.

The ^{14}C ions, emerging from the accelerator, are accompanied by an energy continuum of the other isotopes due partly to ion interactions in the residual gas of the accelerator [21]. These can be removed by the high-energy mass spectrometry and for convenience the ions can be separated into two groups. The first group is that part of the continuum of emerging ions with the same value of E/Q as the 14C ion, where E/Q is the energy divided by ion charge. These ions penetrate all electric analysers (EA). The second group of ions with mass M has the same ME/Q^2 as the ^{14}C ion. These can penetrate all magnetic analysers (MA). Consequently a sequence of electric and magnetic analysers should be sufficient to remove all such isotopes of the wrong mass. The AMS equipment was designed to remove such ions for radiocarbon dating [21] only and not for studies of very low ratios of $^{14}C/C$. Consequently additional analysers must be added. Two electric and magnetic analyser combinations in the series EA, MA, EA MA should suffice to solve this problem [20] provided they have sufficient energy and mass resolution.

The first experiment in 1991 using enriched isotopes [5] encountered a serious ME/Q^2 background due to the fact that the sequence of analysers was EA, MA, MA was designed for radiocarbon dating to about 60,000 years before the present (BP) or a ratio for $^{14}C/C$ of about 10^{-15}. Later versions of radiocarbon dating machines are different and the properties of two versions for low $^{14}C/C$ ratios have been discussed in some detail recently [21].

The tail from the ME/Q^2 interference in each case was first subtracted and then by subtracting the results for the unenriched sample from the results from the enriched sample of CO an estimate of the $^{14}C/C$ level was obtained [5]. The measured values for the ratios of $^{14}C/C$ were $(5.7\pm 1.4)\times10^{-16}$ and $(5.6\pm0.5)\times10^{-16}$ respectively, which after subtraction gives the final result, quoted above, of $< 1.6\times10^{-18}$, after allowing for the isotope enrichment. The errors on these numbers are mainly due to the tail of the ME/Q^2 peak in the ionisation counter, so the elimination of that interference by improved high-energy mass spectrometry would be an advantage. The background due to pulses of the same energy, the E/Q background, was small due to the two MA following the EA. It was estimated to contribute $< 10^{-19}$ to $^{14}C/C$ the ratio. In contrast the peak of the ME/Q^2 interference from $^{13}C^{+3}$ ions, which have 14/13 times the energy that of the $^{14}C^{+3}$ ions, was equivalent to a ratio of near 10^{-14} for $^{14}C/C$ from an unenriched target. However, the resolution of the final detector lowers this by a factor of 10^3 so that background subtraction was feasible.

The measurements of 1991 also clearly showed that background was made up of two major components. Usually CO_2 gas is used as a starting gas for the creation of graphite-like material for the sputter ion source and at the IsoTrace AMS laboratory the CO_2 is first turned into acetylene by a reaction with hot lithium. The acetylene is then converted into graphite-like material by cracking in an electrical discharge onto the surface of an aluminium cylinder, which is then inserted as a target into the ion source. Such graphite-like material gives a good efficiency for converting atoms into ions in a caesium sputter ion source with an efficiency that is greater than about 10%.

Such targets made from the CO_2 from natural gas routinely showed $^{14}C/C$ ratios of $(9.24 \pm 0.60) \times10^{-16}$. That level, however, dropped to $(5.88 \pm 1.4)\times10^{-16}$ when CO from natural gas was used and purified by freezing with liquid nitrogen before converting it to acetylene. It was presumed that some contemporary CO_2 is removed by this process leaving the remainder to come from the lithium metal used to make the acetylene or from the stainless steel bomb used in the process. Undoubtedly both sources are involved. The constancy of these numbers with time allows backgrounds to be subtracted for the radiocarbon dating of younger material. A ratio of 1.2×10^{-15}corresponds to about 57,000 years.

When acetylene generated from calcium carbide is used to create graphite-like targets this gives a lower measured $^{14}C/C$ ratio of $(1.4\pm1.0)\times10^{-16}$. This indicates that the process used to create the acetylene from CO, and especially CO_2, at present introduces ^{14}C atoms. The present system therefore is limited by the method for the preparation of the target material for the sputter ion source. This method will therefore have to be developed further. The creation of acetylene directly from methane is possible [22] so that the step using lithium metal is not essential.

5. The Ultimate AMS Background

First of all it is clear from the recent studies that the addition of electric and magnetic analysers can arbitrarily decrease the ^{14}C background due to the inability of the accelerator mass spectrometry to handle the huge fluxes of ^{12}C and ^{13}C ions from the accompanying mass-14 molecular hydrides. The situation for the heavy element analysis line at IsoTrace, where the ME/Q^2 interference is already at the level of 10^{-21} due to the use of two electric analysers, is already nearly adequate for low-level ^{14}C studies. With the addition of pulse-height spectrometry this level could be reduced further by over a factor of 100. This is to be compared with the ME/Q^2 level of $\sim 10^{-14}$, for the radiocarbon dating equipment, with a further factor of 100 reduction possible, due to the pulse height spectrometry in the final detector. The E/Q interference of the heavy element line will, however, require attention and adding a small magnetic analyser to lower the level to well below 10^{-21} easily does that. Adding a small magnet and electric analyser to the radiocarbon dating line would achieve similar properties to the heavy element line. Consequently both the AMS systems at IsoTrace can have a background lower than the level likely to be encountered with the natural $^{14}C/C$ ratios encountered in nature.

At a level of $^{14}C/C$ lower than about 10^{-18} the negative ion of nitrogen may appear as the predicted metastable ^9S state of N^- [23]. This is analogous to the He^- ion in that all the electrons have their spins parallel. If such an ion appeared at the 10^{-18} level then by the time the level of 10^{-21} was approached the ^{14}N ions would be some 10^3 times as intense as the ^{14}C. The measurement of dE/dx for the ions would allow this exotic ion to be identified and eliminated from the spectrum at 10^{-21}. Below that level it could be a nuisance!

The remaining difficult problems of actually determining the $^{14}C/C$ ratios in natural samples must now be discussed. These have so far been assessed only in a preliminary way [5] by, for example, using acetylene generated from calcium carbide, which gives for the $^{14}C/C$ ratio $(1.4\pm1.0)\times10^{-16}$. Here the error is determined largely by the need to subtract the tail of the ^{13}C peak and to a lesser extent the ^{12}C peak in the final detector. The ME/Q^2 interference was at the level of near $^{14}C/C \sim 10^{-14}$ in the system used for the studies. This measurement of the acetylene created from CaC_2 is consistent with the measurement with the enriched and unenriched CO, but it illustrates that ^{14}C contamination, during aspects of sample preparation, is a formidable foe. However, by eliminating the ME/Q^2 interference the origin of the low levels of ^{14}C will become easier to study and the origins of the contamination more easily determined.

A second useful investigation was the study [5] of the level of carbon and radiocarbon in the standard aluminium support blanks in the ion source. This will provide a measure of the sum of the contamination in the Al, the contamination in Cs^+ sputter ion beam and the hydrocarbons in the residual gas in the ion source. After an initial cleaning period the level of $^{14}C/C$ reached a low plateau that seems to correspond to carbon with an age of between 7000 and 13,000 years but with a carbon content that is apparently near 100ppm, a rather high level for aluminium. This carbon gives an equivalent background level of

$^{14}C/C$ that is near 3×10^{-17}. However, it is possible that this apparent level of carbon in the aluminium target supports is due mainly to vacuum hydrocarbons. This is because a partial pressure of about 1nTorr of a light hydrocarbon could give this result, as the level seems too high for contamination of the aluminium or the Cs^+ ion beam. Further study will be required because even though the measured pressure in the vacuum is usually less than 1μTorr the actual pressure near the target being sputtered must be much lower because of the strong gettering action of the sputtered material. The sputtered material is a mixture of carbon, caesium and aluminium.

Lee [24] observed the effects of a partial pressure of hydrocarbons while studying the counting rate of ^{14}C when acetylene with a $^{14}C/C$ ratio near 1.2×10^{-12} was let into the ion source with graphite and aluminium targets. The counting rate of ^{14}C remained at the background level for graphite until a critical partial pressure of 40 nTorr of C_2H_2 was reached. This was as expected from the gettering action of the material sputtered onto the surroundings by the Cs^+ ions, which included Cs and C atoms. The counting rate from a standard aluminium graphite-backing target remained near the background for all acetylene pressures up to a partial pressure of 150 nTorr. Reactive layers of Cs, Al and C will be present near all sputtered targets in a radiocarbon dating ion source and this should reduce the partial pressure of the hydrocarbons near the sputter target by a large amount. Consequently even a hydrocarbon partial pressure near the sputtered region of 1nTorr, which is needed to obtain the observed counting rates, seems large. The re-creation of hydrocarbons in the nearby active sputtered layer take may place as a result of the interaction with residual water vapour to create hydrocarbons. This problem will have to be investigated further using an ion source that is not used routinely for dating radiocarbon samples.

The caesium-sputtering beam itself is also not above suspicion, as one analysis [25] showed many unexpected elements. However, in this case the C^+/Cs^+ ratio needed would have to $\sim 10^{-4}$. This is considered unlikely. However, the cation spectrum from a Cs^+ ion source has surprisingly yet to be checked completely. For example, the spectrum with masses greater than 133 has not been checked for heavier clusters containing both Cs^+ and carbon. This must now be done if the further development of the Cs^+ sputter ion source for the OCP is to take place.

Finally as pointed out by Raghavan [26] the CO depleted industrially in ^{13}C and ^{14}C could be used for making a lower level neutrino or dark matter scintillation counter. Although the CO made from natural gas is only depleted by about a factor of about three, it could be the gas that has the lowest $^{14}C/C$ value. However, as was pointed out above, the creation of ^{14}C atoms during the enrichment or depletion process would have to be controlled.

6. Conclusions

Studies of the lowest levels of ^{14}C in carbon have been undertaken during recent years on a radiocarbon dating machine continually carrying out routine analyses. These low level studies have had to be fitted in to a tight schedule of analyses using a single ion source, which could only be cleaned partly of the accumulated sputtered carbon. In spite of this handicap it has been possible to demonstrate that underground methane has a very low level of $^{14}C/C$ of less than 1.6×10^{-18}. This required the use of isotope enrichment because of the low-level background of ^{14}C in the natural sample. Recent measurements have demonstrated that the accelerator mass spectrometry can be made essentially free of interference from the residual ^{12}C and ^{13}C. These are created from the destruction of mass 14 molecules injected into the tandem accelerator along with the ^{14}C. However, the region of an ion source dedicated to the problem of looking at the real ^{14}C background will need more attention. The materials used for the ion source will have to be screened for their contamination and that includes the Cs^+ ion beam. The need to prepare samples of material for analysis without introducing contamination and the role of isotope enrichment will also require study. Finally more attention will have to paid to increasing the ion current from the ion sources to speed up the low counting rate measurements. Ion sources with both positive and negative carbon ion beams in excess of 1mA have already been demonstrated elsewhere so that measurements at a low level of ^{14}C can, in principle, be made quickly provided all the other problems are solved. Then it will be possible to study the actual level of ^{14}C in deep underground carbon reservoirs and so assess their potential for lower background measurements with large scintillation counters.

Acknowledgements

This work was supported by grants from the Natural Science and Engineering Research Council of Canada and the National Science Foundation. We also wish to thank the University of Toronto for support. We are indebted to R. S. Raghavan for providing us with the initial impetus for this project, Dr. K. H. Purser for his continuing advice and Y. Kamyshkov for his early advice.

References

1. W. F. Libby, E. C. Anderson and J. R. Arnold, *Age determination by radiocarbon: Worldwide assay of natural radiocarbon.* Science **109** (1949) 227.
2. Mentioned by J. R. Arnold in, *Decay counting in the age of AMS*, Nucl. Instrum. and Meth. **B29** (1987) 424 – 426.
3. G. Alimonti et al., *Measurement of the ^{14}C abundance in a low-background liquid scintillator.* Phys. Letts. **B422** (1998) 349 – 358.
4. A. de Bari et al., *Radiopurity of the Borexino scintillator,* AT&T Bell Laboratories Tech. Memo. 11121-921015 (Oct 1992).
5. R. P. Beukens, *Radiocarbon accelerator mass spectrometry: background and contamination,* Nucl. Inst. and Meth. **B79** (1993) 620 – 623.
6. C. Tuniz, J. R. Bird, D. Fink and G. F. Herzog, *Accelerator Mass Spectrometry,* pp372, 1998, CRC press, Boca Raton, Boston, London, New York and Washington, D.C.
7. J. A. Kuehner, E. Almqvist and D. A. Bromley, 1959 (unpublished).
8. M. Anbar, *The limitations of mass spectrometric radiocarbon dating using CN^- ions,* in the Proceedings of the First Conference on Radiocarbon Dating with Accelerators, Edited by H. E. Gove, University of Rochester April 20 and 21, 1978, p.152 – 156.
9. H. Gnaser, *Formation of metastable N_2^- and CO^- anions in sputtering,* Phys. Rev. A. **56** (1997) R2518 – R2521.

10. R Middleton and J. Klein, *Production of metastable negative ions in a cesium sputter source: Verification of the existence of N_2^- and CO^-*. Phys. Rev. **A59** (1999) 1 – 13.
11. Proceedings of the Symposium of North-East Accelerator Personnel 1972 held at Florida State University at Tallahassee, March 23 and 24, edited by K. Chapman. These proceedings contain many discussions that led to the development of the high-intensity caesium sputter ion source.
12. R. Middleton, *A versatile high intensity negative ion source,* Nucl. Instr. and Meth. **196** (1983) 139 – 150.
13. H. Naylor, D. Elmore, M. R. Clover, L.R. Kilius, R. P. Beukens, B. R. Fulton, H. E. Gove, A. E. Litherland and K. H. Purser, *Determination of ^{36}Cl Isotope ratios*, in the Proceedings of the First Conference on Radiocarbon Dating with Accelerators. Edited by H. E. Gove, University of Rochester, April 20 and 21, 1978, 360 - 365.
14. K. H. Purser et al., *An attempt to detect stable N^- ions from a sputter ion source and some implications of the results for the design of tandems for ultra-sensitive carbon analysis.* Rev. Phys. Appl. **12** (1977) 1487 – 1492.
15. A. E. Litherland, *Fundamentals of accelerator mass spectrometry,* Phil. Trans. of the Roy. Soc. of London, **A 323** (1987) 5 –20.
16. C. L. Bennett et al., *Radiocarbon dating using electrostatic accelerators: Negative ions provide the key,* Science **198** (1977) 508 – 510.
17. The web site of the Isotec Corporation is www.isotec.com
18. A. E. Litherland and K. H. Purser, *The measurement of very old Radiocarbon ages by AMS*, in Proc. Int. Conf. on the Fundamental & Applied Aspects of Modern Physics: Luderitz, Namibia. Edited by S. H. Connell & R. Tegen, pp 457- 473, 2000, World Scientific Publishing Company, New Jersey, London, Singapore and Hong Kong.
19. R. P. Beukens, *Radiocarbon accelerator mass spectrometry: background, precision and accuracy,* in Radiocarbon After Four Decades An Interdisciplinary Perspective, edited by R. E. Taylor, A. Long and R. S. Kra, 1992, pp.596. 210 – 239.
20. R.P. Beukens et al., *The Old Carbon Project: How Old is Old?* Nucl. Instr. and Meth. **B223-224** (2004) 333-338.
21. M-J. Nadeau et al., *E/Q and ME/Q2 interference in the two models of 14C Tandetron systems: Towards the 21st Century,* Nucl. Instr. and Meth. **B 223-224** (2004) 328 – 332.
22. S. Kado, Y. Sekine, K. Fujimoto, Chem. Commun. **24** (1999) 2485.
23. N. A. Piangos and C. A. Nicolaides, *Very highly excited bound states of atomic negative ions having all their electrons with unpaired spins.* J. Phys. B: At. Mol. Opt. Phys. **31** (1998) L147 - L154.
24. H. W. Lee, *Study ofcarbon-14 detection and dating using accelerator mass spectrometry,* Ph.D. thesis, University of Toronto, Canada, 1988.
25. S. Matteson et al., *An ultra-clean microprobe ion source for atomic mass spectrometry,* Nucl. Instr. and Meth. **B52** (1990) 327 – 333.
26. S. Raghavan, private communication.

Trace element analysis of K, U and Th in high purity materials by neutron activation analysis

Ila Pillalamarri

MIT-EAPS Neutron Activation Analysis Laboratory,
Department of Earth Atmospheric & Planetary Sciences,
Massachusetts Institute of Technology, Cambridge, MA 02139.
pila@mit.edu

Abstract. The concept and usage of 'high purity' are explained. Trace element analysis of K, U and Th by neutron activation analysis is described, the radio-isotopes and their corresponding gamma rays used to identify the elements are listed. The interfering elements are described. The advantages and disadvantages of using neutron activation analysis are discussed. Some examples of trace impurity determinations in high purity materials are provided.

Keywords: trace element analysis, K, U, Th, activation analysis, limits, gamma spectroscopy, high purity.

INTRODUCTION

The key words in the title are 'high purity materials', 'trace element analysis' and 'neutron activation analysis'. So the questions that arise are - what is high purity, what are high pure materials, what is trace impurity, what is the necessity and why focus in particular on the trace element analysis of K, U and Th? The purpose of this paper is to address these questions and outline the salient features of trace element determination of K, U and Th in high pure materials by Neutron Activation Analysis.

What is high percent purity?

Purity of metals, compounds, solutions and several different organic and inorganic materials is of great importance to many technological applications and basic research. There are different levels of requirement of the purity of the materials for each and every research and application. The purity of the material, whether compound or metals organic or inorganic, is usually certified by the manufacturer or an organization. Hence, the definition of purity of different grades should always be referred to the source providing the certification. The common usage of purity of analytical solvents is provided by chemical companies, for example with reference to American Chemical Society. (Ref: 1. http://www.reagents.com/products/reagents/grades.html). Similarly, different countries in the world have corresponding certification societies. In this paper, I refer to high purity of materials like metals. High percent purity refers to 99.9%, also referred as three nines (3N) purity. For example the Fe wire used in our laboratory, for routine monitoring of flux variation among irradiated samples, is of 99.9+ % purity. This may also be clearly quoted for example as 3N5 which means 99.95% purity certified. Different grades of high purity have different common usage terms as given in Table 1.

TABLE 1. High percent purity levels and the corresponding trace impurity

High Purity Level	Element	Trace Impurity Level
99.9%	High pure (3N)	0.10%
99.95%	Analytical Grade (3N5)	0.05%
99.99%	High Pure	0.01%
99.998%	High Pure (4N8)	0.002%
99.999%	Spectroscopic Pure	0.001%
99.9999%	High pure (6N)	0.0001%

CP 785, *Topical Workshop on Low Radioactivity Techniques,* edited by B. Cleveland, R. Ford, and M. Chen

The high percent purity levels greater than the 9 nines may be referred as ultra-high purity and 12 nines as hyper pure. The terms ultra-high pure and hyper pure are used according to the context and should always be clarified by providing the quoting source or the source of certification. Also, it should be remembered that not everything is referred in terms of weight %, but may be in terms of volume.

Concepts of trace impurity:

The level of high purity and the corresponding impurity are inversely related. As the purity level goes higher the impurity level becomes trace level. For example, for a 99.95% pure material, the corresponding total amount of impurity would be 0.05%. For 6N pure material, the corresponding total amount of impurity would be 0.0001% which means the impurities will be at sub-ppm level. For the ultra-high pure and hyper pure levels of purity, the impurity levels will be in the trace amounts of 10^{-9} to 10^{-15} weight % or per cc level. The concentration determination of trace impurities requires deep probing nuclear analytical tools. Neutron activation analysis is one of such tools, in practice for more than 50 years.

INSTRUMENTAL NEUTRON ACTIVATION ANALYSIS

A brief description:

Instrumental neutron activation analysis is a well established and versatile technique to determine the trace element concentrations of many elements in the periodic table [Ref. 1]. The principles of neutron activation analysis are described in detail in textbooks [Ref. 1-4]. Instrumental neutron activation analysis does not require any chemical procedure to be performed on the sample. The sample to be analyzed is usually sealed in a polyethylene tube or bag and placed in the core of a nuclear reactor where it is surrounded isotropically by a sea of neutrons. The neutrons used for neutron activation are usually thermal neutrons. A small fraction of each kind of isotope of the elements (called **parent** isotopes) present in the sample material will react with the neutrons, causing neutron reactions, producing (n, gamma)) **product** isotopes. Many of the generated isotopes are radioactive which decay with characteristic gamma energies and half-life to a stable form. The sample is removed from the reactor and carried to the laboratory for radioassay usually by gamma spectrometry. Each element in the sample is uniquely identified by one (or several) gamma-ray energy of the product isotope of the element. The concentration of the parent element is related to the number of decays observed for the corresponding unique product isotope.

The sensitivity of concentration determination for a particular element depends on how many of atoms are available for the neutron interaction to produce detectable and identifiable decays of the product isotope. This in turn depends on 1) neutron flux, cross-section or neutron absorption probability for the parent isotope, 3) irradiation time (limited to time for the production of saturation activity), 4) efficiency of the radiation detector, 5) intensity of the gamma radiations of the product isotope, 6) half-life of the product isotope (it should not be too long or too short).

The concentrations of several elements in a sample can be conveniently determined by comparator method, in which a standard with known concentrations is included along with the sample irradiation, eliminating the need to know the absolute values of parameters like cross-section, intensity and efficiency [Ref. 2].

Gamma activity measurement by Ge detectors – Gamma Spectroscopy:

The gamma activities from the irradiated materials are measured by a gamma spectrometer consisting of HPGe detector and the related nuclear instrumentation like the amplifier, analog to digital converter, multi-channel analyzer and a computer with data acquisition and reduction software and other peripherals. Details of gamma spectrometry are provided in textbooks and references [Ref. 2, 5-9]. The schematic diagram of MIT-EAPS INAA gamma spectrometer is available on the web site http://web.mit.edu/pila/www/INAAHOMEPAGE.HTM

DETERMINATION OF K, U AND Th by INSTRUMENTAL NEUTRON ACTIVATION ANALYSIS

Thermal neutron activations of stable isotopes ^{41}K, ^{238}U and ^{232}Th in the materials create suitable product isotopes for the concentration determination of K, U and Th. ^{233}Th, formed by ^{232}Th (n, gamma)) ^{233}Th (half-life of 22 minutes) decays by beta emission to ^{233}Pa (half-life 27 d). ^{239}U formed by ^{238}U (n, gamma) ^{239}U (half-life 24 minutes) decays by beta emission to ^{239}Np. A summary of parent and product isotopes, the gamma-ray energies and the half-lives of the product isotopes of K, U and Th along with possible interfering gamma-ray energies are listed in Table 2.

TABLE 2. Gamma-ray energies and half-lives of neutron activation product isotopes of K, U and Th

Element	Parent Isotope	Product Isotope	Half-life	Gamma-ray Energy (keV)	Interferences[a] Isotope: Gamma-ray Energy (keV)
K	^{41}K	^{42}K	12.36 h	1524	
U	^{238}U	(^{239}U decays to) ^{239}Np	2.35 d	106, 222, 228, 277	^{153}Sm:103 ^{182}Ta:222 ^{182}Ta:229 ^{132}Te:222
Th	^{233}Th	(^{233}Th decays to) ^{239}Pa	27 d	312	^{239}Yb:308 ^{192}Ir: 316

[a]The interferences listed are considered as first order interferences such as – 1) gamma rays of same energy may be emitted by different isotopes, 2) different gamma-ray energies may be within the detector resolution of the gamma-ray energy of interest. The second order interferences are discussed under limits of detection.

From Table 2. it can be seen that for the determination of K, U and Th abundances by INAA, one irradiation may not be sufficient especially if the trace impurity level is in the sub-ppm range. In such a case, the samples have to be irradiated for longer periods of time and counted for three different delay times. Trace abundances of K, U and Th are determined in materials of different matrices by INAA [Ref: 10, 11, 12 and Bibliography of 1].

LIMITS OF DETECTION

The lower limit of detection of trace level impurities of K, U and Th in high purity materials is set by

1) the matrix activities because of the Compton continuum from higher energy gamma rays than those of interest. Interferences arise from the activities of the elements Cr, Fe, Co, Sc and Na,
2) depends on the major element matrix, and trace elements like Cr, Fe, Co, Sc
   These four elements are contaminants arising from handling of high purity materials (by way of molding, milling, cutting etc.),
3) contamination of Na arises from as much as touching or breathing on the sample.

TABLE 3. Interfering isotopes and their gamma-ray energies for the concentration determination of K, U and Th

K, U, Th Activation product isotopes: Gamma-ray Energies (keV)	Interfering element Activation product isotopes: Gamma-ray Energies (keV)
^{233}Pa: 312 ^{239}Np: 106, 222, 278 ^{42}K: 1524	^{59}Fe: 142, 192, 1099, 1291 ^{51}Cr: 321 ^{46}Sc: 889, 1120 ^{24}Na: 1368, 2754

Best limits in clean materials like acrylic [Ref: 11]
U, Th : 1 ppt; K : 50 ppb

Typical limits of detection
U, Th : 20 ppt, K : 50 ppb

Extending the limits of detection:
Limits can be extended below pg/g by a judicious choice of

1) mass of sample,
2) irradiation time,
3) irradiation flux,
4) INAA and Low Background Counting

If the activities other than from Th and U are negligible, low background counting will be useful to go below pg/g [Ref. 11]. Radiochemical Neutron Activation Analysis (RNAA) is useful in extreme cases, post irradiation separation of Pa and Np activities are necessary and have been done as required [Ref. 2].

APPLICATIONS OF TRACE IMPURITY ANALYSIS IN BASIC RESEARCH AND TECHNOLOGY:

The search engines on the internet list hundreds of references about the industrial applications and basic research of trace impurity analysis of high purity materials. Some names of the current research studies:

1) Impurities of metals define their properties like formability, fracture toughness, corrosion resistance, ultra-pure metal film formation, high temperature oxidation, crystal defect control in high pure metals: "Influence of trace impurities on the high-temperature mechanical properties of Iridium", [Ref. 13];
2) "Consideration of accuracy in the analysis of ultra pure aluminum", [Ref.14];
3) "Detection and characterization of trace element contamination on silicon wafers", [Ref. 15]
3) "High purity metals as primary calibration materials for elemental analysis, their importance and certification", [Ref. 16]
4) "Determination of primordial radioactive trace elements in neutrino detector components", [Ref. 17]

CONCLUSION

The need for high purity materials creates the need for trace analysis of impurities at 10^{-6} (ppm), 10^{-9} (ppb), 10^{-12} (ppt), 10^{-15} (ppf) and 10^{-18} (ppa) level. INAA is a very useful direct tool but RNAA coupled with low background counting becomes very powerful when the impurities to be determined are below ppt level [Ref. 12]. The relative ratio of Th or U to total matrix trace impurities is one issue; the other issue is the major matrix itself for the use of INAA. We cannot use NAA for determining Th and U in Fe, or Ge, or Co or Cu, or any other element where the major element gets activated.

Neutron Activation Analysis is a very important analytical tool and complements many other techniques like Atomic Absorption, X-Ray Fluorescence, Inductively Coupled Plasma Atomic Emission Spectrometry, Inductive Coupled Plasma Mass Spectrometry. Several advantages of NAA are:

1. Nuclear technique that measures the intensity of gamma rays of " characteristic" energy using gamma spectroscopy.
2. Multielement Analysis.
3. Rapid analyses of multiple samples.
4. Sample size can be variable (typically 1 mg to 1 gm).
5. Non-destructive - that is valuable and safe, samples are not destroyed.
6. No Chemical processing; therefore samples are not contaminated during sample preparation, no uncertainty about total dissolution of sample, no need for dilutions of solutions, making the technique valuable and safe. Samples are not destroyed.
7. No need for repeated blank measurements because no memory effects.
8. Gamma ray spectroscopy is largely free from matrix interferences
9. Depending on the sample matrix, elemental concentrations can be determined at parts per million (ppm), parts per billion (ppb) and parts per trillion (ppt) level.
10. Versatile (in use for more than half a century), well established and reliable.

REFERENCES

1. D. De Soete, R. Gijbels, J. Hoste, Neutron Activation Analysis, Wiley-InterScience 1972.
2. J. Dostal and C. Elson, “General principles of neutron activation analysis”, pp. 21- 42; P. Jagam and G. K. Muecke, “Instrumentation in neutron activation analysis”, pp. 73-108; I. L. Gibson and P. Jagam, “Instrumental neutron activation analysis of rocks and minerals”, pp. 109 –132; C. L. Chou, “Radiochemical neutron activation analysis”, pp. 133-166; in Short course in Neutron Activation Analysis in the Geosciences, edited by G. K. Muecke, Mineralogical Association of Canada, 1980.
3. P. J. Potts, “Neutron Activation Analysis”, Handbook of silicate rock analysis, Blackie, 1987, pp. 399-439.
4. S. Parry, “Neutron Activation Analysis”, in Modern Analytical Geochemistry, edited by Robin Gill, Addison Wesley Longman Ltd. 1997, pp. 116–135.
5. K. Debertin and R. G. Helmer, Gamma- and X-Ray spectrometry with semiconductor detectors, North Holland, 1988.
6. G. F. Knoll, Radiation Detection and Measurement. John Wiley & Sons, 1979.
7. P. Ila and F. A. Frey, Atomkernenergie Kern-technik supplement **44**, 710–716 (1984).
8. P. Jagam, J.J. Simpson, J.L. Campbell, B.C. Robertson, H.L. Malm, Nucl. Instr. and Meth**. A 239**, 214 –220 (1985).
9. P. Jagam, J. J. simpson, B. C. Robertson, L. J. Hahn, G. E. Aardsma,, Nucl. Instr. and Meth. **A 267**, 486-491 (1988).
10. P. Ila and F. A. Frey, Jour. Radioanal. and Nucl. Chem. **131**, 37- 42 (1989).
11. P. Jagam, J. –X. Wang, J. J. Simpson, Jour. Radioanal. and Nucl. Chem. **171**, 277–286 (1993).
12. R. J. McDonald, A. R. Smith, D. L. Hurley, E. B. Norman, M. C. Vella “Sub-parts-per-quadrillion sensitivity to contaminants in ultra-pure silicon via low-background instrumental neutron activation analysis”, in Proceedings of the Fourth International Symposium on High Purity Silicon, Electrochem. Soc., Pennington, NJ, USA,1996. pp. 554-65.
13. B. Fischer, A. Behrends, D. Freund, D. Lupton, J. Merker, “ Influence of trace impurities on the high-temperature mechanical properties of iridium”, in Iridium Proceedings of International Symposium 2000, 12-16 Mar 2000, TMS - Miner. Metals & Mater. Soc., Warrendale, PA, USA. pp. 15-26.
14. J. L. Genna, “Consideration of accuracy in the analysis of ultra pure aluminum”, in Second Workshop on Purification of Materials for Crystal Growth and Glass Processing 28 Sept.-1 Oct. 1987, Champion, PA, USA, Jour. Crystal Growth **89(1),** 62- 27 (1988).
15. A. Singh, K. Baur, S. Brennan, T. Homma, N. Kubo, P. Pianetta, “Detection and characterization of trace element contamination on silicon wafers”, in AIP Conference Proceedings 652 , 2003, pp. 472-80.

16. R. Matschat, M. Czerwensky, S. Pattberg, H. -J. Heinrich, S. Tutschku, "High purity metals as primary calibration materials for elemental analysis-their importance and their certification", in Eighth International Conference on Ultra-High Purity Base Metals (UHPM-2001), 25-28 June 2001, Berlin, Germany, Mat. Trans. **43(2)**, 90-97 (2002).
17. R. von Hentig, F. von Feilitzsch, "Determination of primordial radioactive trace elements in neutrino detector components" in Proceedings of the 2nd International Workshop on Low Energy Solar Neutrino Detection LowNu2, 4-5 Dec 2000, World Scientific, Singapore, 2001, pp. 162-171.

III – SCREENING FACILITIES AND LOW BACKGROUND DETECTORS

Background Reduction in Cryogenic Detectors

Daniel A. Bauer

Fermi National Accelerator Laboratory

Abstract. This paper discusses the background reduction and rejection strategy of the Cryogenic Dark Matter Search (CDMS) experiment. Recent measurements of background levels from CDMS II at Soudan are presented, along with estimates for future improvements in sensitivity expected for a proposed SuperCDMS experiment at SNOLAB.

Keywords: Backgrounds, Cryogenics, Dark Matter, WIMPS
PACS: 95.30.Cq, 95.35.+d, 95.55.Vj

THE CRYOGENIC DARK MATTER SEARCH EXPERIMENT

There are a large number of experiments trying to detect interactions of dark matter particles in terrestrial targets. One class of such experiments cools Ge and Si detectors to cryogenic temperatures (< 50 mK) in an attempt to discriminate the low energy (< 100 keV) nuclear recoils caused by dark matter WIMPS from the electron recoil interactions of conventional backgrounds. This paper discusses the background reduction strategy and measurements in one such experiment, the Cryogenic Dark Matter Search (CDMS II). The goal of the present incarnation of this experiment, CDMS II at Soudan, is direct direction of as few as 5 WIMP interactions per year, in the face of ambient backgrounds exceeding 10,000 events per year.

The CDMS II experiment operates at the Soudan Underground Laboratory (2070 mwe) in northern Minnesota. The rock overburden at Soudan reduces the cosmic ray muon flux to approximately 1 per minute hitting the experimental setup. Figure 1 shows elements of the CDMS II facility. The experiment is housed within an rf-shielded, class-10,000 cleanroom, with an auxiliary clean preparation area (anteroom), a room for electronics and data acquisition, and an area for cryogenic systems.

FIGURE 1. The CDMS II infrastructure at Soudan. Clockwise from upper left: the upper-half of the class-10,000 cleanroom which houses the experiment seen from the mezzanine level; the cryogenics system, with the dilution refrigerator and detector cold volume (icebox) connected via the cold stem; Towers 1 & 2 installed in the inner can of the icebox; a partial assembly of the shielding layers in the cleanroom.

CP 785, *Topical Workshop on Low Radioactivity Techniques,* edited by B. Cleveland, R. Ford, and M. Chen

The experimental apparatus is hermetically surrounded by active shielding, consisting of two-inch-thick plastic scintillator panels. Inside of this "veto" is a 200-cm-tall by 200-cm-diameter hollow cylindrical polyethylene shield with a 40-cm-thick top, bottom and wall, which moderates neutrons from surrounding materials. Inside of that is a 130-cm-tall by 130-cm-diameter hollow Pb shield with 22.5-cm walls, which stops gamma rays produced by radioactivity in the rock. The inner 5 cm of this Pb is 'ancient', meaning that it contains very little of the long-lived ^{210}Pb isotope. Farther inside is a second 10-cm-thick polyethylene shield, which moderates neutrons that penetrate the Pb. Finally, at the center is the Icebox, a mostly-copper cryogenic chamber mapped to the thermal layers of the dilution refrigerator via a nested set of copper stems, (or cold stem). There is space in the base temperature (< 50 mK) chamber for seven six-detector "towers" arranged in a hexagonal pattern. The detector signals are brought out through another set of nested copper stems to a vacuum bulkhead and from there to front end electronics.

The CDMS ZIP (Z-dependent Ionization and Phonon) detector technology, using simultaneous athermal phonon and ionization measurement in Ge and Si crystals, has demonstrated outstanding rejection of gamma and beta backgrounds. These background particles scatter off electrons in the detectors, while WIMPs (and neutrons) scatter off nuclei. The ZIP detectors allow discrimination between electron and nuclear recoils through two effects. First, recoiling electrons are more ionizing than recoiling nuclei, resulting in a higher ratio of ionization to phonon signal, called "ionization yield." Figure 2 shows the discrimination based on ionization yield. Second, the athermal phonon signals due to nuclear recoils have longer rise times and occur later than those due to electron recoils, as is shown in Fig. 3. For recoils within a few μm of a detector's surface (primarily from low-energy electrons), the charge collection is less complete, making discrimination based on ionization yield less effective. Many of these events can be rejected by phonon timing cuts because they have even faster phonon signals than those from bulk electron recoils.

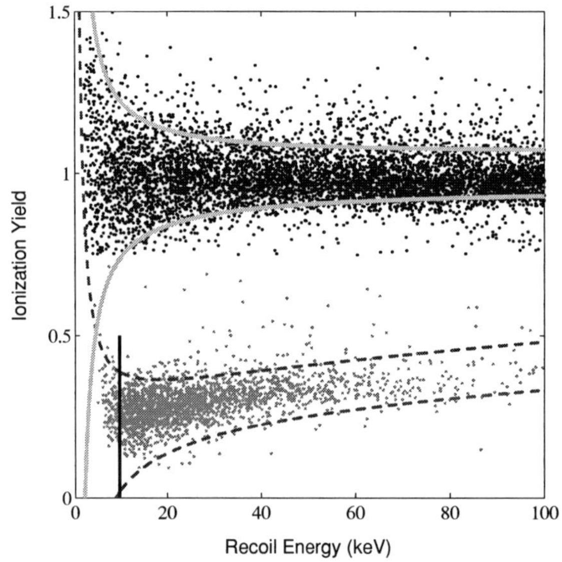

FIGURE 2. Ionization yield versus recoil energy for calibration data with a ^{252}Cf gamma and neutron source for Ge ZIP detectors showing the $\pm 2\sigma$ gamma band (solid curves) and the $\pm 2\sigma$ nuclear-recoil band (dashed curves). Events with ionization yield < 0.75 are shown only if they pass the phonon-timing cuts. The vertical line is the 10 keV analysis threshold.

While neutrons cannot be distinguished from WIMPs on an event-by-event basis, because both produce nuclear recoils, they can be statistically separated in two ways. First, neutrons will frequently scatter in more than one detector, whereas WIMPs will not. Second, the cross sections for neutron interactions in Ge and Si are similar, but the (spin-independent) WIMP cross sections are coherent, scaling as A^2, and are thus about 5 times larger in Ge than in Si.

BACKGROUNDS

All dark matter experiments must employ some form of shielding against backgrounds resulting from small amounts of radioactivity from the three naturally occurring chains ^{238}U, ^{232}Th, and ^{40}K. These chains produce a mixture of gammas, betas, alphas, and neutrons with typical energies of a few MeV. Trace quantities of these isotopes are always present in the materials surrounding the experiment and great care must be taken to reduce the amounts present

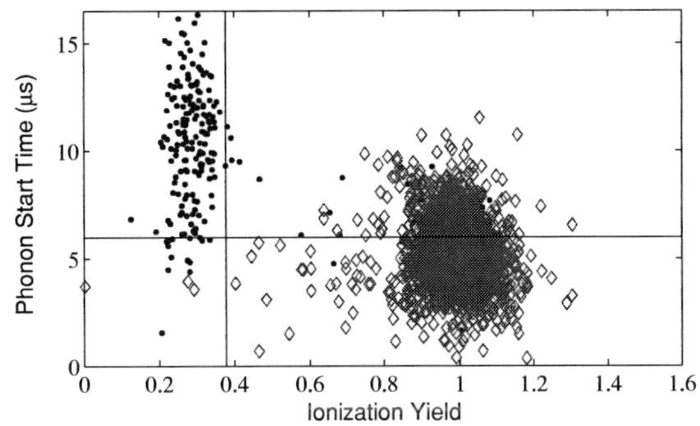

FIGURE 3. Phonon start time versus ionization yield for ^{133}Ba gamma-calibration events (diamonds) and ^{252}Cf neutron-calibration events (dots) in the energy range 20–40 keV in a Ge detector. Lines indicate typical timing and ionization-yield cuts, resulting in high efficiency for nuclear recoils and a low rate of misidentified electron-recoil events.

in the shielding materials themselves. Neutrons from radioactivity arise primarily from fission decays and (alpha,n) reactions. In addition to radioactivity, cosmic ray interactions provide a significant source of potential background particles, requiring that the experiment be located deep underground to reduce the cosmic ray muon flux. The most difficult backgrounds are due to high energy neutrons from cosmic ray interactions, since neutrons lead to nuclear recoil interactions just as WIMPs do. Most experiments surround their detectors with an active veto, to detect cosmic ray muons that would interact in the shielding materials and produce neutrons. However, if the cosmic ray interacts in the surrounding materials, the active veto may not detect the resulting high energy neutron. These neutrons are very difficult to moderate with shielding and form an irreducible background at a given depth. Figure 4 illustrates these background sources, and the following sections detail their measured levels in CDMS.

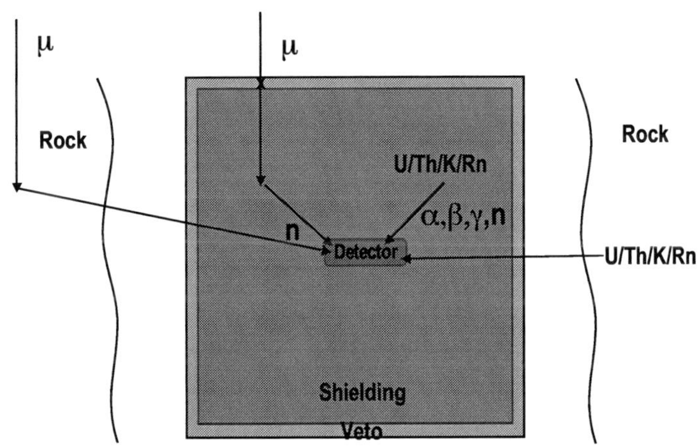

FIGURE 4. Conceptual diagram of background sources for dark matter experiments

Electromagnetic Backgrounds

The term "electromagnetic backgrounds" is used to refer to photons and electrons. Electromagnetic backgrounds divide into two subgroups, "bulk" and "surface" backgrounds, owing to the fact that the ZIP detectors possess a thin surface "dead layer" in which ionization-yield-based rejection is poor compared to that for bulk events. Events falling within the first 35 μm suffer some ionization-yield suppression and events in the first 1 μm lose so much ionization as to be misidentified as nuclear recoils. Phonon-timing-based rejection provides rejection of 97% of surface events, but the overall rejection of such events remains poorer than for bulk events.

Bulk electromagnetic backgrounds consist primarily of high-energy photons, as well as electrons emitted from radioisotopes in the bulk of the detectors. The photon component is determined by bulk contamination levels in the apparatus and shield and by radon penetration into the shield. Because the electron component is usually negligible in comparison to the photon component, bulk events are generically referred to as "photons." They are identified empirically as events with ionization yield ≈ 1 as expected for bulk electron recoils.

Surface electromagnetic backgrounds include (1) "photon-induced" photons or electrons that interact in the dead layer and (2) electrons (and possibly low-energy X-rays) emitted by radioactive contaminants on detector or nearby surfaces. The photon-induced component scales directly with the continuum photon rate and detector surface area/volume ratio. Surface contamination depends on the detailed fabrication and handling history of the detectors. Surface events are generically referred to as "electrons." Surface events are identified empirically as all events with ionization yield below the "bulk" event band and above the nuclear-recoil band.

The rates of bulk and surface electromagnetic background events for Towers 1 and 2 are shown in Table 1.

TABLE 1. Average bulk and surface electromagnetic background rates in Tower 1 and Tower 2 at Soudan, 15-45 keV deposited energy, inner detectors.

event type	Ge	Si
Bulk total [/kg/day]	115.2 ±1.6	501.9 ±4.6
Bulk singles [/kg/day]	30.2 ±0.8	194.5 ±2.3
Surface total [/day]	1.57 ±0.09	1.53 ±0.09
Surface singles [/day]	0.41 ±0.05	0.51 ±0.06

Detector rejection for bulk electromagnetic backgrounds is extremely good. Rejection of events where a photon ejects an electron from the detector surface is similarly excellent, except for a negligible fraction of such events induced in the first μm of the surface. On the other hand, "electron" events from surface radioactive contaminants have a very shallow interaction depth profile and thus suffer far worse misidentification. About 0.6% of the surface single electrons in Table 1 survive to the nuclear-recoil signal region, contributing the dominant Soudan background of 1×10^{-2}/kg/day.

Photon Backgrounds at Soudan

We have simulated the contributions to the photon background from the U/Th/K content of the Cu cans, of the inner and outer polyethylene shields, and from ^{210}Pb in the inner lead shield, with the simulations normalized by high purity Ge (HPGe) spectrometer screening results. The sum of these normalized contributions yields a photon background smaller by about a factor of 4 than the observed background, as shown in Figure 6.

A likely additional source of photons are radon decays outside the currently purged volume. The measured radon levels in the Soudan laboratory in the CDMS experimental room show a mean of about 500 Bq/m^3, with large seasonal variations, as shown in Figure 5. During early running at Soudan, before the radon purge was established, photon rates were a factor of 4-6 higher than shown in Table 1 and the ^{214}Bi radon daughter line at 609 keV was clearly identified in the Ge detectors.

To reduce radon gas in the space directly surrounding the icebox, we are purging this volume with medical grade breathing air stored at least two weeks in metal cylinders to allow its radon to decay ($T_{1/2}$ = 3.2 days). However, the purge system in use when the data presented in Table 1 were taken did not penetrate the full volume, so significant radon buildup was possible outside the purged volume. Monte Carlo simulations show that about 35 Bq/m^3 of radon

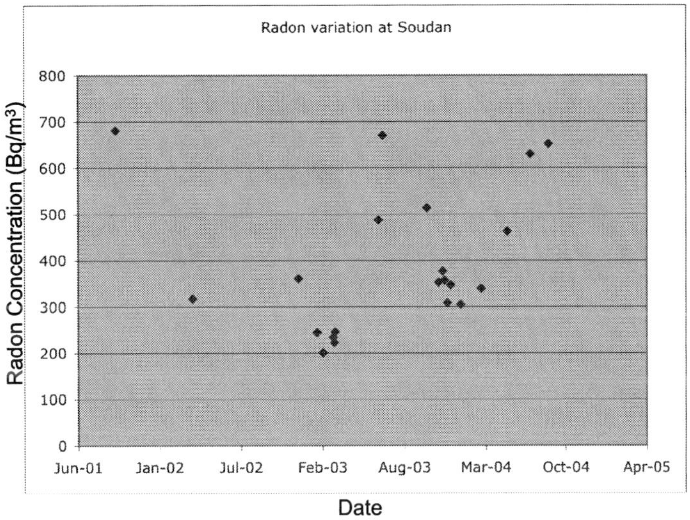

FIGURE 5. Measurements of Radon concentration in the air of the Soudan Underground Laboratory using a Rad7 counter made by Durridge. Radon concentration peaks in the summer and is lowest in the winter.

gas outside the purged volume is sufficient to explain most of the remaining photon background, as shown in Figure 6. In preparation for further running at Soudan, the purge system has been improved greatly.

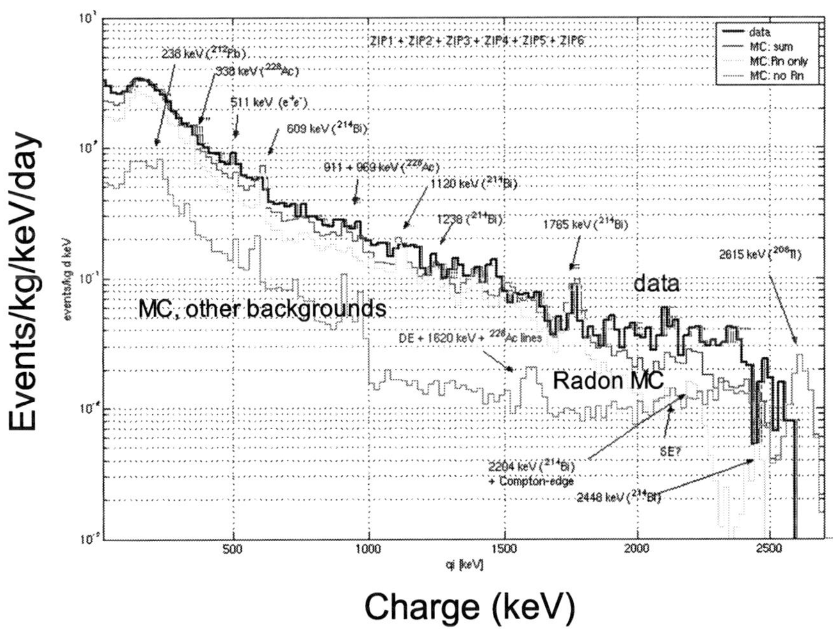

FIGURE 6. Measured gamma backgrounds in CDMS II detectors at Soudan compared with GEANT4 simulations incorporating known levels of radioactive elements in materials surrounding the detectors. Agreement between data and simulations is only achieved if significant Radon gas is present within the inner shielding of the experiment.

Rejection of photon-related backgrounds in the CDMS II experiment has been measured and simulated. Measurements are done by exposing the entire apparatus to high-energy photon sources via penetrations of the lead shield. Simulations are performed using GEANT4, including tracking of low-energy electrons created by photon interactions and conversion of interaction depth to ionization yield via the a yield vs. depth profile measured by exposure of a detector to a ^{109}Cd electron source.

The simulations indicate that of all photon-related events (15-45 keV), 0.31% are single-scatter surface events and 1.8% of those (or 0.005% of all photon-related events) suffer enough ionization-yield suppression to be misidentified as nuclear recoils. Thus, we have 99.995% rejection based on number of scatters and ionization yield alone. The single-scatter fraction is confirmed by photon-exposure calibration runs and by a comparison of pre/post-radon-purge rates. The overall misidentification is confirmed by seeing 1 single-scatter misidentified event in 25,000 photon-exposure events. The phonon-timing cut rejects low-yield events with 97% success, so the current overall rejection of photon-induced events is 99.9999%. (The cuts have 70% exposure efficiency for nuclear recoils).

Beta Backgrounds at Soudan

The dominant CDMS background at Soudan consists of electrons from radioactive contamination on the detector surfaces, which we refer to as "betas," reflecting the likelihood that most of these events arise from beta decay. We include internal-conversion and Auger electrons as betas, and perhaps some soft x-rays as well. The total rate of surface electrons in the energy range 15-45 keV is $\sim$ 1.6/detector/day, of which 0.31/detector/day are beta "singles," having a hit in one ZIP detector alone.

With a half-life $T_{1/2}$ = 22.3 years, ^{210}Pb is implanted into surfaces by a decay chain intiated by airborne ^{222}Rn, which itself evolves from ^{238}U. A total of 63 keV is available in the transitions from ^{210}Pb to the ^{210}Bi ground state, which liberates nearly one internal-conversion electron per decay with energy near 40 keV. The ^{210}Bi daughter with half-life $T_{1/2}$ = 5.01 days undergoes β^- decay to ^{210}Po with 1.2 MeV endpoint, but the β-electron energy distribution is non-standard, and peaks at the lowest energies. The ^{210}Po then decays with a half-life $T_{1/2}$ = 138 days to the stable ^{206}Pb isotope, emitting an α particle with an energy of 5.3 MeV.

Our detectors are exposed to radon during fabrication, mounting and testing. We have measured the ambient radon levels at all the locations where detectors travel and have taken precautions to store them in evacuated vessels or in cabinets purged with nitrogen gas for most of the time. The observed rate of α particles in our detectors is $\sim$ 0.4/detector/day and suggests a similar contribution to the total rate of beta events. Further simulation work is in progress to determine what precise fraction of these events result in single-scatter events in the 15-45 keV range.

Another beta-emitting candidate is ^{40}K, produced in the atmosphere by cosmic-ray interactions. Natural potassium may be deposited through human contact. The lack of a visible 1460 keV photopeak or its Compton edge in the Soudan data sets an upper limit of 15 betas/day per ZIP over the full energy range up to the 1.4 MeV endpoint, which corresponds to a limit on the ^{40}K contribution to the total rate of betas of about 0.15 betas/day per ZIP in the 15-45 keV energy region.

Stricter limits on the amount of ^{40}K were set using materials-science surface-analysis techniques. Assays of test wafers produced at intermediate and final stages of processing were performed at campus facilities at Case Western Reserve University and the University of Minnesota using Secondary-Ion Mass Spectroscopy, Auger spectroscopy, RBS, and PIXE. We found that Rutherford backscattering (RBS) combined with particle-induced X-ray emission (PIXE) techniques, done simultaneously for each test wafer, provided the most sensitive normalized test for the presence of light elements such as potassium. RBS is capable of measuring the depth profile of elements and PIXE resolves ambiguities.

Our preliminary upper limit on natural K is $1-2\times10^{14}$ atoms/cm^2; confusion with Ar makes interpretation of the tests challenging. Further studies are in progress to determine whether this upper limit holds generically for witness wafers at all stages of fabrication.

This contamination level would yield a beta rate of about 1–2/detector/day over all energies. Simulations indicate that about 8% of these betas produce energy depositions in the 15–45 keV range and 15% of those (1.4% overall) are single scatters, resulting in 0.02–0.04 single scatters/detector/day, a negligible contribution.

Our work with the above technique has convinced us we can detect buried and surface K layers as small as $1-2\times10^{13}$ atoms/cm^2 when there is no confusion with Ar. Such confusion can be avoid by replacing the Ar atmosphere used during metal deposition with He or Kr.

A third possibility is ^{14}C, also produced by cosmic-ray induced reactions. Natural carbon could be introduced during processing, and traces of carbon are easily measurable by surface analysis methods. Auger and RBS both indicate that there are 2-3 monolayers of C present on the ZIP surfaces, consistent with exposure to air. Depth profiles with RBS at the carbon nuclear resonance show that there is no buried carbon at this level. This yields 0.3 betas/day per ZIP with a

156 keV endpoint. The contribution to the single-scatter rate in the 15-45 keV range is thus likely small, no more than 15%.

Overall, then, while we have quantitative estimates or limits for contamination by many possible emitters, we do not have a full accounting. We summarize our current knowledge in Table 2.

TABLE 2. Surface-event contributions estimated as indicated in the text. Although photons induce about 25% of the surface singles, they induce less than 1% of the events misidentified as nuclear recoils.

source	origin	rate (#/day)	% of singles
^{210}Pb	airborne radon	< 0.4 all	$< 100\%$
^{14}C	natural carbon	< 0.05 all	$< 15\%$
^{40}K	natural potassium	$< 0.02 - 0.04$ singles	$< 15\%$

To progress toward identifying untested contaminants, we have assembled a list of 79 beta-emitting and electron-capture isotopes and performed a preliminary analysis of the possible contribution of these isotopes to our background of surface electrons. Contamination by these isotopes can be potentially determined by mass spectrometry, γ detection, α detection, or of course by detecting the electron itself.

With a commonly achieved ICP-MS sensitivity of 1 ppb, 17 beta-emitting isotopes could be detected. A better ICP-MS sensitivity of 1 ppt, which has been achieved for some isotopes, could allow detection of 20 additional isotopes. We will work with ICP-MS labs to test for these 37 isotopes on the detectors and in substances used during fabrication. In particular, ICP-MS routinely provides sensitivity to ^{238}U at the 100 part-per-trillion (ppt) level, sufficient for our needs.

Forty-three isotopes may be detected by low-level γ-counting, including 25 that cannot be detected by ICP-MS or α-counting. γ detection is advantageous for screening certain materials where high sensitivity can be achieved by using a large mass of sample. An upgrade of the HpGe (SOLO) γ counting facility at Soudan, combined with a lower-background environment provided as part of the Soudan Low Background Counting Facility, will provide sensitive γ screening.

A small number of isotopes, 12, can be detected by their emission of α's. Although few in number, a crucial candidate, ^{210}Pb, is among these isotopes. We have already operated a small Si(Li) detector for α detection, but we believe that detector has a background level that is higher than desired for detecting α's at the small flux now required. We plan to establish an α screening facility at the Soudan lab capable of detecting α's of a flux $< 5 \times 10^{-4}$/cm^2/d.

Finally, there are 12 to 21 isotopes, depending on ICP-MS sensitivity, that cannot be screened in any manner except by their emission of beta electrons. Direct counting of beta particles from large numbers of witness samples is a feasible method by which to detect unscreened isotopes. A sufficiently large screening chamber (or multiple chambers) can observe beta emission from a large surface area, of order 1 square meter, which is equivalent to 100 detectors. Witness samples can be run through the same fabrication and handling process as full detectors at a fraction of the manpower and cost. Rates can be measured quickly and feedback provided in a timely fashion.

However, such a large-area, low-background beta-screening device does not exist. We are beginning to develop it by prototyping two different technologies. The first is a neon-gas multi-wire proportional chamber, the second a cloud chamber read out by digital video. Both are optimized for identification and meausurement of sub-200 keV electrons. With appropriate construction materials, the limiting background in these detectors is ejection of electrons from the sample surface by Compton-scattering photons. In an environment that would yield a 1/kg/keV/d photon rate in Ge, the rate of such events is about 0.3/m^2/keV/day.

A beta screener is not just useful for detecting betas, but makes an excellent alpha-particle counter as well. This ability provides another handle on many of the isotopes that could give rise to a beta background. Also, we can use the beta screener to check in detail our simulations of surface events. A contaminated detector could be run in a test facility to measure the observed spectrum of beta events in energy, yield, and phonon timing, and then the same detector could be placed in a beta screener.

Our current level of beta rejection is estimated to be 99.4% rejection. The combination of ^{109}Cd calibration and WIMP-search data and simulations indicates that the ionization-yield-based rejection is about 80% for such events because of their shallow depth. The phonon-timing based rejection is 97% at 70% cut efficiency. The phonon-timing analysis is under active development and we are confident that at least a factor of five improvement in the beta rejection is possible. It would also be possible for future running to add 'ancient' Pb shielding around the detectors in the icebox to reduce both gamma and beta rates, as was done in CDMS I at the Stanford Underground Facility. Finally, we are considering the use of ionization-only 'endcap' detectors on our towers, to veto gammas and betas from outside materials.

Neutron Backgrounds

A neutron of kinetic energy approximately 2 MeV can cause a Ge recoil that is indistinguishable from a recoil caused by a WIMP. A principal motivation for deploying WIMP searches at deep underground sites is the reduction of ambient cosmic-ray-induced neutron flux. Neutron background at Soudan arises from environmental radioactivity, primarily from processes in the Uranium/Thorium (U/Th) decay chain, and from particle cascades induced by muons that penetrate to Soudan's depth of approximately 2000 mwe. Neutrons from either process can originate in the material immediately surrounding the ZIP detectors ("internal") or from the relatively distant rock in the mine ("external"). Indeed, the radioactivity in the rock necessitates the massive CDMS II shielding consisting of 50 cm-thick polyethylene to degrade the energy of neutrons from external radioactivity, and of 22.5 cm of lead to stop gammas from the same source. The polyethylene suppresses the rate of nuclear recoils due to neutrons from external radioactivity to insignificance. Selection of clean materials for the inner shield and icebox keeps the background from internal contaminations of U/Th small.

The concentration of shielding material near the ZIPs forms a target in which penetrating particles initiate showers of particles, including neutrons. We estimate the rate of Ge nuclear recoils (in the 15-45 keV region) from neutrons in muon-induced showers in the shielding to be 18×10^{-3} /kg/day, with an uncertainty of a factor of 2. Currently, we suppress this background with a hermetic, active, plastic scintillator veto system that surrounds the Soudan ZIP detectors. The Soudan veto system tags the incoming muon that initates showers in the shielding, with efficiency well in excess of 99%, reducing this background to insignificance. In our published exposure [1], no neutrons were observed, where we would have expected < 0.05 unvetoed neutrons and 1.9 vetoed neutrons. The full CDMS II exposure should constrain the rate of the neutron background with a fractional statistical error of 25%.

The dominant source of neutrons that cause nuclear recoils in Ge, and are anticoincident with the veto, come from showers produced by the ~ 220 GeV muons that penetrate to and interact in the rock that surrounds the CDMS II experimental hall. This muon-induced neutron background was estimated independently using two Monte Carlo techniques (GEANT4 and FLUKA) and checked for consistency.

These simulations indicate that associated shower particles produce signals in our existing veto for $60 \pm 10\%$ of events containing neutrons. This implies a rate for the veto-anticoincident neutron-induced recoils from showers in the rock at Soudan of 3×10^{-4} /kg/day; small contributions from radioactivity raise the estimated total rate of background neutron-induced recoils to 4×10^{-4} /kg/day, with an uncertainty of a factor of 2.

This background rate is sufficiently low that neutron background should be unimportant for the exposures currently planned at Soudan, but eventually they would limit further sensitivity improvements there. That is the principal motivation for proposing a larger version of CDMS (called SuperCDMS) to be located at SNOLab, where the dominant neutron backgrounds are suppressed compared to Soudan by over two orders of magnitude (see Figure 7).

One method to improve neutron rejection would rely on the detection of the neutrons themselves. Studies done five years ago for the CDMS II proposal, as well recent studies by the UKDMC [2], indicate that up to one order of magnitude suppression of neutron-induced Ge recoils can be achieved by an internal veto, just outside the Icebox. A schematic of an internal veto is shown in Fig. 8. The main role of the inner veto is to detect dangerous neutrons in the MeV energy range, via the recoil protons, with threshold of order 100 keV. Our current simulations indicate that upgraded veto systems at Soudan could extend background-free CDMS II running there to as much as 5000 kg-d.

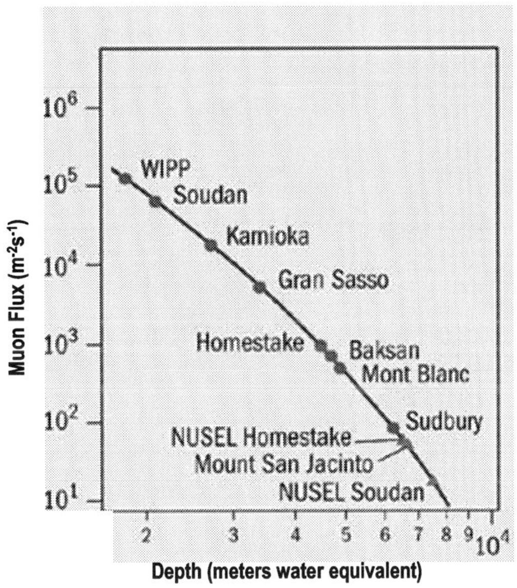

FIGURE 7. Muon flux as a function of depth in meters-water-equivalent, showing existing underground facilties.

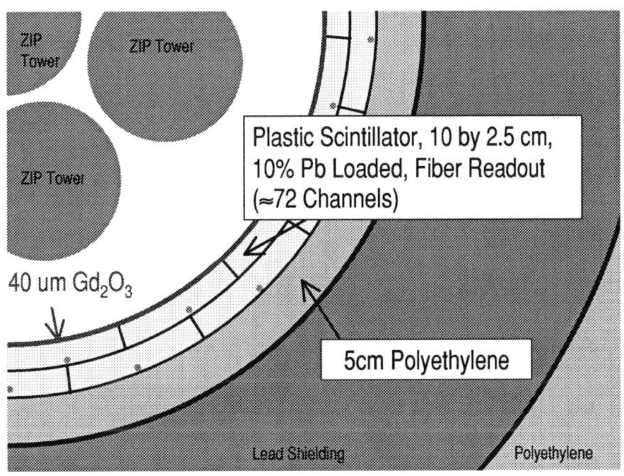

FIGURE 8. Possible Soudan Inner Veto. A double layer of plastic scintillator, read out by waveshifting fiber, would replace the inner 5 cm of the existing polyethylene shielding. Recoils of protons in the scintillator allow detection of MeV neutrons. A layer of Gd_2O_3 would capture slow neutrons, and emit about 8 MeV in several photons, which would be detected in both the ZIP detectors and the scintillator. Only the "barrel" portion surrounding the ZIP detectors is shown; other, probably unsegmented "endcap" portions would go above and below the ZIP detectors.

SCIENTIFIC REACH OF CDMS

In the absence of backgrounds, the search sensitivity of a detector array is directly proportional to the mass $\times$ exposure time (MT). In a mode where subtraction of an estimated background becomes necessary, the sensitivity improvement becomes proportional to $\sqrt{MT}$. Ultimately, the subtraction becomes limited by the systematics of calibrating the detector response to the background and no further improvement in sensitivity is possible. Thus, it is desirable to operate the experiment in "zero-background" mode to obtain the proposed large sensitivity gains.

The initial CDMS II results at Soudan [1] set an upper limit on the WIMP-nucleon cross-section of 4×10^{-43} cm^2 at the 90% C.L. at a WIMP mass of 60 GeV for (spin-independent) coherent scalar interactions and a standard WIMP halo. In Figure 9 we show the expected progression of scientific reach through the end of running at Soudan and the possible reach of two possible SuperCDMS experimental phases at SNOLab. The dash-dot straight line corresponds

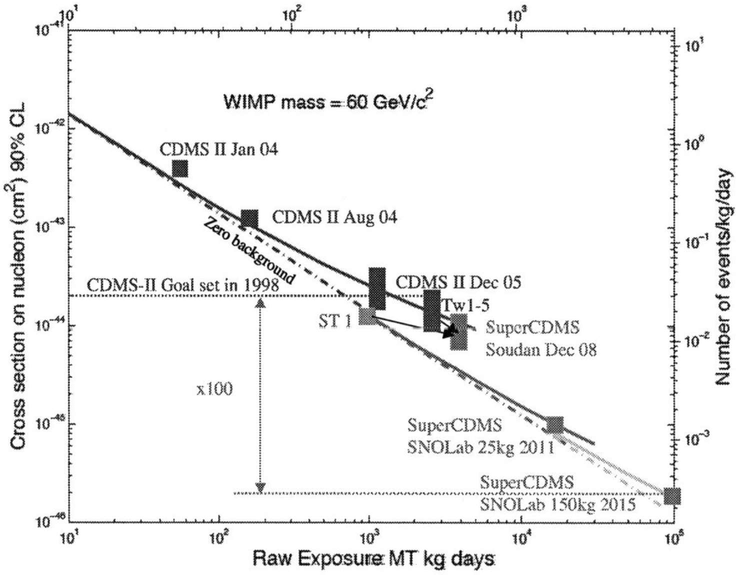

FIGURE 9. Limiting sensitivity (cm^2) as a function of exposure (kg-d) for CDMS II goals and SuperCDMS goals. These estimates use background subtraction and 10 keV thresholds.

to the linear improvement in sensitivity for a zero background experiment. The curved lines show the effect of backgrounds which are subtracted after careful identification, and the rectangles are the projected reach at the end of each phase. Clearly the specific challenge for cryogenic detector experiments such as ours is to continue to improve beta rejection so as to stay near the zero background line.

ACKNOWLEDGMENTS

I would like to acknowledge all of my colleagues from the CDMS collaboration whose work is represented in this article. I would also like to thank all of the organizers of LRT2004 for a very interesting workshop and Richard Ford and Doug Hallman, in particular, for a fantastic tour of the SNO experiment.

REFERENCES

1. D.S. Akerib et al. (CDMS Collaboration), astro-ph/0405033, accepted by *Phys. Rev. Lett.*
2. P.F. Smith, Private communication.

Soudan Low Background Counting Facilty (SOLO)

Michael Attisha*, Luiz De Viveiros*, Richard Gaitksell* and John-Paul Thompson*

*Brown University, Providence RI 02912

Abstract. The Soudan Low Background Counting Facility (SOLO) has been in operation at the Soudan Mine, MN since March 2003. In the past two years, we have gamma–screened samples for the Majorana, CDMS and XENON experiments. With individual sample exposure times of up to two weeks we have measured sample contamination down to the 0.1 ppb level for ^{238}U / ^{232}Th, and down to the 0.25 ppm level for ^{40}K.

Keywords: low-background gamma screening, HPGe, Radon
PACS: 82.80.Ej

INTRODUCTION

The SOudan LOw Background Counting Facility (SOLO) has been in operation at the Soudan mine in Minnesota since March 2003. The SOLO screening facility was constructed by Brown University, in conjunction with PNNL, the University of Minnesota and the Soudan Mine, and has been, and is currently being used to screen materials for the Majorana, CDMS and XENON experiments.

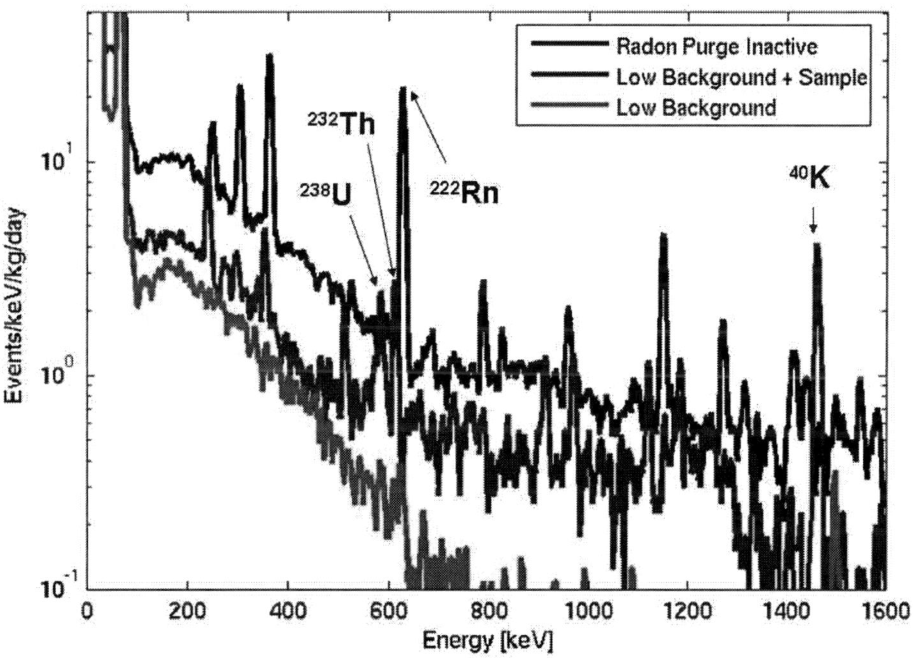

FIGURE 1. SOLO data taken using the 0.6 kg HPGe "Diode M" detector. From top to bottom: data taken without the Radon gas purge active, background data taken with a screening sample present (Hamamatsu R9288 PMTs) plus Rn purge, and background data (no sample) with Rn purge active.

The facility uses HPGe detectors and shielding previously deployed in Double Beta Decay Experiments at the Homestake Mine. Two independent detectors are housed in separate chambers, each surrounded by at least 12" of Pb, the innermost 2" of which is low ^{210}Pb-activity (German) Pb. A 2" wall of German Pb separates the chambers, each of which houses a HPGe detector; 'TWIN' (1kg Ge) and 'Diode M' (0.6kg Ge). The entire enclosure is sealed up with two airtight layers of 50μm thick white mylar, and radon gas is purged using boiloff gas from a LN dewar fed

CP 785, *Topical Workshop on Low Radioactivity Techniques,* edited by B. Cleveland, R. Ford, and M. Chen

in through the shield. Figure 1 shows the spectra with and without the Rn purge active. At present, when samples are introduced no airlock is employed. Gamma lines due to Rn disappear less than 3 hours after the shield is closed, and the N_2 gas purge, at a rate of 3 cfh, is started.

A simple data acquisition system is used, consisting of Spectroscopy Amplifiers with $2\mu s$ shaping times, and pulse height ADCs connected to a PC running acquisition software. Spectra are integrated in 4 hour intervals, and transferred automatically to a web accessible archive.

The TWIN detector is currently offline due to a failure in its front–end FET. A low background/low noise FET is currently being prepared by PNNL as a replacement. In conjuction with the University of Florida, a 2kg HPGe detector was installed in the shield in June 2005, and will perform future screening alongside Diode M.

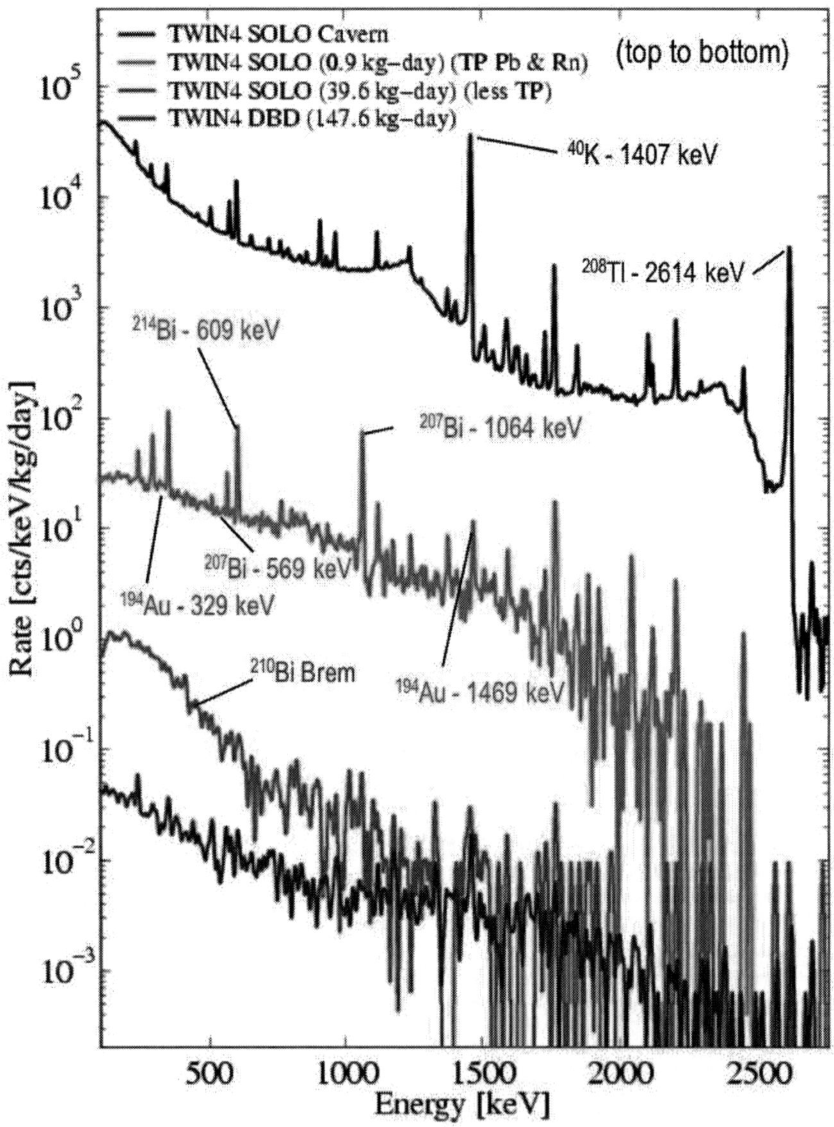

FIGURE 2. SOLO data taken with the TWIN detector. From top to bottom: cavern background rates, initial spectrum including TP brick, final spectrum without TP brick, and background taken during Double-Beta operation at Homestake.

ANECDOTES

We relate two unusual incidents that occurred during the commissioning of the facility, in the hope that it may provide guidance to others in the future.

Upon first operation in June 2003, the TWIN spectrum showed >10 events/keV/kg/day up to 1 MeV, the same detector having operated at <1 event/keV/kg/day in the past (see Fig. 2). At the same time, Diode M observed a background rate ~1 event/keV/kg/day up to 500 keV, suggesting that the TWIN detector was subject to a source that Diode M was not. Analysis of the TWIN spectral lines in the data indicated the presence of two unexpected isotopes: ^{207}Bi (33 years half-life) and ^{194}Hg (400 years half-life). The relative strengths of the lines at different energies appeared to indicate the presence of 3.5-4.5cm of Pb between the source contaminant and the detector. Both isotopes are very rare, but are known to be spallation products arising from high energy protons on Pb.

In order to test the hypothesis that contaminated bricks were causing the high background levels on TWIN, the upper section of the SOLO Pb shield was disassembled to examine the bricks surrounding the chambers. Bricks stamped with the letters 'TP' were found directly above the TWIN chamber, behind the low activity German Pb layer, in a location consistent with our calculations. We understand that these bricks came from Fermilab, where they may have been used as a proton beam stop. Unfortunately, at some stage they must have been combined with the gamma shielding stock of Pb at Soudan. Removal of these 'TP' bricks substantially reduced the background on the TWIN detector, as can be seen in Fig. 2.

In May 2004, SOLO was counting several blocks of polyethylene for the CDMS experiment, in order to establish the intrinsic levels of U/Th/K. The polyethylene data showed a large number of unexpected gamma lines. The polyethylene spectrum was ~5 events/keV/kg/day above background across all energies. (On closer inspection the no–sample backgrounds also appeared elevated.) While the polyethylene lines could not be matched to the patterns from any known combination of radioactive contaminants, we realized they were characteristic of gamma cascades arising from thermal neutron capture in Ge. The low energy neutron flux coming from the rock at Soudan was insufficient to be generating such pronounced features, however, it became clear that the polyethylene was moderating neutrons from a calibration source (10^5 n/s) had been placed in storage 20ft from the SOLO experiment. Needless to say, the neutron source was moved to another storage location.

SAMPLE SCREENING AND REMAINING BACKGROUNDS

Over the past two years SOLO has screened many samples. For sample runs of two weeks, sensitivities better than 0.1 ppb level for ^{238}U/^{232}Th, and 0.25 ppm level for ^{40}K have been achieved.

As Fig. 2 shows, the SOLO background below 1 MeV is 5-10 times greater than when these detectors were in operation at Homestake, while the background above 1 MeV is very similar to the levels achieved previously. The shape of the excess background is consistent with that expected from ^{210}Pb bremsstrahlung. As a general rule of thumb, a brem spectrum with a rate at low energies ($<$ 100 keV) of 1 event/keV/kg/day would arise from Pb with an intrinsic ^{210}Pb level of 1 Bq/kg. The German Pb liner has an intrinsic activity of <50 mBq/kg so it is unlikely that this is the cause of the elevated background. We have systematically altered the shielding geometry, and have also cleaned the surfaces of the inner Pb liner (May 2005) using dilute nitric acid. This intervention has had little effect on the level of this background. We now believe that the brem could be arising from Rn plating on the surface of the Cu housing on the HPGe detectors. The detectors were stored in non–airtight packing cases for over 5 years in the Soudan Mine where the airborne Rn levels fluctuate between 100–500 Bq/m^3. Back of the envelop calculations indicate that Rn daughters plating onto the Cu housing could indeed lead to sufficient activity to explain the residual brem spectra. We intend to try acid etching the Cu during the next major overhaul of the system.

CONCLUSION

Photographs of the SOLO facility can be seen in Figs. 3 and 4. We would like to thank our colleagues from PNNL, UMinn, and Soudan Mine for their collaboration establishing and operating this facility, and also our new collaborators from UFlorida. Further details of the facility can be obtained at **http : //particleastro.brown.edu/SOLO/**. Please contact us if you would like to make use of the facility.

FIGURE 3. SOLO Gamma Screening Facility at Soudan Mine, MN. The Pb shield is surrounded by white mylar providing a Rn barrier. The LN dewars of the two HPGe detectors can be seen to the left and right of the main shield.

FIGURE 4. The two HPGe detectors on the ends of their Cu cold fingers can be seen, during shield assembly.

Beta Cage: A New, Large-Area Multi-Wire Screening Detector For Surface Beta Contamination

T. Shutt[a], C.E. Dahl[a,b], L. DeViveiros[c], R.J. Gaitskell[c] and R.W. Schnee[a]

[a]*Department of Physics, Case Western Reserve University, Cleveland, OH 44106, USA*
[b]*Department of Physics, Princeton University, Princeton, NJ 08544, USA*
[c]*Department of Physics, BrownUniversity, Providence, RI 02912, USA*

Abstract. We report on a design study for a new, dedicated, low-radioactive-background screening detector for low-energy beta-emitting contamination on surfaces. The detector, a large-area gaseous detector with multi-wire proportional readout, has sufficient sensitivity to screen materials at the level needed for the proposed SuperCDMS 1 ton dark matter experiment. Such capability does not currently exist. This detector would have additional applications in low-radioactive-background physics experiments, and radioactive trace analysis, such as ^{14}C dating.

Keywords: low radioactive background screening, beta decay, dark matter.
PACS: 29.40.Cs, 95.35.+d

INTRODUCTION

This paper describes the design of "Beta Cage" - a new detector for low-radioactive-background screening of low-energy beta-emitting contamination on large-area surfaces. Though it has broader applications, which we discuss below, it is primarily motivated by the CDMS dark matter experiment. The detector was proposed, and a preliminary design studies were completed, in 2003. Shifting commitments have so far prevented the authors from further developing this system. The LRT2004 workshop offered a good opportunity to present this idea to the broader community, and this paper serves as a record of this work.

Low-energy beta contamination on or near detector surfaces, most likely Rn daughters, is the current dominant background in CDMS[1], and is likely to remain the most challenging background for CDMS-style dark-matter detectors. Future scaling of detector mass (SuperCDMS) will require that any contamination of detector surfaces (during processing of the detector thin films or during handling) be minimized and screened. To achieve a background of less than one event for a 100-kg or ton phase of SuperCDMS, the rate of beta decays below 200 keV should be limited to 1-100 events/m^2/day, depending on the level of detector discrimination. Fast screening techniques are needed to identify the contaminants and the fabrication procedures by which they are introduced. Chemical assaying techniques are useful for many long-lived possible contaminants, but direct measurement of betas is necessary for 20 isotopes, and removes ambiguities for all contaminants.

There is currently no screening capability that we are aware of that approaches this level of sensitivity for low energy betas. However we note that double-beta decay searches based on gaseous TPCs[2,3] and commercially available alpha screening detectors[4] are similar, though they work at much higher energies than we consider. We do not attempt to survey these prior systems or compare them in detail with this detector, though such an effort would doubtless be fruitful.

CP 785, *Topical Workshop on Low Radioactivity Techniques,* edited by B. Cleveland, R. Ford, and M. Chen

DETECTOR OVERVIEW

Beta Cage is a large area gaseous detector for surface contamination of low-energy beta emitters. Three goals have guided the design:

1. Use only the minimum amount of detector mass necessary to stop electrons of interest. This minimizes the background from ambient penetrating gammas interacting in the detector.
2. The detector should have the minimum possible surface area, which itself can be a source of betas.
3. Have sufficient spatial information about events to distinguish those coming from the sample surface from backgrounds arising elsewhere.

Our solution to these requirements is a gas drift-chamber with multi-wire proportional-gain readout. A sketch of the detector is shown in *Figure 1.* Samples are placed directly in the gas. As sketched, only the top of the sample is actively measured; a two-sided chamber could be readily envisaged. Three sets of readout wire grid structures are employed. One sits directly above the sample surface and tags events at or near the sample. Above this is a drift region that provides the stopping power for the betas. Electrons generated here are drifted to a top set of readout grids. The depth of the drift region sets the energy of betas whose full energy is contained in the detector. A field shaping structure will be needed at the outside of the drift region. The mechanics must allow opening of the gas chamber, and separation of the grids structures to access samples.

Very low energy events at (or near) the sample will deposit energy only in the trigger region. At higher energies events from the sample will also deposit energy in the drift region; these can be confused with events of the same energy originating in the drift region that extend to the sample. It might be possible to distinguish these types of events based on their track geometry using the measured energy partitioning between trigger and top grids, and/or the spatial profile in the z direction as measured by the time profile of the signal. In the background estimates below we assume no such discrimination has been performed.

Events originating below the sample are identified by a lower active veto region and grid structure. Events above the top grid but still intersecting the trigger grids could be rejected based on an energy cut (below some energy the event size cannot extend from the sample to the top of the chamber), or additional veto grids could be employed. Crossed grids will be used for x-y position determination. The granularity of the x-y readout, which will determine the DAQ costs, could be ~mm. High x-y segmentation will allow measurement of variations across a sample, and simultaneous measurement of multiple samples, which could be particularly important for a large area detector.

By design, the only surface of the detector contributing backgrounds comes from the wires, which are a few % of the sample surface. Each grid structure consists of a center anode grid, surrounded by two sense grids. The grid structures will have a grid spacing and wire pitch of a few mm. Anode wires of 20 μm Ø should allow gas gain at a few hundred volts. We note that in principle, a photon sensor could give t_0, thus allowing a time-projection-chamber readout of the z position, eliminating the need for the trigger grid, but this does not seem practical for the small amount of light from ≈ keV betas.

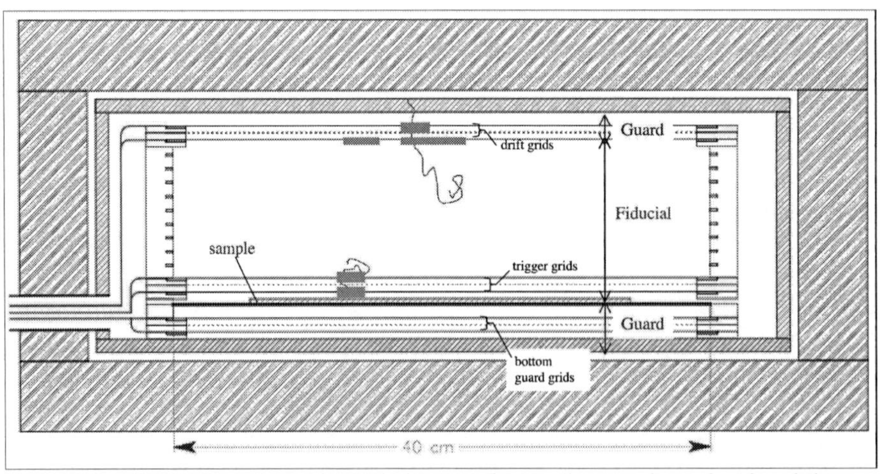

FIGURE 1. Side view of the Beta Cage. Two schematic beta tracks are shown, along the signal they would create on various grids.

The gas choice, and detector thickness are determined by electron stopping. Only He, Ne and Xe are the only obvious gas choices that are (mostly) free of radioactive contaminants. We performed a Monte Carlo study, using

Geant4, of electron tracks. Tracks below < 100 keV have completely curled up on themselves, so their size is not simply related to their total track length. We find that Ne gas at STP makes a good choice for low energy betas, with most 150 keV events fully contained with a ~15 cm drift region. Xe gas should be used to measure alphas or higher energy betas.

The demand for a low radioactive background environment requires that the chamber be constructed of copper or plastic. Gas handling is simplified, and non-atmospheric pressures can be utilized, if a vacuum tight, and, possibly, pressure-capable copper box is constructed. However this will be expensive and technically difficult, especially for large detector sizes. A plastic, ambient-pressure chamber is simpler, but will outgas. For Ne an initial purging might be economical; alternatively a recirculating purification system could be employed for Ne or Xe.

BACKGROUNDS

Several sources of potential background have been considered, and are summarized in Table 1. The dominant background is expected to come from of penetrating gamma rays the shield and possible detector chamber that interact (primarily via Compton scattering) in the gas or sample. In shielded Ge detectors, for example, typical backgrounds at low energy are roughly 1 cnt/kg/keV/day[5]. We studied the gamma background with Geant4, using a an external background tuned to match the gamma background in CDMSII, and scaled to give 1 count/kg/keV/day in Ge at low energy. The resulting background in our detector for all events coincident with the trigger grid are shown in figure 2. The surface area is 700 cm^2, so 2e-2 cnts/keV/day corresponds to 3e-5 cnts/cm^2/keV/day. With sufficient care, it may be possible to achieve background levels an order of magnitude better than this.

Other backgrounds have been roughly estimated as follows. In Xe gas, Kr, which has the beta-decay isotope ^{85}Kr, is thought to be at a level of ≈5 ppb for the best commercially available Xe[6]; this should dominate over Ar (with ^{39}Ar). Kr and Ar should be readily removed from Ne with a trap. ^{14}C in the quench gas use for the multi-wire proportional charge amplification is potentially more serious. The organic liquid scintillator used in Borexino's CTF facility[7] had a $^{14}C/^{12}C \approx 2x10^{-18}$, and it seems reasonable to achieve methane at this level. A significantly higher rate would be needed before this became a problem. If needed, a non-organic gas such as SF_6 could be explored.

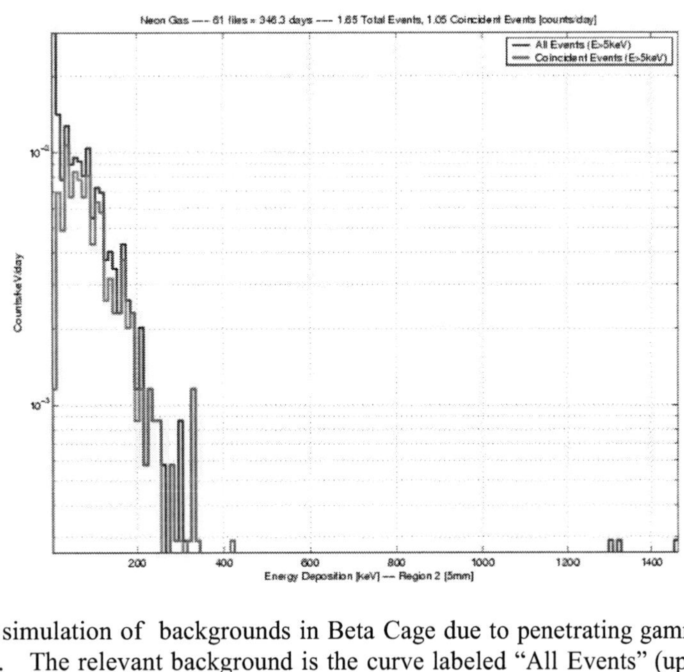

FIGURE 2. Monte Carlo simulation of backgrounds in Beta Cage due to penetrating gamma rays from the shield and outer detector regions. The relevant background is the curve labeled "All Events" (upper curve). The higher labeled rate is 10^{-2} counts/keV/day over 700 cm^2; the x scale is keV.

Most of the materials of the detector could be made of plastic or Cu, which will lead to backgrounds at the level expected for the shield. The grid wires will not be made of copper, but have a small mass; their dominant background is betas from their bulk. The wire surfaces, on the other hand, must be carefully cleaned, and may represent an ultimate background. The wire grids could be fabricated by soldering (using a Pb-free solder) to circuit boards made of Cirlex (a laminated Kapton) which is used in CDMS. The solder and Cirlex should give a small background. Surface-mount, ceramic electrical components, though small, could be problematic – we show a nominal value for 30 resistors. Electrical components could, with some inconvenience, be placed outside the shield.

TABLE 1. Summary of several expected backgrounds in Beta Cage.

Source	Rate (cnts/keV/cm^2/day)
Gammas, equivalent to 1 cnt/kg/keV/day in Ge.	3×10^{-5}
Kr at 5 ppb in Xe.	1×10^{-6}
5% methane quench gas (assume 10^{-16} C^{14}/C^{12})	5×10^{-6}
Wires - emergent beta rate < 100 keV.	$\approx10^{-8}$
Wires: Rate from gammas in full volume.	$\approx 10^{-13}$
30 resistors: Rate from gammas in full volume.	$\approx10^{-6}$

PROJECTED SENSITIVITY

The sensitivity of the detector depends on the sample area and final background. In *Figure 3* we show the 3 σ (99.8 % CL) sensitivity for 30 days counting time, assuming gaussian statistics. For large enough exposures, there is non-zero background due to photons, especially at low energy. It is standard practice in low-background counting to subtract the background after having measured the blank background with no sample present. We do this here, but caution that background photons will also interact in the sample leading to escaping electrons (and X-rays) not measured in a blank run. Thus it may be necessary to use external gamma sources to calibrate the gamma-related background, and/or develop blank samples that are determined to have low surface contamination. However it is clear that the 1-100 cnts/m^2/day goal of the 1-ton SuperCDMS experiment could be readily met with a modest sized detector in a short time.

OTHER APPLICATIONS

A large-area low-mass screening facility such as this will undoubtedly have wider applicability. In a very obvious way it would be ideal for alphas. It is likely that the only background that could contaminate an alpha measurement, especially at a deep site, would come from contamination on and in the wires themselves, which could be made very low. An alpha counter similar to this was considered during construction of SNO[8], and alpha backgrounds are important for dark matter searches[9], solar neutrino experiments such as Borexino[10], and double beta decay experiments.

Besides these rare-interaction physics experiments, this technology could be useful for trace radioactive counting applications. A potentially very promising application is ^{14}C measurements. It is already a common practice to convert organic material for ^{14}C analysis to CO_2, and such a gas could be readily used as the detector medium in this proportional counter. For a background corresponding to the middle curve of *Figure 3*, a sensitivity to a few $\times10^{-16}$ seems possible, depending on how well background subtraction works. With an aggressive effort on the cleanliness of the shield and detector, a sensitivity near 10^{-17} might not be out of the question.

CONCLUSIONS

We have proposed Beta Cage, a large area gas drift chamber with multi-wire proportional readout for low background screening of low-energy surface electrons. Although based on similar existing technologies, it would be the first dedicated facility for this specific background. We find that a modest size counter would allow rapid screening of samples at the needed level for a 1 ton SuperCDMS-style experiment. This counter would also have applications to other low background physics experiments, and low radioactivity counting applications, such as ^{14}C

dating, though we have not fully explored the potential for these applications. It is our hope that this design study will prove to be a useful first-draft for an eventual working detector.

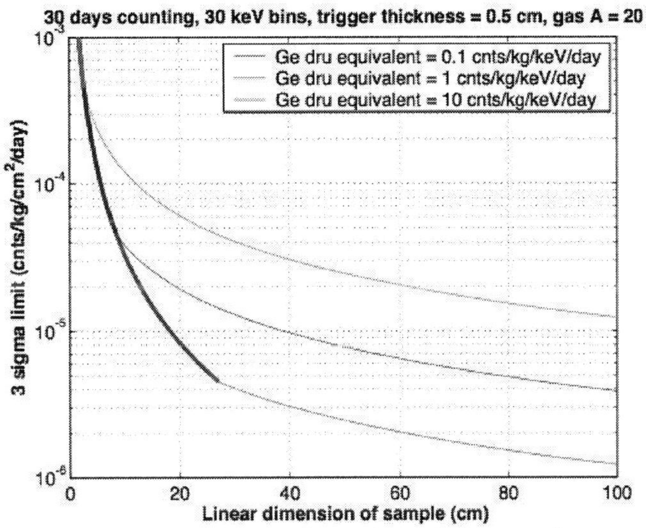

FIGURE 3. Sensitivity as a function of linear dimension of the sample area, and for several photon backgrounds. Our target background is at least 1 cnt/kg/keV/day (Ge equivalent), and ultimately 0.1 dru. The heavy lines indicates no counts in 30 keV energy bins for 30 days counting.

REFERENCES

1. D.S. Akerib, *et al.*, "First Results from the Cryogenic Dark Matter Search in the Soudan Underground Lab", Phys. Rev. Lett. **93**, 211301 (2004).
2. R. Arnold, *et al.*, "Technical design and performance of the NEMO 3 detector", NIM **A**, 536, 79-122 (2005);
3. J.-C. Vuilleumier, "Search for neutrinoless double-beta decay in ^{136}Xe with a time projection chamber", Phys Rev **D** 48, 1009-1020 (1993).
4. For example, X-Ray Instrument Associates, Newark CA, USA, www.xia.com.
5. See, for example, G. Heusser, *"Low-Radioactivity background techniques"* Annu. Rev. Nucl. Part. Sci., 45, 543-590 (1995)
6. S. Moriyama, presented at IDM 2004, http://www.shef.ac.uk/physics/idm2004.html (2004).
7. G. Alimonti *et al.*, "Science and Technology of BOREXINO: a real-time Detector for Low Energy Solar Neutrinos", *Astropart.Phys* **16**, 205 - 234 (2002).
8. T. Bowles, private communication.
9. V.A. Kudryavtsev, *et al.*, "Study and suppression of anomalous fast events in inorganic scintillators for dark matter searches", Astropart. Phys. **17** 401-408 (2002).
10. A. Pocar, Ph.D. Thesis, Princeton University (2004).

IV – LOW RADIOACTIVITY DETECTOR COMPONENTS

Radioactive contamination of $CaWO_4$, $ZnWO_4$, $CdWO_4$, and Gd_2SiO_5:Ce crystal scintillators

F.A. Danevich*, A.Sh. Georgadze*, V.V. Kobychev*, B.N. Kropivyansky*, S.S. Nagorny*, A.S. Nikolaiko*, D.V. Poda*, V.I. Tretyak*, S.Yu. Zdesenko*, Yu.G. Zdesenko[1]*, P.G. Bizzeti†, T.F. Fazzini†, P.R. Maurenzig†, I.M. Solsky**, V.B. Brudanin‡ and F.T. Avignone III§

*Institute for Nuclear Research, MSP 03680 Kiev, Ukraine
†Dipartimento di Fisica, Universitá di Firenze and INFN, 50019 Firenze, Italy
**Institute for Materials, 79031 Lviv, Ukraine
‡Joint Institute for Nuclear Research, 141980 Dubna, Russia
§University of South Carolina, Columbia, South Carolina 29208

Abstract.
Radioactive contaminations of $CaWO_4$, $ZnWO_4$, $CdWO_4$, and Gd_2SiO_5:Ce crystal scintillators were measured in the Solotvina Underground Laboratory. The Monte Carlo simulation of contributing radioactive sources, the time-amplitude and pulse-shape analyses of the experimental data were applied. $CaWO_4$ and GSO scintillators are considerably polluted by uranium and thorium, while the total α activity in $ZnWO_4$ and $CdWO_4$ scintillators does not exceed the level of a few mBq/kg. Particular radioactivity was observed in $CaWO_4$ (α active ^{180}W), $CdWO_4$ (β active ^{113}Cd, $2\nu2\beta$ decay of ^{116}Cd, ^{180}W), and GSO (α active ^{152}Gd) detectors. The radioactive contamination of the studied scintillation crystals is compared with that of commonly used detectors.

Keywords: Scintillation detectors; Radioactive contamination
PACS: PACS numbers: 29.40.Mc

INTRODUCTION

There is class of experiments on searches for rare events (as for instance, low-level gamma-, beta-, alpha- spectrometry, double beta decay (2β) and dark matter particles search, measurements of solar neutrino flux) which demand ultra-low background of detector. After the main sources of background (such as environmental radioactivity, cosmic rays, radioactivity of shield) are suppressed, internal radioactive contamination of detector begins to be the most important source of background.

Many experiments to search for rare α [1, 2, 3] and β decay [4, 5, 9], 2β decay [6, 8, 10, 11, 12, 13, 14, 15] and dark matter particles [16, 17, 10, 18] were performed using crystal scintillators. Some of scintillation crystals are also used as low-background scintillating bolometers [19, 20, 21, 22]. Here we analyze number of low-background measurements to estimate radioactive contamination of calcium, zinc, cadmium tungstate ($CaWO_4$, $ZnWO_4$, $CdWO_4$), and cerium-doped gadolinium orthosilicate (Gd_2SiO_5:Ce, GSO) crystal scintillators as promising detectors for rare events experiments.

EXPERIMENTAL METHODS

$CaWO_4$, $ZnWO_4$, $CdWO_4$, and GSO scintillation crystals studied in the present work were grown by Czochralski method. The main properties of the scintillators are presented in Table 1. All these crystals are non-hygroscopic and chemically resistant.

[1] Deceased.

CP 785, *Topical Workshop on Low Radioactivity Techniques,* edited by B. Cleveland, R. Ford, and M. Chen

TABLE 1. Properties of $CaWO_4$, $ZnWO_4$, $CdWO_4$, and GSO crystal scintillators

	$CaWO_4$	$ZnWO_4$	$CdWO_4$	GSO
Density (g/cm^3)	6.1	7.8	8.0	6.71
Melting point (° C)	1570 – 1650	1200	1325	2173
Hardness (Mohs)	4.5 – 5	4 – 4.5	4 – 4.5	5.7
Wavelength of emission maximum (nm)	420 – 440	480	480	440
Refractive index	1.94	2.1 – 2.2	2.2 – 2.3	1.85
Effective average decay time (μs)*	8	24	13	0.05
Photoelectron yield [% of NaI(Tl)]*	18%	13%	20%	20%

* For γ rays at 20° C.

Radioactive contaminations of the crystals were measured in the Solotvina Underground Laboratory built in a salt mine 430 m underground ($\simeq$1000 m w. e., cosmic muon flux 1.7×10^{-6} cm^{-2} s^{-1}, neutron flux $\leq 2.7 \times 10^{-6}$ cm^{-2} s^{-1}, radon concentration in air < 30 Bq m^{-3}) [23].

To estimate radiopurity of cadmium tungstate scintillators, the data of the ^{116}Cd and ^{106}Cd double beta decay experiments [24, 8, 25, 12] were used. The radioactive contamination of GSO scintillator was determined on the basis of the experiment [26, 11].

The $CaWO_4$ ($40 \times 34 \times 23$ mm) was measured during 1734 h in the low background set-up. The $CaWO_4$ scintillator was viewed by the special low radioactive 5" PMT (EMI D724KFL) through the high pure quartz light-guide 10 cm in diameter and 33 cm long. The $CdWO_4$ crystal ($\oslash 59 \times 21$ mm) was measured during 37 h in the same set-up. The $ZnWO_4$ scintillation crystal ($\oslash 14 \times 4$ mm, 429 h) was viewed by the 3" PMT (FEU-137) through the high pure quartz light-guide 4.9 cm in diameter and 25 cm long. The detectors were surrounded by an passive shield made of teflon (3 – 5 cm), plexiglass (6 – 13 cm), high purity copper (thickness 3 – 6 cm), lead (15 cm) and polyethylene (8 cm). Two plastic scintillators ($120 \times 130 \times 3$ cm) were installed above the passive shield to generate a cosmic muons veto. For each event in the detector the amplitude of a signal and its arrival time were recorded. In addition, scintillation pulse shape of the $CaWO_4$ and $CdWO_4$ scintillators were digitized with a 20 MHz sampling frequency.

Pulse shape analysis was applied to $CaWO_4$ and $CdWO_4$ data as described in [27, 1, 28, 15]. Technique of time-amplitude analysis to recognize a presence of the short-living chains from ^{232}Th, ^{235}U and ^{238}U families is described in [24, 11]. The Monte Carlo simulation of contributing radioactive sources was performed with the help of GEANT3 and GEANT4 packages [29], and event generators DECAY0 and DECAY4 [30].

RESULTS AND DISCUSSION

The energy spectrum measured with the $ZnWO_4$, $CaWO_4$ and $CdWO_4$ crystal scintillators in the low background set-up is presented in Fig. 1 (the spectra are normalized by the measurement times and the detector masses).

The background of the $ZnWO_4$ detector is substantially lower than that of the $CaWO_4$ and is comparable with that of the $CdWO_4$ above ≈ 0.5 MeV. Note that below 0.5 MeV the counting rate of the $ZnWO_4$ detector is one order of magnitude lower than that of $CdWO_4$. Obviously, it is due to presence of the β active ^{113}Cd isotope (natural abundance of $\approx 12\%$) in the $CdWO_4$ crystals.

Activity of thorium and uranium α active daughters in $CdWO_4$ and $CaWO_4$ crystals was determined with the help of pulse shape analysis. In addition, the pulse shape analysis allows to estimate activity of ^{228}Th and ^{226}Ra due to selection of double pulses associated with the fast sequences $^{214}\text{Bi} \rightarrow {}^{214}\text{Po}$ ($T_{1/2} = 164$ μs) $\rightarrow {}^{210}$Pb and $^{212}\text{Bi} \rightarrow {}^{212}\text{Po}$ ($T_{1/2} = 0.3\mu$ s)$\rightarrow {}^{208}$Pb from ^{235}U chain.

More precise data on radioactive impurities associated with the daughters of ^{232}Th, ^{235}U and ^{238}U were obtained with the help of the time-amplitude analysis. For example, the fast sequence of two α decays from the ^{232}Th family was searched for: ^{220}Rn ($Q_\alpha = 6.41$ MeV, $T_{1/2} = 55.6$ s) $\rightarrow {}^{216}$Po ($Q_\alpha = 6.91$ MeV, $T_{1/2} = 0.145$ s) $\rightarrow {}^{212}$Pb (which is in equilibrium with ^{228}Th). In the same way, the activity of ^{226}Ra (^{238}U family) in the scintillators was measured by selection of the fast sequence ^{214}Bi ($Q_\beta = 3.27$ MeV, $T_{1/2} = 19.9$ m) $\rightarrow {}^{214}$Po ($Q_\alpha = 7.83$ MeV, $T_{1/2} = 164$ μs) $\rightarrow {}^{210}$Pb. The presence of ^{227}Ac (^{235}U family) was estimated by time-amplitude analysis of ^{219}Rn ($Q_\alpha = 6.95$ MeV, $T_{1/2} = 3.96$ s) $\rightarrow {}^{215}$Po ($Q_\alpha = 7.53$ MeV, $T_{1/2} = 1.78$ ms)$\rightarrow {}^{211}$Pb from ^{235}U chain.

The background spectrum measured with the GSO crystal ($\oslash 47 \times 57$ mm) during 13949 h [11] is depicted in Fig.

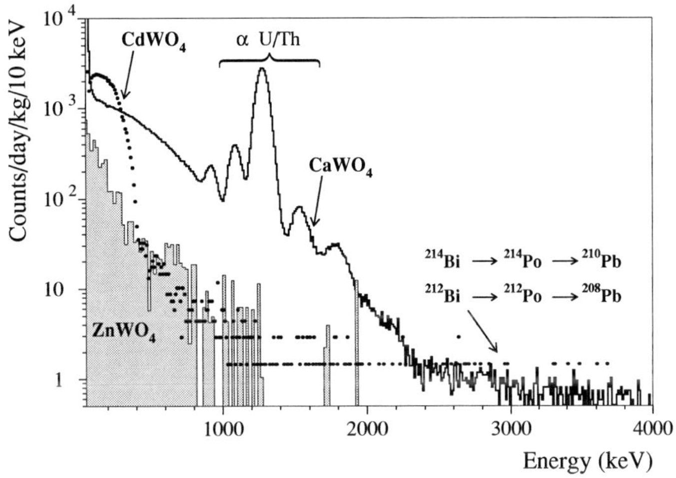

FIGURE 1. Energy spectra of $ZnWO_4$ (4.5 g, 429 h), $CaWO_4$ (189 g, 1734 h), and $CdWO_4$ (448 g, 37 h) scintillation crystals measured in the low background set-up

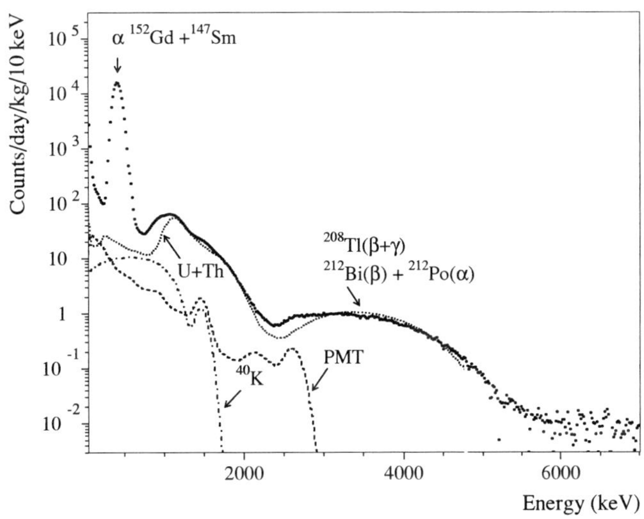

FIGURE 2. The background spectrum of the GSO crystal for 370.5 yr kg of exposure and the model of background.

2, where the following peculiarities exist: the clear peak at the energy 0.42 MeV, the comparatively wide peak at the energy around 1.05 MeV and two broad distributions dropped down at the energies 2.4 and 5.5 MeV. The first peak is attributed to α particles of ^{152}Gd ($T_{1/2} = 1.08 \times 10^{14}$ yr; $E_\alpha = 2140$ keV; abundance $\delta = 0.20\%$) and ^{147}Sm ($T_{1/2} = 1.06 \times 10^{11}$ yr; $E_\alpha = 2233$ keV; $\delta = 15\%$; samarium can be present as impurity of the GSO crystal at the level of $\approx$8 ppm). The peak near 1050 keV as well as the broad distribution up to the energy 2.4 MeV is mainly due to the radioactive contamination of the crystal by the nuclides from the ^{232}Th, ^{235}U and ^{238}U families. The distribution up to the energy 5.5 MeV is caused by decays of ^{232}Th daughter isotopes: a) β decay of ^{208}Tl ($Q_\beta = 5.00$ MeV); b) β decay of ^{212}Bi ($Q_\beta = 2.25$ MeV) followed by fast α decay of its daughter ^{212}Po ($T_{1/2} = 0.3$ μs; $E_\alpha = 8.78$ MeV or $\approx$2.7 MeV in β scale). The estimation of radioactive contamination of the GSO scintillator was performed by using the time-amplitude analysis and by analysis of the energy spectrum shape. The model of background was built with the help of Monte Carlo simulation. It should be stressed that no surface concentration of radionuclides was observed for the GSO crystal.

The summary of the measured radioactive contamination of the $CaWO_4$, $ZnWO_4$, $CdWO_4$ and GSO crystal scintillators (or limits on their activities) is given in Table 2.

As one can see from Table 2, the secular equilibrium in uranium chain is substantially broken in $CaWO_4$ crystal.

TABLE 2. Measured radioactive contaminations in $ZnWO_4$, $CaWO_4$, $CdWO_4$, and GSO crystal scintillators (mBq/kg).

Chain	Source	$CaWO_4$*	$ZnWO_4$†	$CdWO_4$**	GSO‡
^{232}Th	^{232}Th	0.69(10)	≤ 3.3	0.053(5)	≤ 6.5
	^{228}Ra				≤ 9
	^{228}Th	0.6(2)	≤ 0.2	$\leq 0.004 - 0.039(2)$	2.287(13)
^{235}U	^{227}Ac	1.6(3)	≤ 0.2	0.0014(9)	0.948(9)
^{238}U	^{238}U	14.0(5)	≤ 3.2	≤ 0.6	≤ 2
	^{230}Th		≤ 4.5	≤ 0.5	≤ 9
	^{226}Ra	5.6(5)	≤ 0.4	≤ 0.004	0.271(4)
	^{210}Pb	≤ 430	≤ 1	≤ 0.4	≤ 0.8
	^{210}Po	291(5)			
	^{40}K	≤ 12	≤ 12	0.3(1)	≤ 14
	^{90}Sr	≤ 70	≤ 1.2	≤ 0.2	
	^{113}Cd			580(20) §	
	^{113m}Cd			1-30	
	^{116}Cd			1.0(1) ¶	
	^{137}Cs	≤ 0.8	≤ 20	$\leq 0.3 - 0.43(6)$	
	^{138}La				≤ 55
	^{147}Sm	0.49(4)	≤ 1.8	≤ 0.04	700-1100
	^{152}Gd			≤ 0.04	1200
	^{180}W	0.05(2)		0.04(2)	

* [28, 31]
† [15]
** [25, 32, 8, 12, 1]
‡ [26, 11]
§ In crystals produced from natural cadmium
¶ In crystals produced from Cd enriched in ^{116}Cd

The equilibrium in thorium family is broken in $CdWO_4$ scintillators too. More likely it is a rule for different materials. On the other hand, measurement of particular sub-chains with comparable precision is quite complicate problem. For that reason radioactive contamination by uranium and thorium is more properly characterized by the value of the total α activity inside a detector.

Particular radioactivity, i.e. associated with elemental composition of a detector, was observed in $CaWO_4$ (α active ^{180}W), $CdWO_4$ (β active ^{113}Cd, $2\nu2\beta$ decay of ^{116}Cd, ^{180}W), and GSO (α active ^{152}Gd) detectors.

Let us mention the effect of U/Th traces concentration on surface layer of the $CdWO_4$ crystal grown from enriched ^{116}Cd. Contribution from U/Th α active daughters was decreased in ≈ 50 times after removing of surface layer (≈ 1 mm) of the crystal [24].

In Table 3 the radioactive contamination of the studied scintillation crystals is compared with that of commonly used detectors.

CONCLUSIONS

We present the radio-purities of $CaWO_4$, $ZnWO_4$, $CdWO_4$ and GSO crystals and compare them with other detectors. $CaWO_4$ and GSO scintillators are considerably polluted by uranium and thorium (particularly, $CaWO_4$ by ^{210}Po at the level of 0.3 Bq/kg, GSO by U/Th at the level of 0.04 Bq/kg). Radiopurity of $ZnWO_4$ and $CdWO_4$ crystals is significantly better. The total alpha activity of U/Th daughters in $ZnWO_4$ and $CdWO_4$ scintillators does not exceed the level of a few mBq/kg. The contaminations of $CdWO_4$, enriched in ^{116}Cd, by ^{232}Th and ^{226}Ra are at the level of 53 μBq/kg and 4 μBq/kg, respectively. Particular, i.e. associated with elemental composition of the detectors, radioactivity was observed in $CaWO_4$ (α active ^{180}W), $CdWO_4$ (β active ^{113}Cd and ^{113m}Cd, $2\nu2\beta$ decay of ^{116}Cd,

TABLE 3. Radioactive contamination of different detectors (mBq/kg).

Detector	Total α activity U + Th	^{228}Th	^{226}Ra	^{40}K	Particular contamination	Reference
$CaWO_4$	20 – 400	0.6	5.6	≤ 12		[28, 31, 22]
$ZnWO_4$	≤ 20	≤ 0.2	≤ 0.4	≤ 12		[15]
$CdWO_4$	$\leq 0.7 - 2.3(3)$	≤ 0.003	≤ 0.004	0.3	580 (^{113}Cd)	[25, 32, 8, 12, 1]
GSO	40	2.3 – 107	0.3	≤ 14	1200 (^{152}Gd)	[26, 11, 33]
NaI(Tl)	≤ 0.2	0.0004 – 0.014	0.005 – 0.39	1.5		[34, 35, 36]
CsI(Tl)		≤ 0.3	≤ 0.3	≤ 20		[37, 38]
BGO	≤ 0.4	≤ 1.2			$(0.5-3)\times 10^3$ (^{207}Bi)	[39, 2]
CaF_2(Eu)		0.04 – 40	0.05 – 75	5		[14, 10]
CeF_3	3400	1100	≤ 60	≤ 330		[13]
BaF_2		400	1400			[40]
Liquid scintillator	10^{-6}	1.2×10^{-6}	6.3×10^{-6}		0.3 (^{14}C)	[41, 42]
HPGe		$\leq 2\times 10^{-5}$	$\leq 2\times 10^{-5}$			[43, 44]

^{180}W), and GSO (α active ^{152}Gd) detectors. It was found that secular equilibrium in uranium chain is substantially broken in $CaWO_4$ crystals. Uranium and thorium traces were concentrated mainly in the surface layer of the $CdWO_4$ crystals enriched by ^{116}Cd.

REFERENCES

1. F. A. Danevich, A. Sh. Georgadze, V. V. Kobychev, S. S. Nagorny et al., *Phys. Rev. C*, **67**, 014310, 8 p (2003).
2. P. de Marcillac, *Nature*, **422**, 876–878 (2003).
3. C. Cozzini, G. Angloher, C. Bucci, F. von Feilitzsch et al., *Phys.Rev. C* **70**, 064606, 6 p (2004).
4. A. Alessandrello, C. Brofferio, D. V. Camin, O. Cremonesi et al., *Nucl. Phys. B (Proc. Suppl.)*, **35**, 394–396 (1994).
5. F. A. Danevich, A. Sh. Georgadze, V. V. Kobychev, B. N. Kropivyansky et al., *Phys. At. Nucl.*, **59**, 1–5 (1996).
6. F. A. Danevich, Yu. G. Zdesenko, A. S. Nikolaiko, S.F. Burachas et al., *Instr. Exp. R.*, **32**, 1059–1064 (1989).
7. F. A. Danevich, Yu. G. Zdesenko, A. S. Nikolaiko, V. I. Tretyak, *JETP Lett.*, **49**, 476–479 (1989).
8. F. A. Danevich, A. Sh. Georgadze, J. Hellmig, M. Hirsch et al., *Z. Phys. A*, **355**, 433–437 (1996).
9. C. M. Cattadori, M. De Deo, M. Laubenstein, L. Pandola, V. I. Tretyak, *Nucl. Phys. A*, **748** 333–347 (2005).
10. P. Belli, R. Bernabei, C. J. Dai, F. Grianti et al., *Nucl. Phys. B*, **563**, 97–106 (1999).
11. F. A. Danevich, V. V. Kobychev, O. A. Ponkratenko, V. I. Tretyak, Yu. G. Zdesenko, *Nucl. Phys. A*, **694**, 375–391 (2001).
12. F. A. Danevich, A. Sh. Georgadze, V. V. Kobychev, B. N. Kropivyansky et al., *Phys. Rev. C*, **68**, 035501, 12 p (2003).
13. P. Belli, R. Bernabei, R. Cerulli, C. J. Dai et al., *Nucl. Instr. Meth. A*, **498**, 352–361 (2003).
14. I. Ogawa, R. Hazama, H. Miyawaki, S. Shiomi et al., *Nucl. Phys. A*, **730**, 215–223 (2004).
15. F. A. Danevich, V. V. Kobychev, S. S. Nagorny, D. V. Poda et al., nucl-ex/0409014; *Nucl. Instr. Meth. A*, (2005) in press.
16. C. Bacci, P. Belli, R. Bernabei, C. Dai et al., *Astropart. Phys.*, **2**, 13–19 (1994).
    R. Bernabei, P. Belli, V. Landoni, F. Montecchia et al., *Phys. Lett. B*, **389**, 757–766 (1996).
    R. Bernabei, P. Belli, F. Montecchia, W. Di Nicolantonio et al., *Phys. Lett. B*, **424**, 195–201 (1998).
    R. Bernabei, P. Belli, F. Montecchia, W. Di Nicolantonio et al., *Phys. Lett. B*, **450**, 448–455 (1999).
    R. Bernabei, P. Belli, R. Cerulli, F. Montecchia et al., *Phys. Lett. B*, **480**, 23–31 (1998).
    R. Bernabei, M. Amato, P. Belli, R. Cerulli et al., *Phys. Lett. B*, **509**, 197–203 (2001).
    R. Bernabei, P. Belli, R. Cerulli, F. Montecchia et al., *Eur. Phys. J. C*, **23**, 61–64 (2002).
    R. Bernabei, P. Belli, F. Cappella, R. Cerulli et al., *Riv. Nuovo Cim.*, **26**, 1–73 (2003).
17. C. Bacci, P. Belli, R. Bernabei, C. J. Dai et al., *Astropart. Phys.*, **2**, 117–125 (1994).
18. I. Ogawa, T. Kishimoto, R. Hazama, S. Ajimura et al., *Nucl. Phys. A*, **663&664**, 869c–872c (2000).
19. A. Alessandrello, D. V. Camin, O. Cremonesi, E. Fiorini et al., *Nucl. Phys. B (Proc. Suppl.)*, **28A**, 233–235 (1992).
    C. Arnaboldi, C. Brofferio, C. Bucci, S. Capelli et al., *Phys. Lett. B*, **557**, 167–175 (2003).
20. C. Bobin, I. Berkes, J. P. Hadjout, N. Coron et al., *Nucl. Instr. Meth. A*, **386**, 453-457 (1997).
21. M. Bravin, M. Bruckmayer, C. Bucci, S. Cooper et al., *Astropart. Phys.* **12**, 107-114 (1999).
22. S. Cebrián N. Coron, G. Dambier, P. de Marcillac et al., *Phys. Lett. B*, **563**, 48–52 (2003).
23. Yu. G. Zdesenko, B. N. Kropivyansky, V. N. Kuts, A. S. Nikolaiko et al., in *Proc. 2nd Int. Symp. Underground Physics, Baksan Valley, USSR, August 17–19, 1987*, edited by G. V. Domogatsky, Nauka, Moscow, 1988, pp. 291–295.
24. F. A. Danevich, A. Sh. Georgadze, V. V. Kobychev, B. N. Kropivyansky et al., *Phys. Lett. B* **344**, 72–78 (1995).
25. A. Sh. Georgadze, F. A. Danevich, Yu. G. Zdesenko, V. V. Kobychev et al., *Prib. Tekh. Exp.* **2**, 45–51 (1996) [*Instrum. Exp. Tech.* **39**, 191–198 (1996)].

26. S. F. Burachas, F. A. Danevich, Yu. G. Zdesenko, V. V. Kobychev et al., *Phys. Atom. Nucl.*, **58**, 153–157 (1995).
27. T. Fazzini, P. G. Bizzeti, P. R. Maurenzig, C. Stramaccioni et al., *Nucl. Instr. Meth. A*, **410**, 213–219 (1998).
28. Yu. G. Zdesenko, F. T. Avignone III, V. B. Brudanin, F. A. Danevich et al., *Nucl. Instr. Meth. A*, **538**, 657–667 (2005).
29. CERN Program Library Long Write-up W5013, CERN, 1994;
S. Agostinelli, J. Allison, K. Amako, J. Apostolakis et al., (GEANT4 Collaboration), *Nucl. Instr. Meth. A*, **506**, 250–303 (2003), URL http://geant4.web.cern.ch/geant4/.
30. O. A. Ponkratenko, V. I. Tretyak, Yu. G. Zdesenko, *Yad. Fis.*, **63**, 1355–1360 (2000) [*Phys. At. Nucl.*, **63**, 1282–1287 (2000)].
31. Yu. G. Zdesenko, F. T. Avignone III, V. B. Brudanin, F. A. Danevich et al., *Astropart. Phys.*, **23**, 249–263 (2005).
32. S. Ph. Burachas, F. A. Danevich, A. Sh. Georgadze, H. V. Klapdor-Kleingrothaus et al., *Nucl. Instr. Meth. A*, **369**, 164–168 (1996).
33. S. C. Wang, H. T. Wong and M. Fujiwara, *Nucl. Instr. Meth. A*, **479**, 498–510 (2002).
34. R. Bernabei, P. Belli, F. Montecchia, W. Di Nicolantonio et al., *Il Nuovo Cim. A*, **112**, 545–564 (1999).
35. J. C. Barton, J. A. Edgington, *Nucl. Instr. Meth. A*, **443**, 277–286 (2000).
36. S. Umehara, K. Fushimi, N. Koori, N. Kudomi et al., *Nucl. Instr. Meth. A*, **490**, 271–275 (2002).
37. A. Baranov, V. Bashkirov, G. Bondarenko, A. Strifutkin, *Preprint MEPI 040-89*, Moscow Engineering Physics Institute, Moscow, 1989.
38. H. T. Wong, J. Li, C. Y. Chang, C. C. Chang et al., *Astropart. Phys.*, **14**, 141–152 (2000).
39. A. Balysh, A. Gurov, A. Demehin, I. Kondratenko et al., *Prib. Tekh. Exp.* **1**, 118–122 (1993).
40. R. Cerulli, P. Belli, R. Bernabei, F. Capella et al., *Nucl. Instr. Meth. A*, **525**, 535–543 (2004).
41. Borexino Collaboration, *Borexino Status Report*, LNGS, September 30, 2004 (unpublished).
42. G. Alimonti, G. Angloher, C. Arpesella, M. Balata et al., *Phys. Lett. B*, **422**, 349–358 (1998).
43. H. V. Klapdor-Kleingrothaus, L. Baudis, A. Dietz, G. Heusser et al., *Nucl. Instr. Meth. A*, **481**, 149–159 (2002).
44. C. Dörr, H. V. Klapdor-Kleingrothaus, *Nucl. Instr. Meth. A*, **513**, 596–621 (2003).

Production of low radioactivity CsI(Tℓ) crystals for WIMP searches

Y. D. Kim*, I. S. Hahn†, M. J. Hwang**, Li Jin‡, W. G. Kang*, H. J. Kim§, S. C. Kim¶, S. K. Kim¶, S. Y. Kim¶, T. Y. Kim¶, J. W. Kwak¶, Y. J. Kwon**, E. K. Lee†, H. S. Lee*, J. I. Lee*, J. Y. Lee*, M. J. Lee¶, S. S. Myung¶, H. Park‖, Alexander Schiedt††, H. Y. Yang¶ and J. J. Zhu‡‡

**Department of Physics, Sejong University, Seoul, 143-747, Korea*
†Department of Science Education, Ewha Woman's University, Seoul 120-750, Korea
***Physics Department, Yonsei University, Seoul 120-749, Korea*
‡Institute of High Energy Physics, Chinese Academy of Sciences, Beijing, 100039, China
§Physics Department, Kyungpook National University, Daegu, 702-701, Korea
¶School of Physics and DMRC, Seoul National University, Seoul,151-742, Korea
‖Division of Chemical Metrology and Materials Evaluation, Korea Research Institute of Standards and Science, Yuseong, Daejeon, 305-600, Korea
†† Special Metals Division, Chemetall GmbH, Innerstetal 2, D-38685 Langelsheim, Germany
‡‡ Department of Engineering Physics, Tsinghua University, Beijing,100084, China

Abstract. We have extensively studied the mechanism of the ^{137}Cs and ^{134}Cs contamination in CsI compounds. The nuclei ^{137}Cs and ^{134}Cs in various samples have been studied using a HPGe detector at deep underground. The main source of the ^{137}Cs contamination in CsI powder was desolved ^{137}Cs in the water used in the chemical process of cesium extraction at company side. We also found that ^{134}Cs was produced mainly from ^{133}Cs by neutron capture. We have successfully produced CsI powder containing no more than 2 mBq/kg of ^{137}Cs by using only ultra-pure water for whole extraction process from raw ore(pollucite).

Keywords: Dark Matter, WIMP, CsI, Radioactivity, HPGe, Water, Background, ^{137}Cs
PACS: 95.35.+d, 29.40.Mc

INTRODUCTION

Scintillation crystals are commonly used to search for extremely rare events of Weakly Interacting Massive Particles(WIMPs) [1] or neutrino interactions[2]; NaI(Tℓ) [1], CaF_2(Eu)[3], $CdWO_4$[4] being some examples. A thallium doped cesium iodide crystal, CsI(Tℓ) , is another interesting scintillator, since it has a high scintillation efficiency and relatively good pulse shape discrimination capability[5, 6, 7, 8, 9]. To use a CsI crystal in ultra low background experiments, however, it is essential to make sure that the potentially severe internal background of the crystal, specially from ^{137}Cs and ^{134}Cs , is low. The KIMS group has been preparing a WIMP search experiment using CsI crystals and explored the background reduction of CsI crystals[10, 11].

The main internal backgrounds in CsI crystal were found to be due to the cesium and rubidium radioisotopes, ^{137}Cs , ^{134}Cs , and ^{87}Rb [12]. In this report, we will concentrate only on ^{137}Cs and ^{134}Cs radioisotopes. The nucleus ^{137}Cs ($t_{1/2}$ = 30.07 y, Q=1175.6 keV) is a beta-emitter decaying with a 95% branching ratio to the meta-stable state($t_{1/2}$ = 2.55 minutes) of ^{137}Ba at 661.7 keV. The expected background spectra for a 10cm × 10cm × 30cm CsI(Tℓ) detector show that there are about 0.35 cpd (1cpd = 1 counts$/(\mathrm{kev}\cdot\mathrm{kg}\cdot\mathrm{day})$) from 1 mBq/kg of ^{137}Cs and 100 mBq/kg of ^{134}Cs in the energy region, $E < 10$ keV, if a Compton suppressor surrounds the detector[9].

Almost all of the present ^{137}Cs nuclei on earth have been produced artificially by the nuclear bomb tests, nuclear reactor accidents, and fuel reprocessing. The isotope has spreaded globally through air and water circulation. For example, the specific soil activity of ^{137}Cs in the northern hemisphere, e.g. Canada, was 38 Bq/kg from a measurement performed in 1994[13]. Surface sea water is typically contaminated at the level of a few mBq/liter[14]. There are very few measurements on the contamination level of ^{137}Cs and ^{134}Cs in CsI crystal(or powder). The TEXONO[2] group reported ^{137}Cs contamination in a CsI(Tℓ) crystal less than 4×10^{-18}g/g (12.8 mBq/kg) using a high purity

CP 785, *Topical Workshop on Low Radioactivity Techniques,* edited by B. Cleveland, R. Ford, and M. Chen

Germanium(HPGe) detector.

We have previously studied commercially available CsI powders[12]. In this paper, we report a systematic study on the presence of cesium radioisotopes in CsI compounds and how we reduced the ^{137}Cs contamination to a few mBq/kg level for the first time.

THE MECHANISM OF ^{137}CS AND ^{134}CS CONTAMINATION

Cesium is extracted from the pollucite ore, which is in the form of $Cs_2Al_2Si_4O_{12}(H_2O)$. A typical elemental composition of Canadian pollucite from Tanco, Canada, is O(36.7%), Cs(30 %), Si(22 %), Al(8.5 %), Na(1.3 %), Rb(0.7 %), H(0.2 %), Li(0.2 %), and K(0.1 %). First the ore from Tanco as well as Bikita, Zimbabwe, is ground to fine powder at the mine site and delivered to chemical companies for the extraction of cesium. Figure 1 shows the extraction processes of the cesium from pollucite in one of major companies of cesium production, Chemetall GmbH, Special Metals Division in Germany[16]. "High purity" CsI powder is 99.9999% [17]. One notes that the extraction process needs about 70 liters of water in the various stages to produce 1 kg of CsI powder[17].

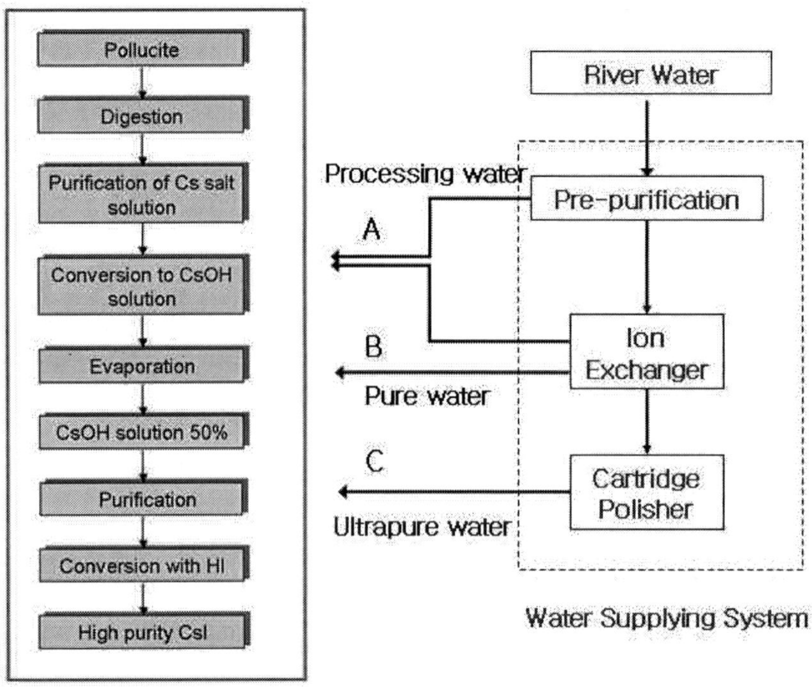

FIGURE 1. The extraction processes and water supplying system at Chemetall GmbH in Langelsheim, Germany. For A, we used 50% of "processing" water and 50% of "pure" water, and used only "pure" water for B. For C, we used only "ultrapure" water.

Extraction processes of cesium from pollucite

We have measured the ^{137}Cs and ^{134}Cs contamination levels in various samples by directly measuring the 661.7 keV, 604.7 keV, and 795.8 keV gamma rays with HPGe detectors. The HPGe detector(Ortec GEM100-S) was a 100% efficient, low-background co-axial detector, and it was installed in a laboratory 750 meter underground, Yangyang Laboroty(Y2L) in South Korea. Shielding includes a 10 cm thickness of oxygen free high conductivity(OFHC) copper, and a 15 cm thickness of lead with the inner 10cm composed of low-background Boliden lead. At present, there is a gap of about 10cm between the HPGe crystal and the inner copper shield in order to house a large volume of samples. The whole detection system, including liquid nitrogen cryogenics bottle was encapsulated, within an acryl box to protect against radon activities. Nitrogen gas from a liquid nitrogen tank flows continuously at a level of 4 liters/hour inside the acryl box. ORTEC PC-based ADC card was used for data acquisition. The background count rate in the 660 keV region was 3 counts/(keV day). The detector has been continuously operated for more than 2 years.

Underground measurements

We have measured the ^{137}Cs and ^{134}Cs contamination levels in various samples by directly measuring the 661.7 keV, 604.7 keV, and 795.8 keV gamma rays with HPGe detectors. The HPGe detector(Ortec GEM100-S) was a 100% efficient, low-background co-axial detector, and it was installed in a laboratory 750 meter underground, Yangyang Laboroty(Y2L) in South Korea. Shielding includes a 10 cm thickness of oxygen free high conductivity(OFHC) copper, and a 15 cm thickness of lead with the inner 10cm composed of low-background Boliden lead. At present, there is a gap of about 10cm between the HPGe crystal and the inner copper shield in order to house a large volume of samples. The whole detection system, including liquid nitrogen cryogenics bottle was encapsulated, within an acryl box to protect against radon activities. Nitrogen gas from a liquid nitrogen tank flows continuously at a level of 4 liters/hour inside the acryl box. ORTEC PC-based ADC card was used for data acquisition. The background count rate in the 660 keV region was 3 counts/(keV day). The detector has been continuously operated for more than 2 years.

Expected contents of ^{137}Cs from spontaneous fission

We have previously measured several CsI powders from various companies[12]. The amount of ^{137}Cs in commercial CsI powder ranges from 30 mBq/kg to several hundred mBq/kg. Since cesium is extracted from pollucite ore, we first studied the amount of ^{137}Cs in pollucite powder. The pollucite powder contained in a Marinelli beaker of 1 liter(about 1.4 kg) was measured directly with a HPGe detector. Since pollucite contains 1.0-1.6 % K_2O, a relatively high amount of potassium, the background level is too high to be sensitive to the gammas from ^{137}Cs and ^{134}Cs . In a previous measurement with a HPGe detector surrounded with a Compton suppressor of NaI(Tℓ) crystals, we gave an upper limit of 5.8 mBq/kg for ^{137}Cs at the 90% confidence level[12]. Therefore we concluded that ^{137}Cs in CsI crystals was mainly contaminated during chemical extraction.

The measurement of pollucite shows the amount of U,Th nuclei as listed in Table 1. Since the half life of ^{137}Cs is much shorter than these primordial nuclei, the activity of ^{137}Cs will be in secular equilibrium, and the expected activites of ^{137}Cs in 1 kg of CsI powder can be calculated as,

$$A_i(^{137}Cs) = \frac{A_i F_i Y_i(^{137}Cs)}{2\alpha M(kg)}. \tag{1}$$

Here, A_i is the activity, F_i is the fission fraction, and $Y_i(^{137}Cs)$ is the yield of ^{137}Cs in each U,Th nuclei. α is the relative amount of cesium in pollucite, and M is the mass of the pollucite sample in kg unit.

TABLE 1. The measured activities of ^{238}U and the expected activites of ^{137}Cs . See the text for descriptions on items.

Nuclei	$T_{1/2}$ (year)	F_i %	$Y_i(^{137}$Cs) %	A_i Bq/kg	$A_i(^{137}$Cs) μBq/kg
^{238}U	4.47×10^9	5.45×10^{-7}	0.058	1.93	0.073
^{235}U	7.04×10^8	7.0×10^{-11}	≈ 0.06	0.086	4.32×10^{-7}
^{232}Th	1.41×10^{10}	1.8×10^{-11}	≈ 0.06	0.24	3.09×10^{-7}

The expected amount of ^{137}Cs from spontaneous fissions of U nuclei in the pollucite powder is estimated to be less than 0.1 μBq/kg as shown in Table 1, which is negligible compared to the amount present in typical CsI powder. The contributions from spontaneous fissions of ^{235}U and ^{232}Th are much smaller than for ^{238}U due to smaller fission fractions of these nuclei. Therefore spontaneous fission is not the main source of ^{137}Cs in CsI crystals, and not a problem even when we consider a highly sensitive WIMP signal search experiment.

Water Samples

Naturally environmental anthropogenic ^{137}Cs should be the source of ^{137}Cs in CsI powder. There are many studies on ^{137}Cs and ^{134}Cs levels in environmental water before and after the Chernobyl accident[14, 15]. The average level of ^{137}Cs in Pacific and Indian Ocean surface water was 1-3 mBq/liter[14] in the year 2000. The ^{137}Cs level of Finland rivers was 100-1000 mBq/liter just after the Chernobyl accident, and decreased to 10-100 mBq/liter in the year 1996 depending on the location of the river[15]. At present, no information on the level of ^{137}Cs in river water used by the company is available. Since 70 liters of water is used in the chemical process of 1kg CsI powder production, even an extremely small amount of ^{137}Cs in the water could be responsible for the whole ^{137}Cs in commercial CsI powder.

To understand the contamination mechanism of ^{137}Cs and find out how to rcducc thc ^{137}Cs in CsI compounds, we have asked the company to produce CsI powder with different water supplied as shown in Figure 1. In the normal production procedure, mixed water from "processing" and "pure" water was supplied with the ratio of 1:1(A). Secondary, CsI powder was produced with only "pure" water for the whole processes(B). Thirdly, high quality ion exchangers(cartridge polisher)[1] was installed after the usual ion exchanger, and only "ultrapure" water from the cartridge polisher was supplied(C). The three kinds of water, "processing", "pure", and "ultrapure" water were supplied from the company to us in the amount of 20 liters each for analysis. In addition to the water samples, we prepared two more water samples of "ultrapure" water (18 MΩ) in the amount of 100 liters produced in Korean laboratories.

To measure the extremely small activity of ^{137}Cs in the water samples, we used the standard method of applying AMP(Ammonium Molybdophosphate) to precipitate the cesium ions in the water. The water samples were supplied in Polyethylene(PE) bottles, and the precipitation was done in the same bottle. First, we added 0.5 cc of $HNO_3(70\%)$ per one liter of water to the water container. Then we stirred the water container for about 10 minutes, and let the water stabilize several hours. Then we added 0.5 gram of AMP per one liter of water to the container, and stirred the whole container for about 30 minutes. After leaving the container over night, the water was decanted slowly, and the remaining yellow liquids were moved to a one-liter beaker. The beaker was left for one day, and the upper water was decanted again. The remaining liquids were moved to a PE bottle(ϕ=5cm). We dried the PE bottle in a drier at about 80 degree. After complete drying, the PE bottles were moved to the HPGe detector for measurement. The detection efficiencies were simulated with GEANT3. All the samples were treated in the same way.

RESULTS AND DISCUSSIONS

^{137}Cs

Figure 2 shows the measured spectra of the three water samples in the energy region of ^{137}Cs γ's. One of the "processing" water samples, (a), shows clearly large ^{137}Cs activity, and the "pure" water, (b), shows very little.

The three powders, A,B, and C were produced as described in previous section(see also Figure 1) in a large quantity of several hundred kgs with different water. The powder samples were measured in a Marinelli beaker of 1 liter, which contains 2.9 kg of CsI powder. Figure 3 shows the ^{137}Cs contamination in measured spectra of powder samples. The peak around 665 keV are due to the gammas from ^{132}Cs and ^{126}I produced by high energy neutrons during the transportation by airplane. However these isotopes have halflives of 6.479 days and 13.11 days respectively, and disappear after a month or so.

Table 2 lists the measured specific activities of ^{137}Cs and ^{134}Cs in the water and powder samples. The activities were 0.50 - 1.66 mBq/liter and 0.061 - 0.084 mBq/liter for the "processing" and "pure" water samples respectively. The "ultrapure" water didn't show a peak, so only an upper limit is obtained, 0.036 mBq/liter. As shown in the Table 2, ^{137}Cs contamination is reduced by an order of magnitude from the best "normal" powder by supplying purer water.

The measured ^{137}Cs in the powder produced with "ultrapure" water only(C) was 1.89 ± 0.5 mBq/kg, which is close to the expected upper limit of ^{137}Cs contamination from the water only. Figure 4 shows the expected ^{137}Cs activites in CsI powders and measured activities of CsI powders for 3 different production procedures, assuming all ^{137}Cs in water was adsorbed in the final CsI powders. With the good agreements between the expected and measured activities,

[1] Two cartridges were installed, and the volume of each was 100*l*. The throughput maximum was 2000 *l*/h for each. The resin was a mixed bed of 40 % cation-exchanger and 60% anion-exchanger, a product from Rohm & Haas, Amberlite MB20.

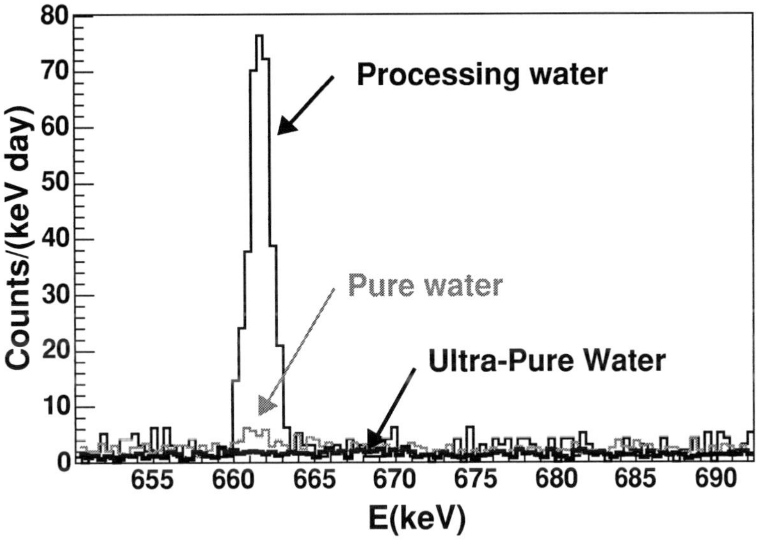

FIGURE 2. The measured spectra of water samples with HPGe detector for ^{137}Cs . (a)"processing" water, (b) "pure" water, and (c) "ultrapure" water samples from Chemetall GmbH.

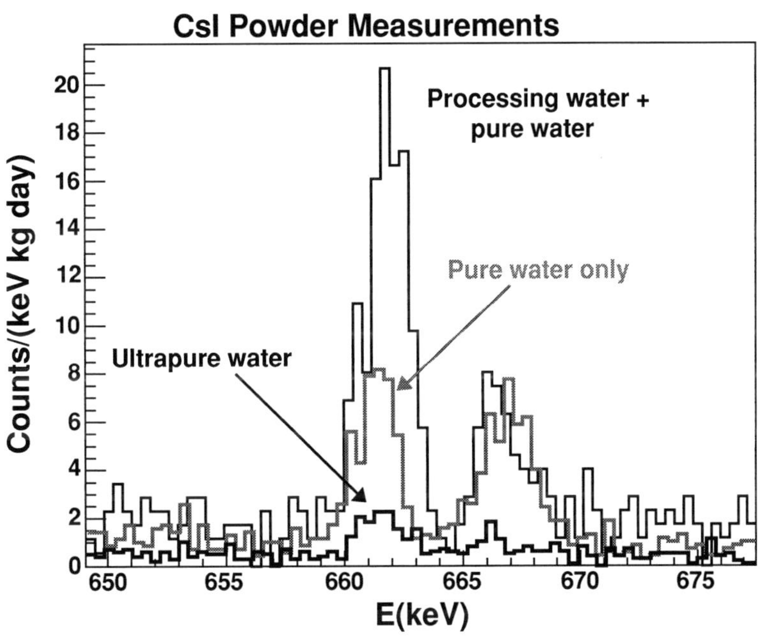

FIGURE 3. The measured spectra of three CsI powder samples.

it may be possible to eliminate the contamination further to the sub mBq/kg level by continuing to purify the water.

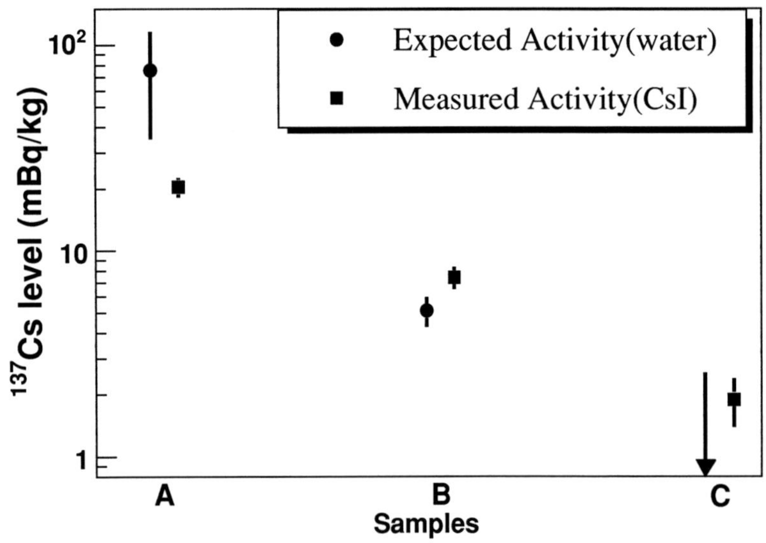

FIGURE 4. The expected activities in CsI powders from water measurements and measured activities in CsI powders. For A, we used 50% of "processing" water and 50% of "pure" water, and used only "pure" water for B. For C, we used only "ultrapure" water.

TABLE 2. Measurements of ^{137}Cs and ^{134}Cs in water and CsI powder samples.

Samples	Description	^{137}Cs (mBq/kg)	^{134}Cs (mBq/kg)
water	"processing" 1	1.66 ± 0.08	< 0.07
water	"processing" 2	0.50 ± 0.07	< 0.06
water	"pure" 1	0.084 ± 0.031	< 0.011
water	"pure" 2	0.061 ± 0.018	< 0.053
water	"ultrapure"	< 0.036	< 0.012
powder	"processing" + "pure"	20.5 ± 2.2	37.2 ± 2.7
powder	"pure" water only	7.41 ± 0.7	15.9 ± 1.5
powder	"ultrapure" water only	1.89 ± 0.5	30.1 ± 1.5

^{134}Cs

The nucleus ^{134}Cs is also an anthropogenic nucleus, but the half life is only 2 years. It can be also produced by the ^{133}Cs(n, γ)^{134}Cs reaction. The spectra of all water samples measured didn't positively show peaks from ^{134}Cs decay. The upper limits of ^{134}Cs in the water samples were a few tens of μBq/kg, which can contribute only a few mBq/kg in CsI powder at most. The ^{134}Cs level in CsI powder was 16-37 mBq/kg, and there was no correlation with water quality as expected. The ^{134}Cs in CsI powder should have been present in pollucite before cesium extraction or produced in-situ in CsI powder by the neutron capture process of ^{133}Cs. The activity of radioactive nuclei produced by constant thermal neutron flux can be written as

$$A(t) = N_0 \sigma F_n (1 - e^{-\lambda t}), \quad (2)$$

where N_0 is the number of ^{133}Cs nuclei, σ is the Maxwell averaged capture cross section, F_n is the thermal neutron flux, and t is the exposure time. The thermal neutron capture cross section of ^{133}Cs is 25.8 barn, and the typical thermal neutron flux at sea level is reported as 0.0014 neutrons/cm^2/sec[18]. With these values, we can estimate the specific activity of ^{134}Cs to be about 24 mBq/kg for pollucite, and about 41 mBq/kg for CsI powder assuming constant flux and infinite exposure(saturation). For two years of exposure, the values are 49 mBq/kg and 81 mBq/kg in pollucite and CsI powders respectively. The measured level of ^{134}Cs in pollucite powder was 11.7 ± 3.3 mBq/kg, which is about factor two lower than the two-year exposure values. Even if we don't know the detail history of exposure for the specific pollucite powder samples, the measured activities of ^{134}Cs in CsI powder are consistent with the production from thermal neutron capture by ^{133}Cs during a few year exposure to environment thermal neutrons.

Backgrounds of CsI(Tℓ) crystal

A CsI(Tℓ) crystal grown with the CsI powder produced with "pure" water only had a background of about 6 counts/(kev $\cdot$ kg $\cdot$ day) level. The background level is consistent with a background expected with the ^{137}Cs amount measured with the powder. For this crystal the dominant background is ^{137}Cs , and it is anticipated that we can produce CsI(Tℓ) crystals with 2 cpd level even with the current Rb and ^{134}Cs levels of background if we grow a CsI(Tℓ) crystal with the powder produced with "ultrapure" water only. The crystallization work is under progress.

CONCLUSION

We have studied the contamination and inhibition of ^{137}Cs and ^{134}Cs in CsI powder to produce low-background CsI crystals for very low rate physics experiments such as WIMP search. The contamination of ^{137}Cs in CsI powder and other cesium compounds was due to ^{137}Cs in water supplied at the manufacturing sites. Using high quality ion exchanger cartridge polisher, we could decrease the ^{137}Cs contamination down to 2 mBq/kg level from an otherwise a few tens of mBq/kg. The expected and measured ^{137}Cs level in CsI powder agrees well to a few mBq/kg, and further purification can be achieved below 1 mBq/kg. The ^{134}Cs contamination level in CsI powder was 16-37 mBq/kg, unaffected by water purity. ^{134}Cs is found to be produced mainly by environmental(cosmic) thermal neutron capture processes.

ACKNOWLEDGMENTS

This work is supported by Korean Ministry of Science and Technology under a Creative Science Initiative Research Program of KOSEF, and partially supported by the Korea Research Foundation Grant(KRF-2002-070-C00027). Authors would like to appreciate the staffs at the Special Metals Division of Chemetall GmbH. Specially, the kind collaborative work by late Dr. Horst Prinz was essential to achieve the reduction of ^{137}Cs . The authors are very grateful to the Korea Midland Power Co. and their staffs for providing the undeground laboratory space at Yangyang.

REFERENCES

1. R. Bernabei *et al.*, Phys. Lett. B **480**, 23 (2000).
2. H. B. Li *et al.*, Nucl. Instrum. Methods **A459**, 93 (2001).
3. I. Ogawa *et al.*, Nucl. Phys. A. **663**, 869c (2000).
4. G. Angloher *et al.*, Nucl. Instru. Meth. A. **520**, 108 (2004).
5. S. Pecourt *et al.*, Astroparticle physics **11**, 457 (1999).
6. V. A. Kudryavtsev *et al.*, Nucl. Instrum. Methods **A456**, 272 (2001).
7. H. J. Kim *et al.*, Nucl. Instrum. Methods **A457**, 471 (2001).
8. H. Park *et al.*, Nucl. Instrum. Methods **A491**, 460 (2002).
9. T.Y. Kim *et al.*, Nucl. Instrum. Methods **A500**, 337 (2003).
10. S. K. Kim *et al.*, Proceedings of Dark Matter 2002., J.W. Kwak *et al.*, Proceedings of IDM 2004., H.S. Lee *et al.* ibid.
11. H. J. Kim *et al.*, Astroparticle Physics **20**, 549 (2004).
12. Y. D. Kim *et al.*, J. Korean. Phys. Soc. **40**, 520 (2002).
13. R. Blagoeva *et al.*, J. Environ. Radioactivity, **27**, 269 (1995).
14. Pavel P. Povinec *et al.*, Journal of Environmental Radioactivity **76**, 113 (2004)
15. Ritva Saxen and Erkki Ilus, Journal of Environmental Radioactivity **54**, 275 (2001)
16. Special Metals Division, Chemetall GmbH(www.specialmetals.chemetall.com).
17. Private communication with Chemetall GmbH.
18. R. Wordel *et al.*, Nucl. Instrum. Methods **A369**, 557 (1996).

Low radioactivity at the Modane Underground Laboratory

Pia Loaiza

Laboratoire Souterrain de Modane, 90 rue Polset, 73500 Modane, France

Abstract. The Modane Underground Laboratory, with an overburden of 4800 m.w.e, offers an excellent site for experiments requiring a low-background environment. The laboratory is currently hosting experiments in particle and astroparticle physics and low-level germanium gamma-ray spectrometry activities. This paper sketchs the main characteristics, scientific activities and facilities of the laboratory as well as its insertion in the European program ILIAS. Special emphasis is given to the description of the low radioactivity measurements and Germanium detectors.

Keywords: Underground laboratories; Gamma-ray spectrometry
PACS: 01.52.+r, 29.40,-n

INTRODUCTION

The Modane Underground Laboratory (Laboratoire Souterrain de Modane, LSM) is located in the tunnel connecting France and Italy under $\sim$ 1800 m of rock ($\sim$ 4800 m.w.e). The laboratory is jointly operated by DSM (Direction des Sciences de la Matière) of CEA (Commissariat à l'Energie Atomique) and by IN2P3 (Institut National de Physique Nucléaire et de Physique des Particules) of CNRS (Centre National de la Recherche Scientifique).

The operational staff is composed of a team of about 10 people, 1 director and deputy, 1 post doc, 1 technical and administrative manager, 1 administrative assistant and 5 technicians.

Table 1 summarizes the main environmental background in the laboratory:

TABLE 1. Environmental background in the laboratory

U content	about 1 ppm
Muon Flux	4 μ/m^2/day
Neutron flux	1.6 10^{-6} cm^2/s (E 1MeV) [1].
Radon concentration	5 to 15 Bq/m^3

The low radon radioactivity level is kept thanks to a ventilation system ensuring a full renewal of the air twice per hour. Besides, all the surfaces of the laboratory were covered with a low U/Th content concrete.

The gamma-ray fluxes in the laboratory have been studied using an NaI detector. Results can be found in [2].

SCIENTIFIC ACTIVITIES AND FACILITIES

The laboratory is currently hosting experiments in particle and astroparticle physics, dark matter and double beta decay searches, super heavy elements search and low level gamma spectroscopy activities performed with 13 low background Germanium detectors. Recently, a radon trapping facility has been installed. Fig. 1 shows the location of the experiments in the main hall, the gamma spectrometry room and the radon trap towers.

Current status of running experimental programs and low radioactivity measurements are described in the following.

EDELWEISS is dedicated to the search for dark matter in the form of WIMPs. The EDELWEISS collaboration gathers laboratories from France, Germany and Russia. This direct search experiment uses Germanium heat-ionisation bolometers. The simultaneous measurement of phonon and ionisation signals provides an excellent event-by-event discrimination between nuclear recoils (induced by WIMPs and neutrons) and electronic recoils (induced by β's or γ's). The first phase, EDELWEISS-I, with three Germanium detectors of 320 g each has succeeded in putting stringent limits to the spin independent WIMP-nucleon cross-section [3]. The set up of EDELWEISS-I has been dismounted and the second phase, EDELWEISS-II is being installed. It will involve, in a first step, 28 detectors of 320 g each. Data taking is scheduled for the end of 2005. Ultimately, this set-up will be able to accommodate up to 30 kg of detectors. The goal is to test Supersymmetric models predicting a WIMP-nucleon cross section close to 10^{-8} pb [4].

CP 785, *Topical Workshop on Low Radioactivity Techniques,* edited by B. Cleveland, R. Ford, and M. Chen

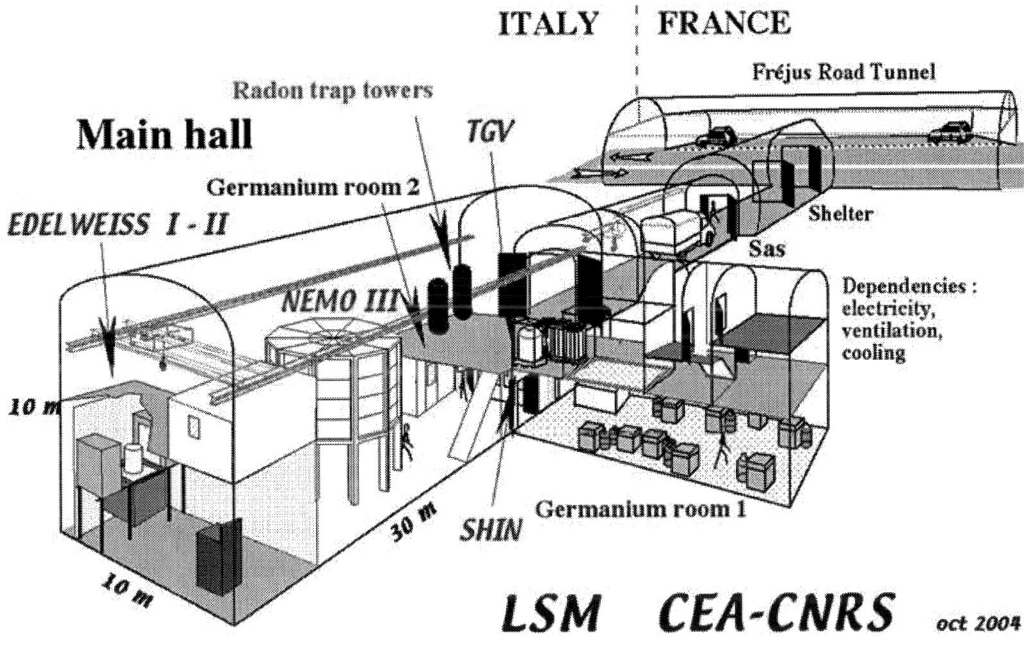

FIGURE 1. Plan of the laboratory

NEMOIII [5] is a neutrinoless double beta decay detector using a tracking-calorimeter technique. The NEMO collaboration involves laboratories from France, Russia, Czeck Republic, UK, Finland, USA and Japan. The detection principle is based on the direct detection of the two electrons emitted from double beta decay isotopes, ^{100}Mo, ^{82}Se, ^{130}Te, ^{116}Cd, ^{150}Nd, ^{96}Zr and ^{48}Ca. The tracking detector, composed of drift wire chambers operated in Geiger mode allows to identify electron tracks while plastic scintillators measure their total energy. The NEMOIII detector has been running since mid-February 2003. Current data is being analysed, reaching a sensitivity on the half-life of neutrinoless double beta decay of 10^{24-25} years.

TGV has been designed to search for rare events like double beta decay or double electronic capture. The TGV collaboration involves Russian, French, Czech and Slovak laboratories. The first phase, TGV1, has measured for the first time the half-life of $\beta\beta 2\nu$ decay with 1 g sample of ^{48}Ca [6]. The second phase, TGV2 has been installed in 2002. TGV2, with 32 Ge detectors is able to accommodate 10 g of ^{48}Ca, the goal being to measure the $\beta\beta 2\nu$ half-life with a precision of few per cent in order to determine the nuclear matrix elements. These elements are necessary for the experiments searching for $\beta\beta 0\nu$ to deduce the neutrino mass limits. Recently, the detector has been upgraded for low energy event detection with the goal of observing for the first time double electron capture process in ^{106}Cd with a 10 g sample.

The SHIN experiment aims at studying the stability of Super-Heavy Elements (Z $\sim$ 110) in nature. The detector has been designed and built by a Russian team from Dubna. As a first step long-lived isotopes of element 108 are being searched by measuring multiple neutron emission accompanying spontaneous fission of the super-heavy nucleus. The neutron detector consists of 60 counters filled with ^{3}He [7]. The detector was installed in the Modane Underground Laboratory in November 2004 and is actually taking data. A sensitivity to the presence of superheavy element 108 of 10^{-22} g/g in earth material can be achieved in one year, assuming that its half life is 10^{9} years.

Low background gamma spectrometry. The low-background Ge facility counts with 13 detectors, belonging to 5 different users. These detectors are used for several purposes :

- Screening of ultra-low radioactivity materials. One of the HP coaxial Ge detectors (called GENTIANE) has been installed by the EDELWEISS collaboration, two other detectors (IRIS and JASMIN) are used to control materials for NEMO and the Ge so-called HELLAZ is used for material selection to develop a low-background Ge spectrometer. Each detector is shielded by a layer of regular lead, copper and an inner shielding of archaeological lead. For ^{222}Rn protection all shields are continuously flushed either with nitrogen gas or with Rn-reduced air. The detectors, built by Eurisys, have a background of the order of 0.1 count/keV/kg/day.

- Environmental studies. Radionuclides through ingrowth or decay, provide useful chronometers for the rate determination of numerous natural processes. Large, low-background well-type detectors are the best choice in spite of several disadvantages, due to their high efficiency. Three of such detectors (MONDEUSE, ROUSSETTE and ABYME) belonging to the Laboratoire des Sciences du Climat et l'Environnement, LSCE (CEA, CNRS), are devoted to the study of numerous processes occurring in the environment [8].
- Monitoring environmental radioactivity levels. Two national institutions perform environmental control using the low background Germanium detectors (HAO, BORA BORA, MOOREA, HUAINE, HERMINE and XXL[9]).
- Various applications such as Bordeaux wine dating or quality control.

Table 2 summarises the detector characteristics and their total background count rate.

TABLE 2. Germanium spectrometers at the LSM

Detector	Type	Vol	Total background (counts/day) 40-2700 keV
JASMIN	P	380 cc	529
HAO	P	360 cc	384
IRIS	P	400 cc	282
HELLAZ	P	204 cc	515
BORA BORA	N	207 cc	530
HERMINE	N	197 cc	313
GENTIANE	N	215 cc	178
MONDEUSE	well	220cc	770
MOOREA	well	300 cc	455
ROUSSETTE	well	430 cc	692
ABYME	well	980 cc	828
XXL	well	844 cc	821
HUAHINE	planar	45 cc	179*

* 40-410 keV

Radon trapping facility. Radon is the main background source for the NEMO III experiment, searching for double beta decay events. A radon reduction device has been installed in September 2004 in the laboratory. The radon trapping facility relies on a well-known procedure consisting of trapping radon onto charcoal. As the air is passed over the charcoal, radon is trapped on the charcoal surfaces by adsorption. Adsorption increases with decreasing temperature. Typically the active carbon in the towers (2 x 500 kg) are cooled down to -50 °C. The facility provides 150 m^3/h of air with a Rn level of $\sim$ 10 mBq/m^3.

THE ILIAS PROGRAM

ILIAS [10] is a common European program to strengthen the coordination between Europe's leading infrastructures in Astroparticle Physics. Part of the ILIAS activities concern the deep underground laboratories, four European underground labs, Gran Sasso (Italy), Canfranc (Spain), Boulby (UK) and the LSM are involved in the coordinated development of underground infrastructures, low level measurements and background reduction techniques. The program has three main activities, networking, research and development and exchange of research teams.

Concerning the research and development, the LSM runs a dedicated program :

- There are many applications for a gamma-ray spectrometer which has a background one order of magnitude lower than the actual low-background Germaniums at LSM. A goal of the program is to improve the sensitivity of a 400 cc Ge diode by testing and carefully selecting construction materials.
- A database with data on radiopurity of materials is being set up. Although all used materials involved in the construction of low background detectors should be screened, we believe the database provides an useful guide for material selection. The database will gather measurements from the different collaborations in the four European labs involved, as well as intercontinental contributions from all collaborations willing to share their data. In the

long term, the aim is to develop methods and techniques to obtain materials extremely free from radioactive impurities.

ACKNOWLEDGMENTS

We acknowledge partial funding from FP6 program ILIAS.

REFERENCES

1. G. Chardin and G.Gerbier, "Neutron background studies for EDELWEISS" in *Proceedings of the 4th International Workshop on Identification of Dark Matter*, edited by N.J. Spooner and V. Kudryavtsev, World Scientific, Singapure, 2003.
2. H. Ohsumi et al *NIM A*, **482** 832 (2002).
3. A. Benoit et al. *Phys. Lett.*, **B545**, 43 (2002).
4. V. Sanglard, "From EDELWEISS-I to EDELWEISS-II" to be published in *Proceedings of the 5th International Workshop on Identification of Dark Matter, IDM2004*, astro-ph/0411629.
5. R. Arnold et al, *NIM A*, **536**, 79 (2005).
6. V. Brudanin et al *Phys. Lett*, **B495** 63 (2000).
7. E. Sokol et al *NIM A*, **400** 96 (1997).
8. J-L. Reyss et al *NIM A*, **357** 391 (1995).
9. R. Gurriaran et al *NIM A*, **524** 264 (2004).
10. http://ilias.in2p3.fr.

Low Radioactivity in CANDLES

T. Kishimoto, I. Ogawa, R. Hazama, S. Yoshida, S. Umehara, K. Matsuoka, H. Sakai, D. Yokoyama, K. Mukaida, K. Ichihara, Y. Tatewaki, K. Kishimoto, Y. Hirano, A. Yanagisawa, S. Ajimura

Department of Physics, Osaka University, Toyonaka, Osaka, 560-0043, Japan

Abstract. CANDLES is the project to search for double beta decay of ^{48}Ca by using CaF_2 crystals. Double beta decay of ^{48}Ca has the highest Q value among all nuclei whose double beta decay is energetically allowed. This feature makes the study almost background free and becomes important once the study is limited by the backgrounds. We studied double beta decays of ^{48}Ca by using ELEGANTS VI detector system which features $CaF_2(Eu)$ crystals. We gave the best limit on the lifetime of neutrino-less double beta decay of ^{48}Ca although further development is vital to reach the neutrino mass of current interest for which CANDLES is designed. In this article we present how CANDLES can achieve low radioactivity, which is the key for the future double beta decay experiment.

Keywords: Double beta decay, neutrino mass, CANDLES
PACS: 13.15.+g, 23.40.-s, 11.30.Fs

WHY ARE WE INTERESTED IN DOUBLE BETA DECAY OF ^{48}CA

It is well known that the Q value of the ^{48}Ca $\beta\beta$ decay is the highest (4.27 MeV) among all $\beta\beta$ decay nuclei. Table1 lists Q values and phase space factors ($G_{0\nu}$) of typical $\beta\beta$ nuclei. The Q value of the ^{48}Ca $\beta\beta$ decay is far above energies of γ rays from natural radioactivities (maximum 2.615 MeV from ^{208}Tl decay), therefore we can expect small backgrounds in the Q-value region. The large Q value also means large phase space factor which enlarges the $0\nu-\beta\beta$ decay rate for a given Majorana mass and nuclear matrix element. Another advantage is the least contribution from the $2\nu-\beta\beta$ decay which is possible in the standard model. Although sum energy of two electrons can never be the same as that of the Q-value, finite energy resolution of detectors makes leak of the $2\nu-\beta\beta$ decay into the $0\nu-\beta\beta$ region unavoidable. The large Q value of the ^{48}Ca $\beta\beta$ decay makes the leak of minor concern for a given energy resolution.

TABLE 1. Q values and phase space factors ($G_{0\nu}$) of $\beta\beta$ nuclei

Nucleus	Q value(MeV)	$G_{0\nu}$ x 10^{-25} (year^{-1} eV^{-2})
^{48}Ca	4.27	2.44
^{76}Ge	2.04	0.244
^{82}Se	3.00	1.08
^{100}Mo	3.03	1.75
^{116}Gd	2.80	1.89
^{136}Xe	2.48	1.81
^{150}Nd	3.37	8.00

However, these nice features of the ^{48}Ca $\beta\beta$ decay are counterbalanced by the low natural abundance of ^{48}Ca (0.187%) because of which a few experiments [1,2,3,4,5,6,7] have been carried out so far.

CP 785, *Topical Workshop on Low Radioactivity Techniques,* edited by B. Cleveland, R. Ford, and M. Chen

STUDY OF ^{48}CA BY CAF$_2$ CRYSTALS

We have been studying ^{48}Ca ββ decay by ELEGANT VI detector system. It has 25 CaF_2(Eu) crystals with a dimension of 45 mm^3 which are placed at the center of the detector. They are surrounded by active and passive shields to reduce backgrounds. The central CaF_2(Eu) crystals has light guides of CaF_2(pure) crystals which are also scintillators. Background signals, which give dominant energy deposit to the light guides, are effectively rejected by this design. The schematic drawing of the detector is shown in figure 1.

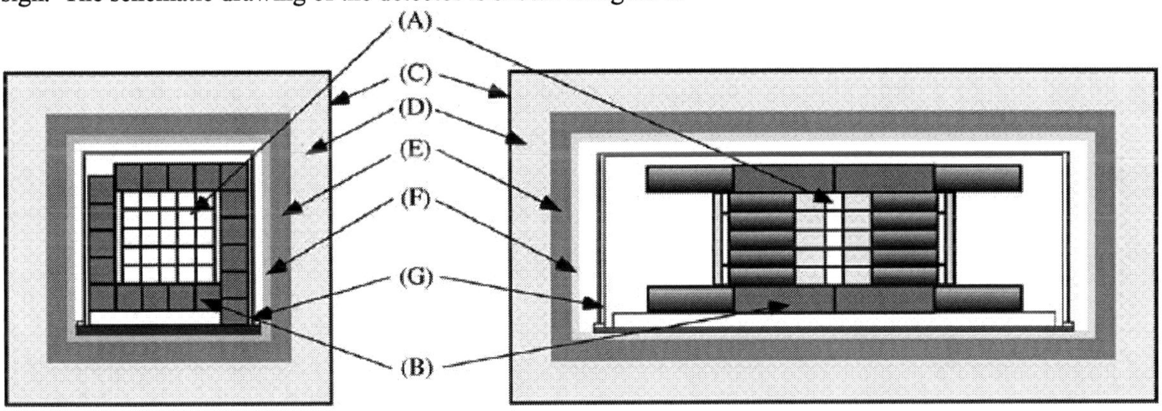

FIGURE 1. A schematic drawing of ELEGANT VI system. Cross sectional view is on the left and side view is on the right. (A) 25-element segmented CaF2 detector. The detectors are numbered from 1 (bottom left in cross-sectional view) through 5 (bottom right) to 25 (top right). (B) 38 CsI(Tl) scintillators. (C) Cd sheet. (D) Pb shield. (E) OFHC Cu shield. (F) LiH-loaded paraffin. (G) Air-tight box. The whole system is surrounded by H3BO3-loaded water tanks.

The CaF_2 scintillators and CsI scintillators veto γ rays from outside. They are in the air-tight box which purges radon by a flow of pure N_2 gas. Oxygen-free high-conductivity (OFHC) copper shield of 5 cm thick and lead shield of 10 cm thick are used to passively shield γ rays. We have layers of LiH-loaded paraffin of 15 mm thick, Cd sheet of 0.6 mm thick and whole system is surrounded by H_3BO_3 loaded water tank. Details of the ELEGANTS VI system have been given elsewhere[8].

4 π Active Shield by Roll-off Ratio

The characteristic of the ELEGANTS VI system is its capability of 4π active shield. Usually scintillators can have only passive shield in the PMT direction for which long light guide have been used. In the ELEGANT VI system the central CaF_2(Eu) is sandwiched by two CaF_2(pure) crystals which act not only as a light guide and passive shield but also as an active shield since the CaF_2(pure) is also a scintillator. Its scintillation light is dominantly in the ultra-violet (UV) region although CaF_2(Eu) emits visible light. Using PMT's sensitive to UV light, a number of photons per energy from the CaF_2(pure) crystal is about 1/3 of that from the CaF_2(Eu). The UV-light from the CaF_2(pure) hardly pass the central CaF_2(Eu) thus gives large pulse height only to either side of PMT's. Thus the background events can be rejected by setting a proper window on the asymmetry parameter (roll-off-ratio) between the signals from 2 PMT's. The roll-off-ratio (R) is defined to be,

$$R = \frac{V_R - V_L}{V_R + V_L}, \quad (1)$$

where V_R and V_L are the pulse heights of right and left PMT's, respectively [9]. It reduces backgrounds from the PMT side thus 4π active shield is effectively achieved. This method is particularly effective for low energy signals.

Limit on the Lifetime of 0ν–ββ Decay of ^{48}Ca

ELEGANT VI has been in operation at Oto Cosmo Observatory (OCO) located in Nara, Japan. It is a 5 km tunnel and its depth is around 500 meters which corresponds to 1.3 km water equivalent shield. The muon flux of OCO was found to follow the relation given as a function of the depth for various underground laboratories [10].

It is an almost straight tunnel and wind is always blowing. This makes OCO ideal for the reduction of background from radon. Reduction of radon is particularly important for detection of low energy signals. Usually underground laboratories have concentration of radon a couple of orders-of-magnitude higher than that in the laboratory at surface of the Earth. We observed no particular radon concentration in OCO.

Here we show a result from the duration of middle of 1998 to end of 1999 (total live time of 5567 hours). Figure 2 shows energy spectrum of CaF_2(Eu) obtained in the duration after the event selection described in ref [11,12]. Total weight of 23 CaF_2(Eu) crystals is 6.66 kg which corresponds to 9.61 x 10^{22} of ^{48}Ca atoms. The statistical significance of the energy spectrum is 4.23 kg yr. No events are observed in the 0ν–ββ decay energy window (4.02-4.75 MeV) which is a 3σ peak interval centered at 4.27 MeV.

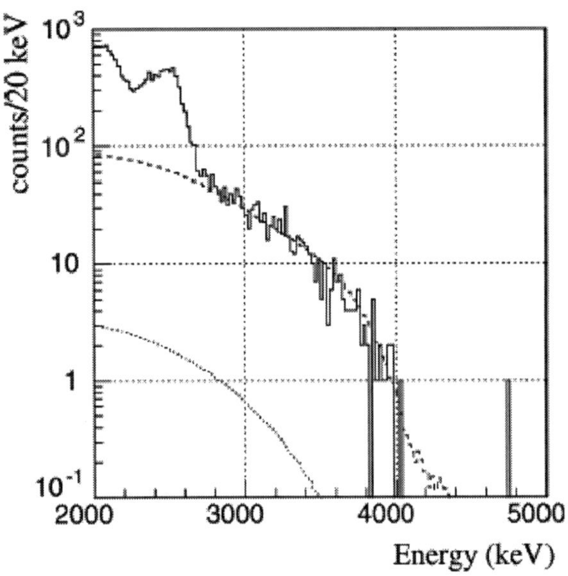

FIGURE 2. A comparison of Monte Carlo simulations from the internal radio-activities (dashed line) and the 2ν–ββ of ^{48}Ca (dotted line) with experimental data (solid line) of a statistical significance of 4.23 kg yr.

Since we observed no event in the Q-value region, we set lower limit on the lifetime of the 0ν–ββ decay. In order to derive the limit we estimated background events in this energy window for which details are given elsewhere [12]. Conceivable radioactivities in the Q value region are ^{214}Bi → ^{214}Po(Q_β=3.27 MeV, $T_{1/2}$=164.3μsec)→ ^{210}Pb (Q_α=7.83 MeV) in the uranium chain, and ^{212}Bi → ^{212}Po(Q_β=2.25 MeV, $T_{1/2}$=0.299μsec)→ ^{208}Pb(Q_α=8.95 MeV) in the thorium chain. Since the electron equivalent energy of α rays in the CaF_2 crystal appear at about ~1/4 of its energy, only simultaneous detection of the β and α rays gives background at around the Q value region. Since decay time of CaF_2 signal is quite long (~1μsec) and we have 4μsec for the ADC gate, two successive decays are in the same ADC gate occasionally. Total simulated energy spectrum is plotted in Fig. 2 which reproduced well shape and absolute yield of the experimental spectrum in an energy range of 2.8-4.0 MeV. The shoulder at around 2.6 MeV in the experimental spectrum is from ^{208}Tl decay in the CsI(Tl) scintillators which is irrelevant for present analysis. The resulting lower limit of half-life or the 0ν–ββ decay is

$$T_{1/2}(0\nu) \geq 8.6 \times 10^{22}\ \text{year}(68\%\ \text{C.L.})$$
$$\geq 1.8 \times 10^{22}\ \text{year}(90\%\ \text{C.L.})$$

This result can be converted to the upper limit for the effective Majorana neutrino mass m_ν neglecting right-handed currents. Depending on the theoretical nuclear matrix elements (see Table 8 and 9 in Ref. [14]), the upper bound at the 90 % C.L. is $m_\nu \lesssim$ 7.2–44.7 eV. Our value is better than the current limit of 9.5 $\times 10^{21}$ year(76 % C.L.) given by Beijing group [3] using 37 kg of Ca_{F2} scintillation crystals. It has to be noted that we have no background in the signal region although Beijing group had 365 events. Therefore our measurement is not limited by backgrounds.

The ^{208}Tl→ ^{208}Pb(Q_β=5.00 MeV) decay could give background in the Q value region. It gives a β ray up to 2.4 MeV and a 2.6 MeV γ ray. The probability that both radiation is confined in a 45mm cube crystal is small thus its contribution is currently negligible.

CANDLES

In order to explore the mass region of $m_\nu \sim 10^{-(1-2)}$ eV, one needs to design detector sensitive to lifetime of 10^{26} year for $0\nu-\beta\beta$ decay of the ^{48}Ca. One then need to prepare several tons of calcium since the natural abundance of the ^{48}Ca is very tiny (0.187 %). So far we have ~5 kg • year of CaF_2 in ELEGANTS VI. We wish to increase amount of CaF_2 crystals more than 4 orders-of-magnitude to sense the region around 50 meV region. In order to achieve this we proposed CANDLES which stands for CAlcium florid for the study of Neutrinos and Dark matters by Low Energy Spectrometer. Conceptual design of the detector is schematically shown in figure 3. Pure CaF_2 crystals with a dimension of ~10cm cube are immersed in a liquid scintillator. Scintillation light is viewed by large PMT's. Use of the CaF_2(pure) crystal is essential to build large detector as described in the following.

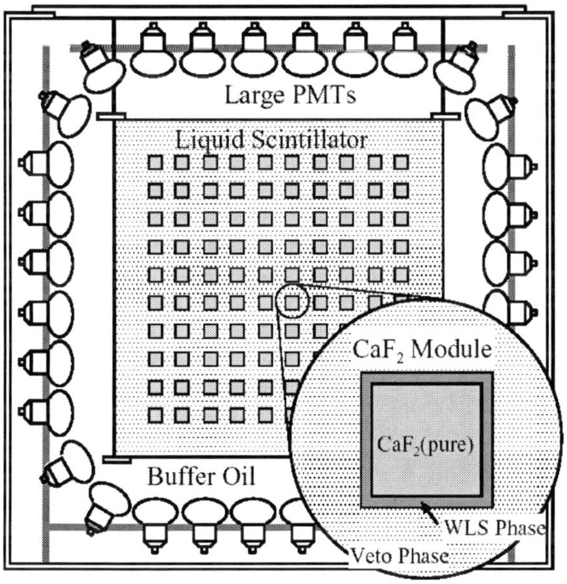

FIGURE 3. Schematic cross sectional view of the proposed CANDLES detector.

The CaF2(pure) has only ~1/3 of light output of that of CaF2(Eu). The CaF_2(Eu) in ELEGANTS VI gives energy resolution of 3 % (FWHM) at the Q value region. However, the CaF_2(Eu) crystal has attenuation length as short as ~10 cm. Therefore if one wants to stick to the CaF_2(Eu) crystal to keep high light output, scale up is possible only in two dimensions which is quite inefficient to build a big detector. On the other hand, the CaF_2(pure) crystal is a scintillator and has long attenuation length of a meter or longer. Therefore there is essentially no limit to construct huge detector. Problem is the low light output of the CaF_2(pure) crystal.

The light emission of scintillation is centered in the UV region. We increased the light output of CaF_2(pure) crystals by using wave-length shifter since the scintillation light is in the UV region. Light output was studied for various wave-length shifter with CANDLES I detector which is a prototype detector and has simple configuration that a 45 mm cube CaF_2(pure) crystal is immersed in liquid scinitillator in 5" cube lucid container. Scintillation lights are viewed by four 5" PMT's. We found that light output can be increased up to about 60% of the CaF_2(Eu) crystal. This light output gives 3% (FWHM) energy resolution in the Q value region if we have PMT's which cover whole solid angle. It is only energy resolution to reduce contribution of $2\nu-\beta\beta$ decay in the Q value region. The contribution in the window of the Q value is small enough for the lifetime of 10^{26} years which enable us to search for the neutrino mass of 50 meV region. CANDLES thus has no problem to build big detector sensitive to the mass region.

Pulse Shape measurement and Background Rejection

In CANDLES detector the liquid scintillator acts as an active shield. We make use of decay time difference of liquid scintillator (~10 nsec) and CaF_2 crystal (~1μsec) to separate backgrounds from signals. In order to demonstrate our background rejection method, we measured pulse heights of prompt (~40 nsec) and total (~4 μsec) component from the CANDLES I detector. Figure 4 shows two dimensional plot of prompt and total pulse heights when we irradiate CANDLES I detector by ^{137}Cs γ source. Background signals that fire liquid scintillator has large fprompt component. On the other hand signals from CaF2 has very small prompt component. We can clearly see two loci corresponding to the backgrounds and signals. Figure 4 clearly demonstrates that we can effectively reduce backgrounds by pulse shape measurement.

We use mixture of paraffin oil (80%) and Pseudo-cumene (20%) for liquid scintillator with wavelength shifter (PPO). This mixture give large scintillation light thus quite effective for active shield. However, increase of Pseudo-cumene decreases CaF_2(pure) scintillation light due to absorption of UV light. We thus use two phase method where CaF_2(pure) has thin outer wavelength shifter region which has no Pseudo-cumene. This design satisfies enough scintillation light from both CaF_2(pure) and liquid scintillator.

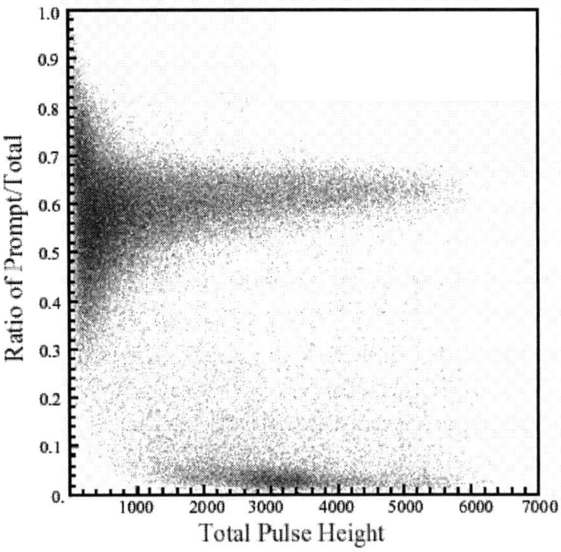

FIGURE 4. Two dimensional plot of pulse height of prompt (~40nsec) and total (~4 μsec) components. The background rejection is clearly demonstrated

We now know that the backgrounds in the Q-value region are solely from the successive β- and α decays from radioactivities contained in a crystal. Sum energy of the two decays can be as large as that of the Q value. Actually the biggest background is from the decay ^{212}Bi. Since the intermediate ^{212}Po has lifetime of 0.299μsec there is a large probability that two successive decays are in an ADC gate of ~4 μsec. This dominant background can be reduced by observing pulse shape of the signal. We measured pulse shape of signals of ELEGANTS VI using 100 MHz Flash ADC and demonstrated that we can detect a rise of second signal if the separation is more than 2 channels (~25 nsec). We can thus reduce the background down to 5%. Since this separation largely dependent on the time resolution of the detection system, we will reduce it down to 1% by introducing 500 MHz Flash ADC in the future CANDLES detector.

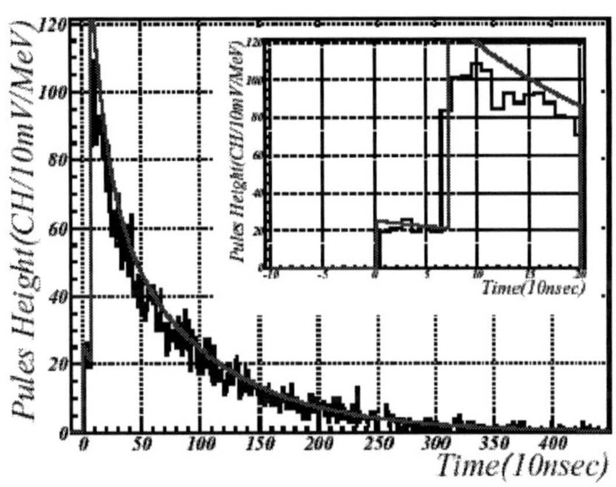

FIGURE 5. Typical pulse shape of the sequential decay events measured by the 100 MHz Flash ADC is displayed.

Measurement of pulse shape by Flash ADC leads us further possibility to reduce background by utilizing a difference in decay times of signals of α rays and β- or γ rays. We found that the decay time difference for CaF_2(pure) crystal although we saw almost no difference for CaF_2(Eu) crystal. Since we will use CaF2(pure) crystal in CANDLES we can use this difference to reduce backgrounds. Quantitative value for the reduction will be given after our analysis although currently we can reduce the background more than an order-of-magnitude.

Further Background Rejection

The β decay of ^{208}Tl has large Q value of 5 MeV. However, it emitts β ray and 2.6 MeV γ rays. The probability that the high energy γ rays are contained in a single crystal of 10cm cube is small. However, if we are interested in a very rare signal, it could be a background seriously considered. The ^{208}Tl decay has a preceding α decay with a lifetime of 3 minutes. If a counting rate is small enough we can reduce the background by identifying the preceding α decay. The counting rate can be reduced if we have good position resolution. The position resolution was tested by CANDLES I detector where two crystals were immersed in a scintillator and four PMT's gave position information. We obtained a few centimeter for the position resolution. The resolution is very good. This, however, may be due to small size of CANDLES I. We will study the position resolution for a larger detector in CANDLES II.

Currently we are assembling CANDLES II detector which has 9 CaF_2 crystals with a dimension of $45\times45\times70$ mm^3. This will demonstrate the design of the CANDLES where we have more than 2 crystals in the liquid scintillator. We are designing the CANDLES III detector which will have around 200 kg CaF_2 crystals. This will be constructed in our laboratory which is at the sea level. For the construction of the detector, low radioactive contamination in a crystal is essential. We are having collaboration with crystal makers to develop crystals with low radioactive contamination.

CONCLUSION

We are developing CANDLES for the study of $\beta\beta$ decay of ^{48}Ca. So far CANDLES has no technical difficulty to construct detector which is huge enough to be sensitive to neutrino mass region of current interest. CANDLES has ways to reduce backgrounds and we will demonstrate by constructing series of CANDLES detectors.

ACKNOWLEDGMENTS

This experiment is financially supported in part by the Grant-in-Aid for Scientific Research 14204026 and 17104003.

REFERENCES

1. E.der Mateosian and M. Goldhaber, Phys. Rev. 146 (1966) 810.
2. R.K. Bardin et al., Nucl. Phys. A 158 (1970) 337
3. K. You et al., Phys. Lett. B 265 (1991) 53.
4. A. Balysh et al., Phys. Rev. Lett. 77 (1996) 5186.
5. V.B. Brudanin et al., Phys. Lett. B 495 (2000) 63.
6. A. Bakalyarov et al., Nucl. Phys. A 700 (2002) 17.
7. R. Bernabei et al., Nucl. Phys. A 705 (2002) 29.
8. R. Hazama et al.,Proc. of the 4th Int. Conf. on Weak and Electromagnetic Interactions in Nuclei (WEIN 95), Osaka, Japan, June 1995, World Scientific, Singapore P.~635; R. Hazama et al., Proc. of the XIV Int. Conf. on Particles and Nuclei (PANIC 96), CEBAF, USA, May 1996, World Scientific, Singapore p.~477; R. Hazama et al., Proc. of the Int. Workshop on the Identification of Dark Matter (IDM 96), September 1996, Sheffield, UK, World Scientific, Singapore p.397; T. Kishimoto et al., Proc. of the 2nd RESCEU Int. Symp. on Dark Matter in the Universe and its Direct Detection, November 1996, Tokyo, Universal Academy Press Inc., Tokyo, p.~71; I. Ogawa et al., Nucl. Phys. A663-664 (2000) 869c; T. Kishimoto et al., Proc. of the Identification of the Dark Matters (IDM2000), September 2000, York, UK.
9. R. Hazama, Doctoral thesis Osaka University (1998).
10. P. F. Smith and J. D. Lowing, Phys. Rep. 187 (1990) 203
11. I. Ogawa et al., Proc. Int conf. Particles and Nuclei, Nucl. Phys. (2002).
12. I. Ogawa et al., Nucl. Phys. A.
13. G.J. Feldman and R.D. Cousins, Phys. Rev. D 57 (1998) 3873.
14. J. Suhonen and O. Civitarese, Phys. Rep. 300 (1998) 123.

NOSTOS: a spherical TPC to detect low energy neutrinos

S. Aune*, P. Colas*, J. Dolbeau†, G. Fanourakis**, E. Ferrer Ribas*, T. Geralis**, Y. Giomataris*, P. Gorodetzky†, G. J. Gounaris‡, I. G. Irastorza [1]*, K. Kousouris**, V. Lepeltier§, T. Patzak†, E. A. Paschos¶, P. Salin†, I. Savvidis‡ and J. D. Vergados‖

*DAPNIA, Centre d'Études Nucléaires de Saclay (CEA-Saclay), Gif-sur-Yvette, France
†APC, Université Paris 7 Denis Diderot, Paris, France
**National Center for Scientific Research "Demokritos", Athens, Greece
‡Aristotle University of Thessaloniki, Greece
§Laboratoire de l'Accélérateur Linéaire, Orsay, France
¶University of Dortmunt, Dortmunt, Germany
‖University of Ioannina, Ioannina, Greece

Abstract. A novel low-energy (∼few keV) neutrino-oscillation experiment NOSTOS, combining a strong tritium source and a high pressure spherical Time Projection Chamber (TPC) detector 10 m in radius has been recently proposed. The oscillation of neutrinos of such energies occurs within the size of the detector itself, potentially allowing for a very precise (and rather systematics-free) measure of the oscillation parameters, in particular, of the smaller mixing angle θ_{13}, which value could be determined for the first time. This detector could also be sensitive to the neutrino magnetic moment and be capable of accurately measure the Weinberg angle at low energy. The same apparatus, filled with high pressure Xenon, exhibits a high sensitivity as a Super Nova neutrino detector with extra galactic sensitivity. The outstanding benefits of the new concept of the spherical TPC will be presented, as well as the issues to be demonstrated in the near future by an ongoing R&D. The very first results of small prototype in operation in Saclay are shown.

Keywords: Spherical Time Projection Chamber; neutrino oscillations; Tritium; Supernovae; Weinberg angle; Neutrino-nucleus interaction; Neutrino magnetic moment
PACS: 29.40.Cs; 12.15.Ff; 14.60.Pq; 13.15.+g; 16.60.st; 95.55.Vj; 95.85.Ry

INTRODUCTION

Nowadays there is a compelling evidence that neutrinos change flavor as they propagate. Appearance or disappearance of neutrinos has been solidly proved in experiments looking at neutrinos of either extraterrestrial [1] (solar and atmospheric) or terrestrial [3] (reactor or accelerator) origin. The neutrino mixing, which provokes the oscillation of the flavor change probability along the propagation of the neutrino, is invoked to explain the observed appearance or disappearance.

The atmospheric neutrino oscillation data [1, 2] strongly suggest that ν_μ oscillate into ν_τ with maximal mixing angle ($\theta_{\rm atm} \sim \pi/2$) and a corresponding mass squared difference of $\Delta m^2_{23} \simeq 3\times10^{-3} eV^2$. Results from accelerator neutrinos support this interpretation [3]. On the other hand, the solar neutrino data [4, 5, 6] could be explained by an oscillation of ν_e into ν_μ and/or ν_τ with a non maximal –but large– mixing angle ($\theta_\odot \sim \pi/3$) and a mass squared difference of $\delta m^2_{21} \simeq 7\times10^{-5} eV^2$, which has been recently supported by evidence of disappearance of reactor antineutrinos [7]. The third mixing angle completing the standard three neutrino oscillation scheme is not known but it is constrained to be quite small $\theta_{13} < \pi/6$ [9]. The determination of this parameter is the remaining question to complete our understanding of the leptonic mixing and, moreover, it will open the way to study the CP-violating effects in the neutrino sector [9].

Such small mixing angle could have measurable consequences in experiments involving electron neutrinos but sensitive to the oscillation length driven by the large squared-mass gap Δm^2_{23}, i.e., the smaller oscillation length. In fact, for a detector close enough to the neutrino source the contribution from the larger oscillation length to the disappearance oscillation probability is negligible, and therefore it is driven only by θ_{13} and Δm^2_{23} in the following way:

[1] attending speaker: Igor.Irastorza@cern.ch

CP 785, *Topical Workshop on Low Radioactivity Techniques*, edited by B. Cleveland, R. Ford, and M. Chen

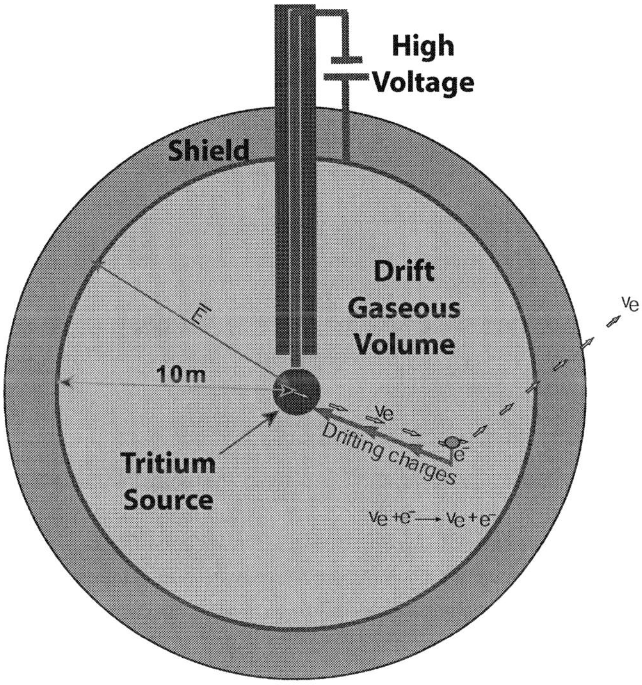

FIGURE 1. Illustration of the proposed NOSTOS experiment, showing the spherical TPC surrounded by an appropriate shield. The neutrino source and the detector are located in the center of the curvature of the sphere. Inside the inner sphere, the tritium source is surrounded by the shield and cooling system and by several Micromegas flat detectors.

$$P(\nu_e \rightarrow \nu_e) = 1 - \sin^2 2\theta_{13} \sin^2 \pi \frac{L}{L_{23}} \tag{1}$$

where L is the distance between source and observed interaction and L_{23} is the oscillation length related to the neutrino energy E in the way: $L_{23} = 2\pi E_\nu/(\Delta m^2_{23})$. While for reactor and accelerator neutrinos, this length is of the order of ~ 1 km or ~ 100 km respectively, we want to point out the fact that in the case of very low energy (few keV) neutrinos, like those emitted by a tritium source, the oscillation length L_{23} is only 13 m.

The NOSTOS concept proposes the detection of such low energy neutrinos by using a large spherical TPC of about 10 m in radius, surrounding the tritium source, so the whole oscillation would occur within the detector volume. That could allow the observation, for the first time, of the space oscillating signature and the determination of the smaller neutrino mixing angle θ_{13}, completing our knowledge of the leptonic mixing scheme [8]. The fact that the whole oscillation occurs inside the detector volume allows us to highly reduce systematic effects due to backgrounds or to bad estimates of the neutrino flux, which is the main worry in most neutrino experiments, where the interaction rate is measured at a single space point.

THE NOSTOS DETECTOR: THE SPHERICAL TPC CONCEPT

The NOSTOS principle of detection is based on the concept of the spherical TPC, which has several interesting potential advantages, as will be shown later on. It consist of 2 concentric spheres, the external one, 10 m of radius, at ground and the inner one, 50 cm of radius, at high potential (Fig. 1). The tritium source is located inside the central sphere. Neutrinos escaping radially from the center of the sphere interact in the gas mainly by electron elastic scattering. The ionization charges produced in the interaction drift towards the center and are collected by an adequate gaseous readout, which covers the surface of the inner sphere (possibly composed by a series of flat segments). The preferred choice for the readout devices are Micromesh Gaseous Structures (Micromegas [10]) due to the high precision, fast response and excellent energy resolution. High efficiency for detecting single electrons have been proved with Micromegas [11] even at high pressures [12]. In addition, Micromegas readout is currently being used for solar axion detection in the CAST experiment [13, 14] where a great stability and ability to reject background events has been achieved.

This novel approach is radically different from all other neutrino oscillation experiments in that the neutrino source and the detector are located in the same vessel; it is then possible to measure the neutrino interactions continuously as a function of the distance source-interaction point, and directly observe the oscillation length that is fully contained in the detector. Indeed we expect a counting rate oscillating from the centre of curvature to the depth of the gas volume, i.e. at first a decrease, then a minimum and finally an increase. Fitting such an observed curve will provide all the relevant parameters of the oscillation by a single experiment.

The radial coordinate of the interaction point is measured by inspecting the time pattern of the charge pulse detected at the center of the sphere which temporal extension is determined by the longitudinal diffusion of the ionization cloud and therefore by the distance drifted. In fact, for our spherical TPC, the electric field at a distance r from the center is given by:

$$E(r) = \frac{R_2 R_1 V_0}{R_2 - R_1} \frac{1}{r^2} \tag{2}$$

where R_1 and R_2 are the inner and outer sphere radius and V_0 is the applied voltage. At low electric fields the drift velocity v_d is roughly proportional to the electric field and the longitudinal diffusion coefficient depends on E as $D_L \sim 1/\sqrt{E}$. Hence the spherical geometry of the electric field enhances the longitudinal dispersion occurred after a given drifted distance r, with respect to the one for a homogeneous field. First calculations show that a precision better than 10 cm, which would largely satisfy the needs of the NOSTOS concept, can easily be achieved. Preliminary results without Micromegas regarding the experimental demonstration of this strategy by the first prototype actually running in Saclay are presented in section 4.

The use of Micromegas will improve those results in two ways. First, the fast pulses then available (rise times of ~1 ns) will allow a better measurement of the temporal dispersion of the signal; and second, information from transversal dispersion will in also be available, by means of an appropriate pixelization of the Micromegas readout.

In general, the use of a spherical TPC detection scheme presents the following advantages for the proposed experiment:

- The spherical geometry naturally focuses a large drift volume into a small amplifying detector with only a few read-out channels. It is the most cost-effective way of instrumenting a large detector volume with a minimum of front-end electronics. Such approach simplifies the construction and reduces the cost of the project.
- The placement of the neutrino source in the center of the detector provides a close to 4π acceptance. The spherical detector geometry optimizes the detection efficiency per unit of volume for a given flux of outgoing neutrinos.
- The spherical drift enhances the relation of longitudinal diffusion with drifted distance, as have been shown above, improving the resolution in the determination of the point of interaction, a key issue of the project.
- The ratio of external surfaces (and therefore external materials) over detector volume is optimized for a spherical geometry, therefore allowing a lower background per unit volume due to external surface or material contaminations. In addition, in a spherical geometry the thickness of material needed to hold the gas is minimum, further reducing possible sources of background.
- Large drift volumes can be built without the use of a field cage, unlike cylindrical TPCs. In addition, the symmetry of the design and the compact placement of the readout in the center of the sphere may provide a lower sensitivity to electronic noise (in fact, the outer sphere acts as a perfect Faraday cage to the inner electrode).

NOSTOS EXPERIMENTAL CHALLENGES AND EXPECTED SENSITIVITY

The NOSTOS concept relies not only on the detection, for the first time, of very low energy neutrinos from a tritium source, but on the measurement of the expected oscillation in that detection rate along the detector volume. We are confident, given the present status of the TPC technology, and the potential of the spherical TPC concept exposed above, on the feasibility of the project. However, an R&D program to assess each of the experimental key issues on which the project relies, as well as to experimentally demonstrate them, is already ongoing. The main experimental issues can be summarized as follows:

- **Energy threshold:** Neutrinos from tritium decay have a maximum energy of 18.6 keV, therefore the maximum energy of the recoiling electron is of only 1.27 keV. An energy threshold of the order of $\sim$ 100 eV is desirable to have a good sensitivity. Single electron efficiency has been achieved by Micromegas readout devices [11], but low threshold operation with large volumes like that of NOSTOS is being checked experimentally, as part of the ongoing development phase.
- **Background:** Although absence of background is not a requisite for the NOSTOS scheme, as runs without the source can be performed to determine the background, it is true that the presence of a given level of background will decrease the sensitivity of the experiment. As a first measure, the experiment should be located underground. In addition, studies are ongoing to determine precisely the loss in sensitivity for a given level of background, in order to assess to which extent are low background techniques necessary, including shielding, vetoing, material radiopurity and gas purification, as well as offline rejection techniques. Ultimately, an experimental test in underground location with a prototype is envisioned to prove that the necessary low background level can actually be obtained.
- **Radial resolution:** The determination of the radial coordinate of the interaction point through the measurement of the time dispersion of the detected charge pulse is a key issue of the NOSTOS concept. Preliminary results concerning this point have been already obtained with the prototype detector which is not equipped with Micromegas and will be presented in the next section. Work is in progress to assess the final radial resolution achievable.
- **Scaling up:** Finally, even if good experimental parameters are demonstrated for small prototypes, work has to be done to assure that they will be maintained after the scaling up to the full size required by the NOSTOS experiment. To this end, an intermediate scale (4 m diameter) prototype is being designed and will be used in a second stage of the ongoing development phase.
- **Electrostatics:** Right now, a metallic rod supports the central sphere. This provokes a distortion of the electric field, so that only a third of the volume (opposite to the stick) is reasonable close to the desired spherical field of (2). Two main ideas are being considered to solve this problem:
  - **a)** Use of field shaping rings along the metallic rod at corresponding potentials. The extremity of the rod, next to the small sphere, would be made of a resistive cone.
  - **b)** A charging system like the one used in electrostatic accelerators, using a series of small metallic balls on an insulator chain. Due to the absence of beam discharging the "terminal", a very small chain would be sufficient.

Using a 20 kg tritium source (=200 MCi) the total number of emitted neutrinos is $6 \times 10^{18} \nu/s$. With the TPC filled with Xe at 1 bar the number of detected neutrinos is about 1000/year, assuming an energy threshold of 100 eV. The use of a less intense source or cheaper gases like Ar or Ne is possible at the expense of operation at higher pressures. High pressure TPC with Micromegas readout have been successfully tested [12] in the past. They are included in the NOSTOS development program in order to study the viability of such mode of operation.

A detailed study based on Monte Carlo simulations with realistic experimental parameters and with the rigorous treatment of the neutrino interaction including atomic effects developed in [15] is currently under way in order to determine more precisely the sensitivity prospects of this proposal.

PRELIMINARY RESULTS FROM THE FIRST SPHERICAL PROTOTYPE WITH A NEW PROPORTIONAL COUNTER

A prototype of spherical TPC has been built as a first step towards the NOSTOS detector, and is currently being used at Saclay to perform demonstration tests. The spherical vessel is 1.3 m of diameter and is made of 6 mm thick copper, allowing to hold up to 5 bar of pressure. The first tests were oriented to the assessment of the tightness of the vessel, so the gas could keep the sufficient level of purity for right operation. The volume was pumped by a primary pump followed by a turbo molecular pump, reaching a level of vacuum below 10^{-6} mbar. The outgassing rate measured was below 10^{-9} mbar/s, which allows us to avoid permanent gas circulation through special cleaning filters and to operate instead in seal mode.

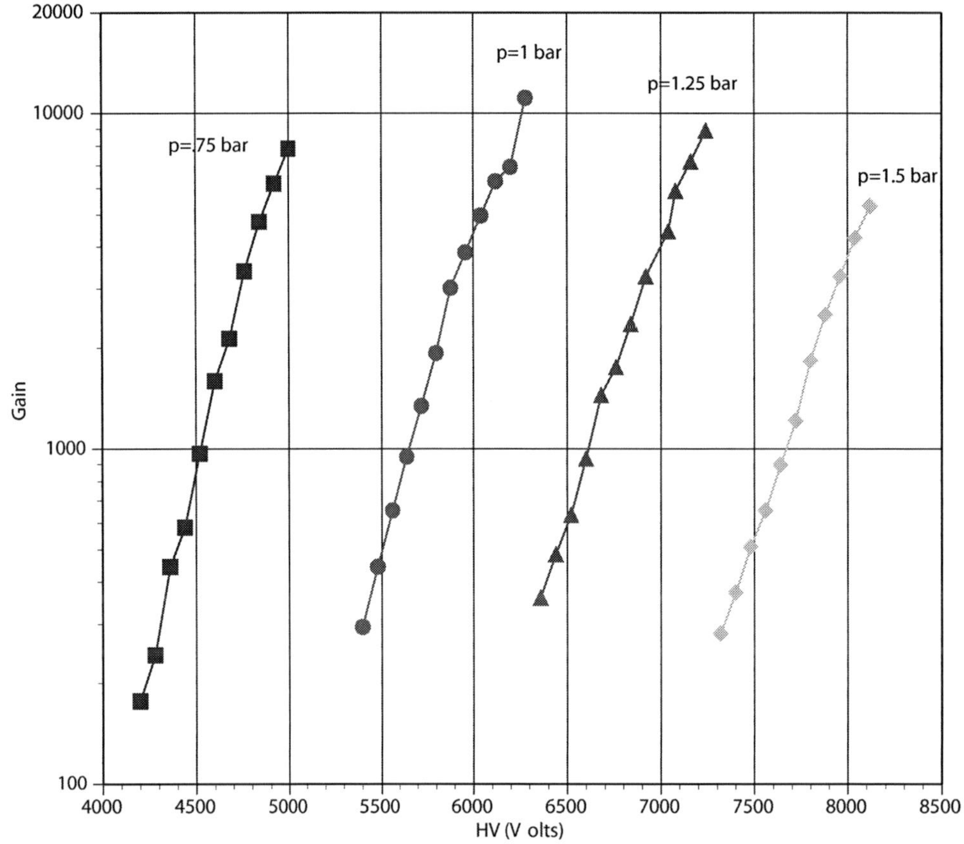

FIGURE 2. Measured gain as a function of the voltage applied, for Ar + 2% Isobutane at four different pressures.

As mentioned before, a Micromegas-type readout is proposed as amplification structure in the center of the TPC. Work is in progress to actually design and build a spherical Micromegas detector with new technologies. A more conventional alternative would be to approximate the spherical geometry by a composition of several flat Micromegas elements. While working independently in the design of the amplification structure, the first tests with the spherical vessel were performed using a small spherical electrode (10 mm diameter) placed in the center of the TPC, working as a proportional counter. The signals from this proportional counter are very slow, for only the ion movement can be observed, and their path is long. The use of Micromegas, where the electron movement in the 100 μm gap will be observed will lead to much shorter pulses. The system shows a very small capacity, $\sim$1pF, and therefore a very low level of electronic noise is expected ($< 1000e^-$) allowing for a potentially very low threshold.

So far the prototype has been operated with two different gas mixtures, namely Ar + 10% CO_2 as well as Ar + 2% Isobutane; and at different pressures up to 1.5 bar. These first tests showed that even with such a simple amplification element, high gains (above 10^4) are achieved. Figure 2 shows the obtained gain versus voltage curves. Stable operation was tested up to 40 days, without gas circulation (seal mode). This results are very encouraging but we are not yet in the tritium configuration. Runs using calibration sources of ^{109}Cd and ^{55}Fe or cosmic rays have been performed. An example of a ^{55}Fe 5.9 keV event is shown in Fig. 3. The remarkable low noise that the baseline of the pulse of fig. 3 shows that thresholds as low as $\sim$100 eV are already at hand.

To study the drift, the ^{55}Fe source is introduced inside the sphere by means of a movable insulant stick which allows us to place the source at any distance from the inner electrode. Data taken at different source distances show no evidence of loss of signal intensity due to electron attachment. In Figure 4 the recorded energy spectrum of one of these runs is shown.

These data allow the study of the drift properties of the chamber. Preliminarily, no appreciable electron attachment has been observed. The main concern at this point of the development phase is to demonstrate whether the time diffusion of the event can be measured and the drift distance extracted from it. To this end, as long as we do not equip

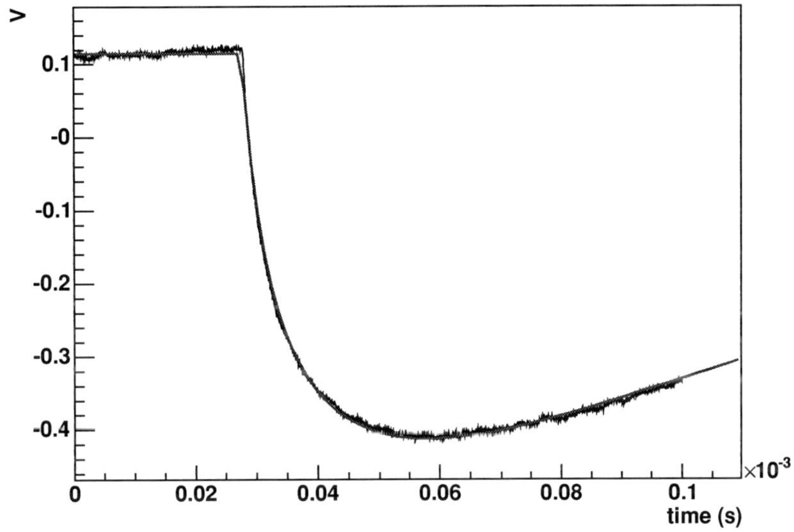

FIGURE 3. Example of a Fe^{55} pulse.

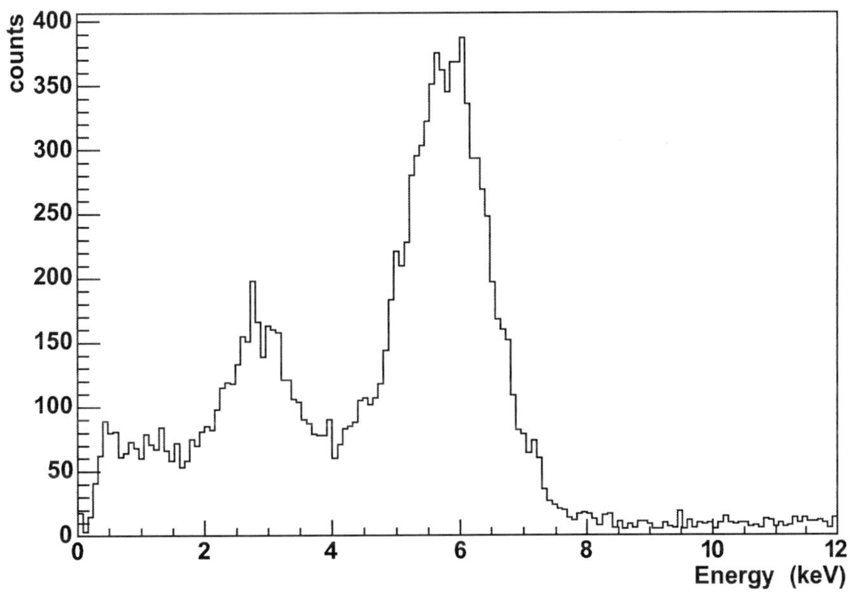

FIGURE 4. ^{55}Fe spectrum.

the anode with Micromegas, a detailed Pulse Shape Analysis is being developed based on deconvolution techniques. Although the work is still under progress, the first results are encouraging, showing that a 10 cm resolution is already achievable as shown in Fig. 5. The technique is supposed to unfold the effect of the electronics and the charge induction from the raw pulse, to arrive to the deconvoluted pulse, a bare reflection of the temporal pattern of the electron cloud arriving at the central electrode. In Fig. 5 various "template pulses", each one obtained by averaging 20 deconvoluted pulses, are shown for different positions of the calibration source. The effect of the diffusion (as the width of the pulse) is clearly visible.

Due to the mean free path of the ^{55}Fe X-rays in the gas, there is an uncertainty in the position of the interactions in those tests. To overcome this difficulty a new way of calibrating this effect is being designed. It consists of the use of

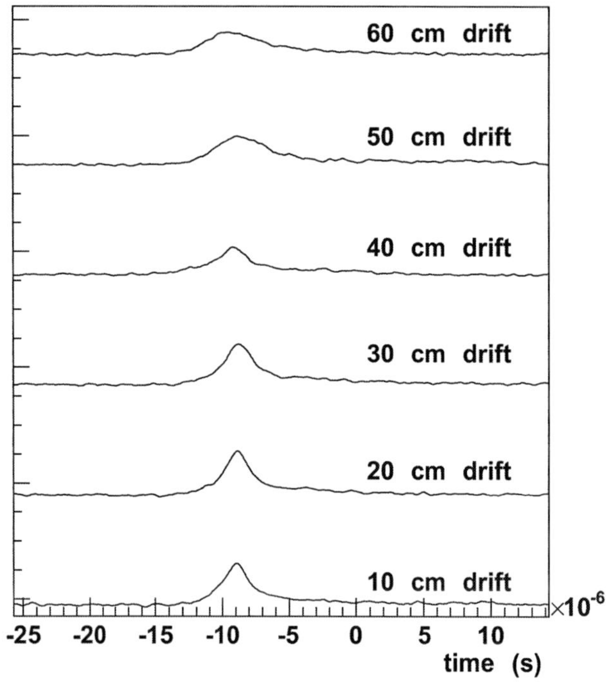

FIGURE 5. Effect of the diffusion on the deconvolved pulses. Each line corresponds to pulses taken with the source at different distances from the inner electrode, from 10 cm up to 60 cm in steps of 10 cm.

a ^{241}Am source, emitting simultaneously an α and a 60 keV γ. The α is detected by a small silicon diode providing the trigger to detect the γ interacting in the sphere. The time delay between the α and the arrival of the electrons at the small sphere provides an accurate measure of the drift time.

ADDITIONAL PHYSICS PROGRAM

We briefly mention some additional application of the NOSTOS detector in the domain of neutrino physics.

Sensitivity to the neutrino magnetic moment

Because of the low energy of the incoming neutrinos and the low energy electron recoils detected in this experiment the sensitivity for the neutrino magnetic moment is high. The cross section of the magnetic moment can be written as[16]:

$$\left(\frac{d\sigma}{dT}\right)_{EM} = \sigma_0 \left(\frac{\mu_l}{10^{-12}\mu_B}\right)^2 \frac{1}{T}\left(1 - \frac{T}{E_\nu}\right) \quad (3)$$

Because of the dependance $1/T$ (T is the electron recoil energy) the sensitivity for the magnetic moment is obviously higher at low energy. Recent measurements from the MUNU experiment [17] show a limit of 10^{-10} μ_B for the neutrino magnetic moment. Our experiment opens the way to improve this value by two orders of magnitude.

Measurement of the Weinberg angle at low energies

Another interesting quantity is the Weinberg angle appearing in standard neutrino cross-sections, which is a function of the momentum transfer and it has not been measured at such low transfers. To this end atomic physics experiments, which utilize the neutral current, have thus far been considered.

By plotting the differential neutrino-electron cross section as a function of the electron energy we obtain a straight line. We hope to construct the straight line quite accurately. Thus we can extract a value of the Weinberg angle both from the slope and the intercept achieving high precision.

Supernova sensitivity. Coherent neutrino scattering

It is generally believed that the core-collapse supernova explosion produces a large number of neutrinos and 99% of the gravitational energy is transformed to neutrinos of all types. The supernova (SN) neutrino flux consists of two main components: a very short (< 10 msec) pulse of ν_e produced in the process of neutronization of the SN matter through the reaction e + p → e + n, which is followed by a longer (< 10 sec) pulse of thermally produced ν_e, ν_μ, ν_τ, and their antiparticles. Only a small fraction, about 1%, of the neutrinos are prompt, while the rest are neutrino-antineutrino pairs from later cooling reactions. It is expected that spectra of thermally produced neutrinos are characterized by the different mean energies: $\nu_e = 11$ MeV, $\bar{\nu}_e = 16$ MeV, $\nu_{e,\mu} = 25$ MeV. Our idea is to use the large cross section offered by the coherent neutrino-nucleus cross section for detecting neutrinos from Super Nova explosions. Coherent scattering occurs when neutrinos interact with more than one particle and the amplitudes from the various constituants of the target add up. The increase of the cross section is proportional to the square of the number of particles in the target leading to increased counting rates:

$$\sigma = \frac{G^2 N^2 E^2}{4\pi} \tag{4}$$

where G is the weak coupling constant, N is the number of neutrons in the target nucleus and E is the neutrino energy. In order to get advantage of the coherent scattering amplification of heavy nuclei gases are needed.

For instance, using Xenon as detector target the coherent cross sections at E=25 MeV, the energy that is relevant for Supernova detection, is quite large ($\sigma = 1.5 \times 10^{-38}$ cm^2). Even at lower energy (11 MeV) where the coherent cross sections decreases quadratically with energy, the cross section is still high. The recoil energy energy is quite low and it takes a maximum value of 1.5 keV for 11 MeV and 9 keV for 25 MeV neutrinos. This implies that detector thresholds must be set quite low with one advantage that backgrounds are highly suppressed given the narrow time window in which the burst takes place. The collected energy may be even lower by a significant factor (quenching factor) and therefore sub-keV detector threshold is required. For a typical galactic SN explosion the detector used for the tritium experiment (10 m in radius, p=10 bar of Xenon) the number of detected neutrinos will exceed 100,000. A possibility to test the efficiency of detecting coherent neutrino scattering will be the nuclear reactor. The expected number of neutrino interactions in the gas volume, from a typical reactor neutrino flux and spectrum (10^{13} cm^{-2} s^{-1}) using a detector filled with Xenon is about 350/day/Kg[18]. The drawback is the very low energy threshold needed since the maximal recoil energy is 185 eV. Therefore single electron counting is required imposing a high gain operation of the detector (and a measurement of the quenching factor of the ionization produced by the low energy recoils). A 4 m diameter prototype, as foreseen as a later stage of the development phase, would be a perfect tool for this measurement, as it could contain 2000 kg of Xe at 10 bar.

CONCLUSIONS

The NOSTOS proposal aims at the detection of low energy (~few keV) neutrinos from a strong tritium source, in order to measure the smaller mixing angle θ_{13}. The new concept of the spherical TPC is proposed as principle of detection. The very first results of small prototype in operation in Saclay have been shown.

REFERENCES

1. Y. Fukuda *et al.* [Super-Kamiokande Collaboration], Phys. Rev. Lett. **81**, 1562 (1998) [arXiv:hep-ex/9807003].
2. S. Fukuda *et al.* [Super-Kamiokande Collaboration], Phys. Rev. Lett. **85**, 3999 (2000) [arXiv:hep-ex/0009001].
3. M. H. Ahn *et al.* [K2K Collaboration], Phys. Rev. Lett. **90**, 041801 (2003) [arXiv:hep-ex/0212007].
4. Q. R. Ahmad *et al.* [SNO Collaboration], Phys. Rev. Lett. **89**, 011301 (2002) [arXiv:nucl-ex/0204008].
5. S. Fukuda *et al.* [Super-Kamiokande Collaboration], Phys. Rev. Lett. **86**, 5651 (2001) [arXiv:hep-ex/0103032].
6. S. Fukuda *et al.* [Super-Kamiokande Collaboration], Phys. Rev. Lett. **86**, 5656 (2001) [arXiv:hep-ex/0103033].
7. K. Eguchi *et al.* [KamLAND Collaboration], Phys. Rev. Lett. **90**, 021802 (2003) [arXiv:hep-ex/0212021].

8. Y. Giomataris and J. D. Vergados, Nucl. Instrum. Meth. A **530**, 330 (2004) [arXiv:hep-ex/0303045].
9. S. Eidelman *et al.*, Phys. Lett. B **592**, 1 (2004)
10. Y. Giomataris, P. Rebourgeard, J. P. Robert and G. Charpak, Nucl. Instrum. Meth. A **376**, 29 (1996).
11. J. Derre, Y. Giomataris, P. Rebourgeard, H. Zaccone, J. P. Perroud and G. Charpak, Nucl. Instrum. Meth. A **449**, 314 (2000).
12. P. Gorodetzky *et al.*, Nucl. Instrum. Meth. A **433**, 554 (1999).
13. S. Andriamonje *et al.*, Nucl. Instrum. Meth. A **518**, 252 (2004).
14. S. Andriamonje *et al.* [CAST Collaboration], arXiv:hep-ex/0411033.
15. G. J. Gounaris, E. A. Paschos and P. I. Porfyriadis, arXiv:hep-ph/0409053.
16. G. C. McLaughlin and C. Volpe, Phys. Lett. B **591**, 229 (2004) [arXiv:hep-ph/0312156].
17. Z. Daraktchieva *et al.* [MUNU Collaboration], Phys. Lett. B **564** (2003) 190 [arXiv:hep-ex/0304011].
18. J. I. Collar and Y. Giomataris, Nucl. Instrum. Meth. A **471** (2000) 254.

V – REDUCED BACKGROUND AIR AND PURIFIED NOBLE GASES

Adsorption techniques for gas purification

Hardy Simgen

Max-Planck-Institute for Nuclear Physics, P.O. Box 10 39 80, D-69029 Heidelberg, Germany

Abstract.
Adsorption techniques offer an easily applicable, efficient way to purify gases to the highest purities. The appropriate choice of the adsorber in combination with the right operational parameters allows a selective removal of impurities. In this work the principles of the single- and multi-component adsorption process are discussed. A simple model of chromatography is applied to predict the breakthrough curves of impurities through adsorption columns. The theoretical results were used to develop purification procedures for nitrogen from radon and krypton and argon from radon. More than 200 m^3 of nitrogen can be purified from krypton with 1 kg of the appropriate adsorber. For radon no breakthrough was observed in nitrogen and in argon and the ^{222}Rn concentration in the final product was less than 0.5 μBq/m^3.

Keywords: Adsorption, chromatography, low-level physics, radon, krypton, argon, nitrogen, gas purification
PACS: 68.43.-h, 82.80.Bg

THEORETICAL OVERVIEW

Gas purification by adsorption is a multi-component adsorption process where the main component is the carrier gas and the other components are represented by the impurities. To predict the purification efficiency the adsorption of single components as well as possible interferences between them must be taken into account. The latter are difficult to predict, but important qualitative relations for multi-component systems can be obtained from the analysis of the single-component adsorption process.

Single-component adsorption

If a gas is exposed to the surface of an adsorber, part of it will be adsorbed on the available adsorption sites. For low partial pressures Henry's law predicts a linear relation between the number of adsorbed particles n and the partial pressure of the gas p (in equilibrium):

$$n = H \cdot p \tag{1}$$

For a given temperature T Henry's constant H can be calculated when the BET-surface area S_{BET} of the adsorber and the binding energy $\phi(z)$ of the gas particle to the adsorber is known. For spherical gas particles H is given by the following expression [1]:

$$H = \frac{n}{p} = \frac{S_{BET}}{RT} \cdot \int_{z=0}^{z_{max}} \left[\exp\left(\frac{-\phi(z)}{kT} \right) - 1 \right] \mathrm{d}z \tag{2}$$

R is the gas constant. The integration runs from the surface of the adsorber ($z = 0$) to an upper limit z_{max}, which depends on the structure of the surface. For a flat surface z_{max} goes to infinity and for porous adsorbers z_{max} equals the pore size b.

A simple model was introduced by Steele [2] to predict the potential $\phi(z)$ for the adsorption of unpolar particles to an unpolar wall. The binding energy is modeled by a Lennard-Jones-like potential between the particle to be adsorbed and any atom of the adsorber. In practice only the first few atomic layers of the adsorber contribute to the interaction because the Lennard-Jones potential vanishes at larger distances. With an approximation proposed in [3] the final expression for $\phi(z)$ is:

CP 785, *Topical Workshop on Low Radioactivity Techniques,* edited by B. Cleveland, R. Ford, and M. Chen

TABLE 1. Relevant parameters for the Steele model to predict Henry's constants for the adsorption of nitrogen, argon and krypton on carbon-based adsorbers (source: [5, 6]).

Element	$\sigma_{Gas,Ads}$ [Å]	$\varepsilon_{Gas,Ads}$ [10^{-22} J]	ρ_{Ads} [10^{19} m^{-2}]	d [10^{-10} m]
Carbon	3.40	3.86	3.81	3.35
Argon	3.41	16.5		
Krypton	3.60	23.6		
Nitrogen	3.70	13.1		

TABLE 2. Critical temperatures, critical pressures and Henry's constants calculated according to equation (7) for argon, krypton, radon and nitrogen.

Gas	T_C [K]	p_C [bar]	$\frac{T_C}{\sqrt{p_C}}\left[\frac{\mathrm{K}}{\sqrt{\mathrm{bar}}}\right]$	$H\left[\frac{\mathrm{mol}}{\mathrm{kg \cdot Pa}}\right]$
Argon	150.7	48.6	21.6	63
Krypton	209.4	55.0	28.2	$4.7 \cdot 10^4$
Radon	377.0	62.8	47.6	$1.3 \cdot 10^{13}$
Nitrogen	126.3	34.0	21.7	70

$$\phi(z) = 2\pi \cdot \rho_{Ads} \cdot \sigma^2_{Gas,Ads} \cdot \varepsilon_{Gas,Ads} \cdot \left[\frac{2}{5}\left(\frac{\sigma_{Gas,Ads}}{z}\right)^{10} - \left(\frac{\sigma_{Gas,Ads}}{z}\right)^{4} - \frac{\sigma^4_{Gas,Ads}}{3d \cdot (z+0,61d)^3}\right] \tag{3}$$

ρ_{Ads} is the 2-dimensional surface density of adsorber-atoms and d is the distance between the adsorber layers. The quantities $\sigma_{Gas,Ads}$ and $\varepsilon_{Gas,Ads}$ are derived from the Lennard-Jones parameters of the gas and the adsorber by the Lorentz-Berthelot mixing rules [5].

$$\varepsilon_{Gas,Ads} = \sqrt{\varepsilon_{Gas} \cdot \varepsilon_{Ads}} \tag{4}$$

$$\sigma_{Gas,Ads} = \frac{1}{2} \cdot (\sigma_{Gas} + \sigma_{Ads}) \tag{5}$$

Equation (5) gives the distance from the wall at which the particle will be adsorbed and equation (4) gives the depth of the potential well. The relevant parameters for the adsorption of argon, krypton and nitrogen on carbon-based adsorbers are summarized in Table 1.

The model can easily be extended to porous adsorbers when a slit-like pore model is applied. In that case a pore is considered as the space between two opposing parallel walls at a distance b. The combined potential is given by:

$$\phi_{Pore}(z) = \phi(z) + \phi(b-z) \tag{6}$$

With the parameters from Table 1 the potential $\phi_{Pore}(z)$ can be calculated. It is shown in Figure 1 for the adsorption of krypton in slit-like carbon pores of different slit widths. Using this potential in equation (2) Henry's constant H can be calculated for any adsorber with known pore size distribution. Activated carbons which are widely used as adsorbers usually have a broad pore size distribution. Therefore their adsorption behaviour can be approximated for most types in a simplified model. Maurer found a semi-empirical formula describing the adsorption of spherical gases on activated carbon to a sufficient accuracy [7]:

$$H\left[\frac{\mathrm{mol}}{\mathrm{kg \cdot Pa}}\right] = \exp\left\{\left(-0.05 + \frac{81}{T[\mathrm{K}]}\right) \cdot \frac{T_C[\mathrm{K}]}{\sqrt{p_C[\mathrm{bar}]}} - 17.5\right\} \tag{7}$$

T_C and p_C is the critical temperature and the critical pressure of the adsorbed gas. Table 2 shows Henry's constants derived using this formula for the adsorption of argon, krypton, radon and nitrogen on activated carbon at 77 K.

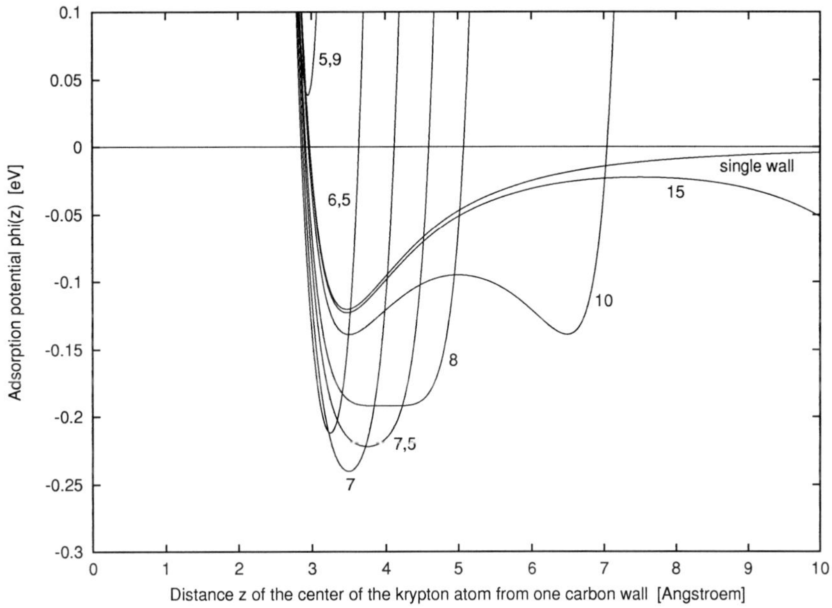

FIGURE 1. The adsorption potential for krypton in slit-like carbon pores according to the potential theory of Steele. The eight curves show the shape of the potential for different pore widths b (number at each curve) and for the case $b \rightarrow \infty$ (single wall). The strongest adsorption potential is obtained for pores of ~7 Å.

It should be emphasized that the results of this section are based on a single-component adsorption model. In general it is not applicable to multi-component systems and Henry's constants may differ significantly from the values that are calculated here.

Chromatography

The purification of gases in a chromatography column is based on different Henry's constants for the adsorption of the gas to be purified (carrier gas) and the contaminants. The contaminants are supposed to be adsorbed stronger and move slower through the column than the carrier gas. That effect can be described quantitatively by the plate theory of chromatography which requires only two parameters: The retention volume V_{Ret} and the number of purification stages (plates) N in the column. The idea of the plate theory is to divide the column in N theoretical stages in which the gases come to equilibrium with the adsorber (described by equation (1)). The height of a single stage (also called *HETP*, Height Equivalent to a Theoretical Plate) depends on many parameters like pressure, temperature, shape of the adsorber, velocity of the gas, viscosity of the gas etc. If the *HETP* is small and consequently N is large ($HETP=L/N$, where L is the column length), the breakthrough curve will be steep. On the other hand if there are only a few theoretical stages the breakthrough of the contaminants almost starts immediately. The retention volume V_{Ret} is defined as the volume of carrier gas passed through the column until the breakthrough happens. It depends on Henry's constant H and on the mass of the adsorber m_{Ads} and is given by the following relation (T is the temperature) [7]:

$$V_{Ret} = H \cdot R \cdot T \cdot m_{Ads} \tag{8}$$

For large N V_{Ret} corresponds to the point where the concentration of the contaminant reaches 50 % of its initial value.

The following equation gives an analytical expression for the breakthrough curve assuming that the entering carrier gas has a constant concentration of contaminants c_0. Its derivation can e.g. be found in [8]:

$$c_N(V) = c_0 \cdot \left(1 - \sum_{r=0}^{N-1} \phi_r^{\frac{N \cdot V}{V_{Ret}}}\right) = c_0 \cdot \sum_{r=N}^{\infty} \phi_r^{\frac{N \cdot V}{V_{Ret}}} \tag{9}$$

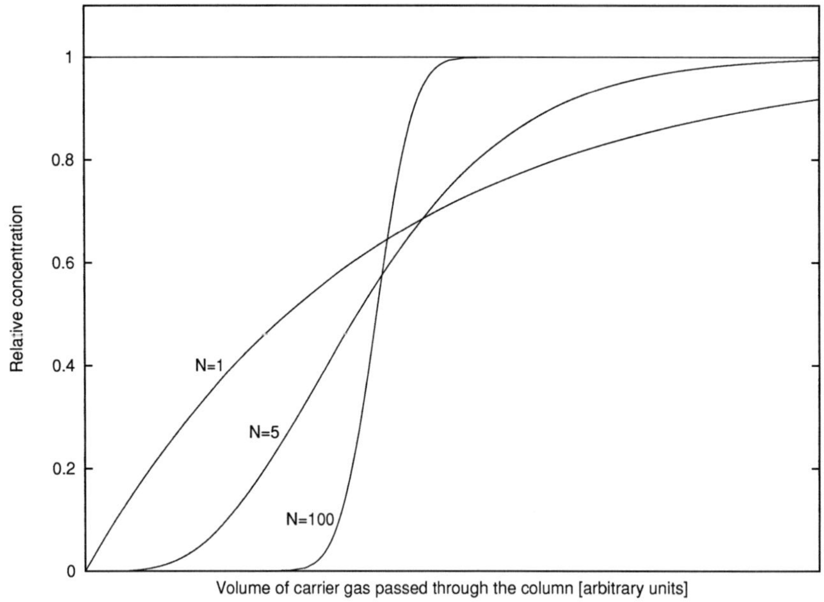

FIGURE 2. The shape of three breakthrough curves for fixed Henry's constant H and three different theoretical stages (N= 1, 5 and 100). For small N the breakthrough starts immediately.

$$\text{with} \quad \phi_r^x = e^{-x} \cdot \frac{x^r}{r!} \quad \text{(Poisson distribution)} \tag{10}$$

V is the volume of the carrier gas passed through the column. For large N the sum can be replaced by an integral and equation (9) becomes:

$$c_N(V) = c_0 \cdot \left(1 - \frac{1}{\sqrt{2\pi}} \cdot \int_t^\infty e^{-\frac{y^2}{2}} \mathrm{d}y\right) \tag{11}$$

$$\text{with} \quad t = \sqrt{N} \cdot \left(\frac{V}{V_{Ret}} - 1\right) \tag{12}$$

Figure 2 shows examples of three breakthrough curves with different N for fixed Henry's constant H. By scanning the breakthrough curves and fitting the obtained results to equation (9) or (11), Henry's constant H as well as the number of theoretical stages N in the used column can be measured.

Multi-component adsorption

As mentioned above Henry's constants for a multi-component system may differ significantly from those which are given in Table 2. Therefore these constants cannot be applied for quantitative evaluations. However the model allows different adsorbers to be compared relative to each other. In particular it can be applied to chose the optimum pore structure of an adsorber.

The nitrogen/radon- and the argon/radon-system

The removal of radon from gases is a common requirement in low-level physics experiments. Due to its size the radon atom has a high polarizability compared to argon and nitrogen and can easily be adsorbed on cold surfaces. The pore size structure of the adsorber is less important for the investigated systems than for systems with components of similar sizes. Therefore activated carbon with a wide pore size distribution is well suited for radon adsorption. Table

2 shows that Henry's constant for the radon adsorption at 77 K is many orders of magnitude larger than for nitrogen and argon. The huge difference suggests that there is only negligible interference between radon and nitrogen/argon. Consequently the removal of radon from nitrogen/argon is expected to be very efficient.

In practice the more serious problem is the re-contamination of the gas by ^{222}Rn which is emanated from the activated carbon by trace contaminations of ^{226}Ra. Several types of activated carbon were tested for ^{222}Rn-emanation [9] and the best results were obtained for the synthetic activated carbon CarboAct. Its ^{222}Rn-emanation of 0.3 mBq/kg is about 2 orders of magnitude lower than for most of the other tested carbons.

The nitrogen/krypton-system

The krypton atom is significantly smaller than the radon atom and is therefore much weaker adsorbed. On the other hand Table 2 shows that the single-component Henry's constant for krypton is still larger than for nitrogen (and argon) thus making a purification in principle possible. However, due to the similar size of the krypton atom and the nitrogen molecule, nitrogen can displace adsorbed krypton and block appropriate pores. Consequently Henry's constant for the krypton adsorption from the binary krypton/nitrogen-system will be much smaller than for the single-component krypton adsorption. Regardless the retention ability can be obtimized by the choice of the right adsorber. Two conditions have to be fulfilled: The pore structure should be optimized for the size of the krypton atom ($\sim$ 7 Å, see Figure 1) and the adsorber should have no internal polarity. The latter condition is important because krypton has to be removed from a bi-atomic gas (nitrogen). If the adsorber is non-polar, only Van-der-Waals forces (induced-dipole/induced-dipole interactions) can cause the binding to the surface (for krypton as well as for nitrogen). If there is a permanent internal polarity of the adsorber, the binding will not only be caused by pure Van-der-Waals forces, but also by stronger permanent-dipole/induced-dipole forces. The strength of the latter is proportional to the polarizability of the gas particle to be adsorbed. In general poly-atomic gases have higher polarizabilities than mono-atomic gases. Consequently, any permanent polarity of the adsorber should weaken the binding of krypton compared to the binding of nitrogen and will therefore be less favorable.

EXPERIMENTAL SETUP

Gas purification techniques in low-level physics must be able to produce ultrapure gases. To prove the purification efficiencies at very low concentrations highly sensitive detection methods are required. For ^{222}Rn measurements low background proportional counters are available. Certain pre-treatments of the samples are required before the radon can be transferred to the counter. The applied technique is described in [9] and has a sensitivity of less than 1 μBq/m^3 (STP) for ^{222}Rn in nitrogen.

For krypton concentration measurements in nitrogen a specially tuned VG 3600 mass spectrometer [10, 11] is available which can detect krypton to a level of 10^{-13} cm^3 [12]. Typically the sample size is 1 cm^3 resulting in a detectable krypton concentration of about 0.1 ppt. Figure 3 shows a sketch of the experimental setup. Nitrogen is evaporated and flows through a cold trap filled with the adsorber to be examined. After the trap there are 4 pipettes to take samples in temporal sequence. In the sample preparation section the nitrogen is removed by a getter pump and the remaining noble gases are transferred to the ion source of the mass spectrometer. Each run starts with a thorough baking of the adsorber and all the lines ($\sim$ 200 °C) for 2–3 days to release contaminations from the surfaces. Next, the initial krypton concentration is measured. Then the trap is cooled in a cryogenic bath and the nitrogen is led through it to the pipettes (flow rate between 120 l/h and 250 l/h). The krypton content in the purified gas was measured in temporal sequence (usually every 3 hours) in order to scan the profile of the breakthrough curve. At the end of the run the krypton concentration was checked again with the adsorber by-passed in order to confirm that it has not changed during the run.

In principle the setup can also be used to test the krypton purification in argon. However, since the argon can not be removed by the getter pump a further treatment of the sample is required. The argon/krypton separation can be done either by gas chromatography or by fractional pumping with a cryo-pump. Presently these techniques are under investigation and first measurements should start soon.

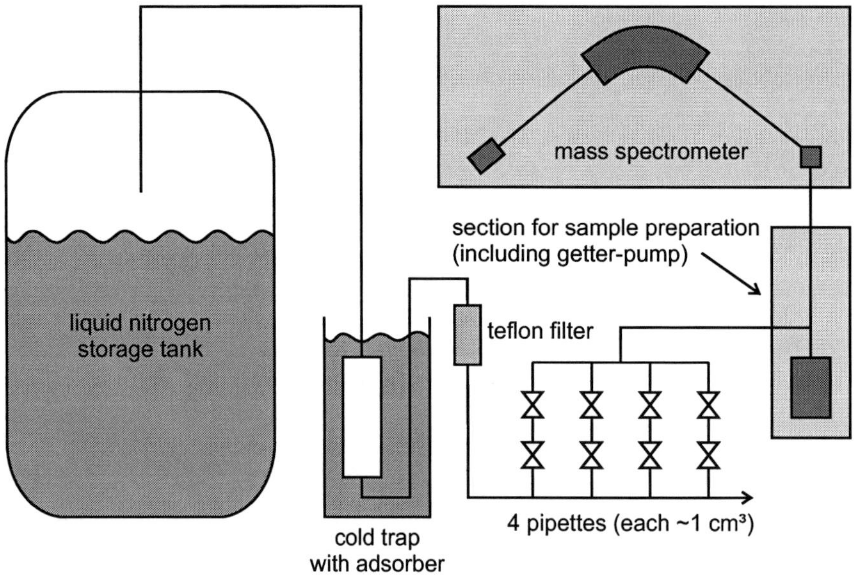

FIGURE 3. Sketch of the experimental setup to measure krypton in nitrogen.

RESULTS

Purification from radon

The theoretical investigations suggest that radon should be adsorbed very efficicently on activated carbon at 77 K - even in the presence of nitrogen or argon. Consequently no breakthrough should be observable in the case of nitrogen purification. Several purification tests were performed with a 2 kg column filled with the pure activated carbon CarboAct. More than 100 m^3 (STP) of nitrogen were processed and the integrated amount of ^{222}Rn in the purified nitrogen was measured. Although pure gas phase adsorption is more efficient, the nitrogen was purified directly in the liquid phase in order to save time and energy. The measurements show that even for liquid phase adsorption no radon was detectable after the purification column. The corresponding upper limit for the radon concentration is 0.5 $\mu Bq/m^3$ (STP) of nitrogen [9]. It also shows that the ^{222}Rn-emanation rate of the activated carbon CarboAct is sufficiently low.

It was also investigated whether argon can be purified from radon. About 140 m^3 (STP) of argon in gas phase were passed through a cold trap filled with 150 g of CarboAct which was cooled in a bath of liquid argon. The main difference of argon compared to nitrogen is the 10 K higher temperature of liquid argon. Although the radon adsorption at 87 K should be less efficient than at 77 K no ^{222}Rn could be observed in the purified argon and an upper limit of 0.5 $\mu Bq/m^3$ (STP) was obtained. Also for argon a purification directly from the liquid phase would be desirable and first liquid phase tests are planned for the near future.

Purification from krypton

For the purification of krypton from nitrogen two types of adsorbers were investigated: Carbon-based adsorbers and zeolites (AlSiO-based). Zeolites usually have a strong internal polarity due to a large fraction of aluminium in the molecular chain. However so-called hydrophobic zeolites are depleted in aluminium and have much smaller internal polarity. The problem is that aluminium is required to stabilize any structure with large pore channels. Consequently hydrophobic zeolites can be made only with rather small pores. Since the krypton atom is relatively large, pore sizes of at least 7 Å are required (see Figure 1). Such a structure contains a non-negligible fraction of aluminium in the molecular chains and thus reveals some internal polarity. Several zeolites of different pore sizes and different Al/Si ratios in the structure were tested. The obtained results confirmed our theoretical predictions. All adsorbers showed very little (if any) abilities to remove krypton from nitrogen. Either due to the pores being too small or due to the high

TABLE 3. The main properties of the investigated carbon adsorbers. The first three products are 'standard' activated carbons with wide pore size distributions. Carbosieve SIII is a carbon molecular sieve with a narrow pore size distribution.

Product	CarboAct F3/F4	C 38/2	FM 1-250	Carbosieve SIII
Manufacturer	CarboAct	CarboTech	Charcoal Cloth Int.	Supelco
Pore size distribution	wide distribution	wide distribution, enhanced fraction around ∼7 Å	wide distribution, enhanced fraction around ∼20 Å	narrow distribution (15 Å – 40 Å)
Composition of product	grains	pressed cylinders	fabric ,	grains
Size	0.85 mm – 2.8 mm	Length: 2 mm – 7 mm ∅: 2 mm	Thickness: 1.2 mm – 1.4 mm	60 mesh – 80 mesh
Density	∼0.1 g/cm^3	0.65 g/cm^3	∼210 g/m^2	0.61 g/cm^3
Surface area	1200 m^2/g	1300 m^2/g	1000 m^2/g – 1200 m^2/g	820 m^2/g

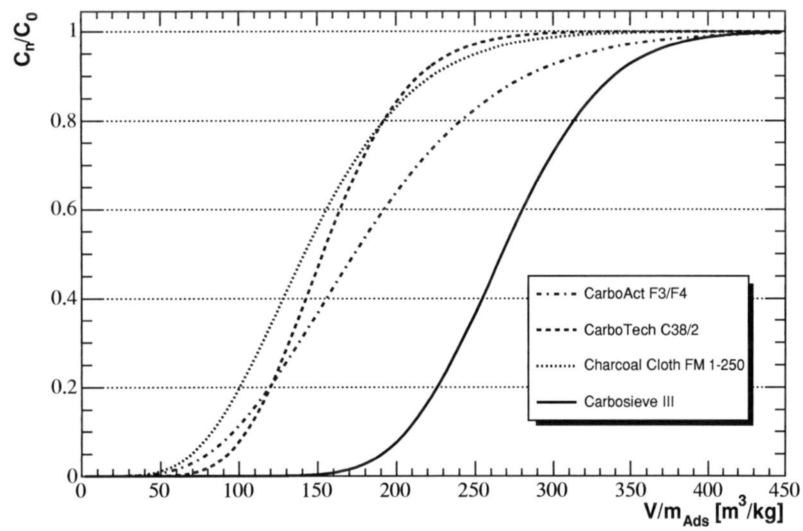

FIGURE 4. Measured breakthrough curves for krypton adsorption from nitrogen on different carbon-based adsorbers cooled by LAr.

polarity. Therefore the future investigations are focused on carbon-based adsorbers only which are free of internal polarity.

The carbon-based adsorbers which were investigated are listed with their main properties in Table 3. C 38/2 is a product mainly used for solvent recovery. According to the manufacturer this product has a larger fraction of pores in the range around 7 Å than 'standard' products and should therefore be favorable for our applications. Charcoal Cloth FM 1-250 is a fabric made out of carbon. The product has an enhanced fraction of pores around 20 Å, but a significant fraction should also be in the range of 7 Å. Finally Carbosieve SIII is a carbon molecular sieve with a narrow pore size distribution between 15 Å and 40 Å. It is mainly used for trapping volatile organic compounds (VOCs). The measurements were performed with different masses of adsorbers so the results cannot be compared directly. However the volume V never appears alone in equation (9) and (11), but only as a quotient V/V_{Ret}. Further there is a linear relation between V_{Ret} and the mass of the adsorber m_{Ads} (see equation (8)). Therefore it is possible to scale the obtained breakthrough curves with the mass.

For the nitrogen purification tests the adsorber was cooled to 87 K (liquid argon temperature). At this temperature the nitrogen to be purified remains always in the gas phase. As explained above a liquefication would block the pores of the adsorber and would reduce the mobility of krypton resulting in a slower kinetics of the adsorption process. For

TABLE 4. Overview of the obtained parameters for krypton adsorption from nitrogen on different carbon-based adsorbers cooled by LAr.

Adsorber	*HETP* [mm]	H [10^{-2} mol/Pa/kg]	V_{Ret}/m_{Ads} [m^3/kg]
CarboAct F3/F4	15	21	183
CarboTech C 38/2	17	19	158
Charcoal Cloth FM 1-250	24	16	125
Carbosieve SIII	10	34	282

krypton this effect is more serious than for the large radon atom. Test measurements at liquid nitrogen temperature have shown little retention abilities thus confirming the predictions. Figure 4 shows the results of the adsorption measurements of krypton from nitrogen with the adsorber cooled to 87 K. The y-axis gives the concentration of krypton in the purified nitrogen normalized to the concentration in the unpurified gas. The curves are fits of the breakthrough curves (9) or (11) to the measured data points. From these fits we can get the retention volume V_{Ret} (and Henry's constant H) and the number of theoretical stages N (*HETP*). The results are summarized in Table 4. It should be mentioned that the *HETP* depends not only on the adsorber, but also on the geometry of the cold trap. Therefore different adsorbers cannot be compared directly by their *HETP*. A good quantity to compare different adsorbers at the same temperature is Henry's constant H.

The results show that 'standard' activated carbons with broad pore size distributions have similar retention abilities. The carbon molecular sieve Carbosieve SIII with the narrow pore size distribution shows a better performance. With 1 kg of this adsorber it is possible to purifiy about 200 m^3 (STP) of nitrogen. This capacity is sufficient for typical demands of low-level physics experiments.

SUMMARY

Adsorption techniques represent a powerful tool to obtain gases with highest purities. In this work single- and multi-component adsorption processes for inert gases on carbon-based adsorbers were studied. The results were used to develop purification techniques for nitrogen and argon. The radon removal ability of standard activated carbons at low temperatures was found to be very efficient - even if the carrier gas is purified in the liquid phase. No radon breakthrough was observable in the used columns. The krypton adsorption efficiency depends strongly on the carrier gas and on the pore size distribution of the adsorber. The removal of krypton from nitrogen works only if the nitrogen remains in the pure gas phase. The best results could be achieved with a carbon molecular sieve (Carbosieve SIII) with a narrow pore size distribution. More than 200 m^3 (STP) of nitrogen can be purified with 1 kg of this adsorber.

ACKNOWLEDGMENTS

I would like to thank all people involved in the presented work. In particular I am grateful to G. Zuzel for his valuable contributions.

REFERENCES

1. Pierotti, R.A. and Thomas, H. E., Surf. Col. Sci. 4, 93-259 (1971).
2. Steele, W.A., Pergamon Press Oxford (1974).
3. Steele, W.A., Surface Science 36, 317-352 (1973).
4. Heusser, G. et al., Appl. Rad. and Isot. 52, 691-695 (2000).
5. Hirschfelder, J.O. et al., Chem. Rev. 44, 205-231 (1949).
6. Maurer, S. et al., Chem. Eng. Sci. 56, 3443-3453 (2001).
7. Maurer, S., Dissertation, Technical University Munich, Herbert Utz Verlag – Wissenschaft (2000).

8. Said, A.S., Hüthig Verlag, Heidelberg (1981).
9. Heusser, G. et al., Appl. Rad. and Isot. 52, 691-695 (2000).
10. Hopp, J. et al.: Max-Planck-Institut für Kernphysik Heidelberg, Annual Report, 299-300 (2000).
11. Kiko, J. et al.: Max-Planck-Institut für Kernphysik Heidelberg, Annual Report, 251 (1992).
12. Zuzel, G. et al., Appl. Rad. Isot. 61, 197-201 (2004).

Distillation purification and radon assay of liquid xenon

Yasuo Takeuchi

Kamioka Observatory, ICRR, Univ. of Tokyo, Kamioka-cho, Hida-shi, Gifu 506-1205, Japan

Abstract. We succeeded to reduce the Kr contamination in liquid xenon by a factor of 1/1000 with a distillation system in Kamioka mine. Then, the remaining radioactivities (Radon and Kr) in purified liquid xenon were measured with the XMASS prototype detector. In this talk, the distillation system and the remaining internal radioactivity levels are reported.

Keywords: xenon, distillation, dark matter, neutrino
PACS: 26.65.+t, 95.35.+d

INTRODUCTION

Liquid xenon is a good scintillator. Table 1 shows a list of properties of liquid xenon. Liquid xenon has a good photon

TABLE 1. Properties of liquid xenon.

item	Liquid Xenon
Scintillation light	$\sim$ 42 photon/keV
Scintillation light wave length	175 nm
Scintillation light width	$\sim$ 40 nsec
Atomic number	54
Atomic weight	131.29 amu
Density	3.0 g/cm^3
Melting (boiling) point	161.4 K (165.1 K)
Chemical series	Noble gases

yield (about 42photon/keV). Therefore, sensitivity against low energy events is good. The wave length is rather short, but it still can be observed by PMTs directly without a wave length shifter. The large atomic number provides a short radiation length for the low energy gamma rays. So, the external low-energy gamma rays would be absorbed in the outer region of the liquid xenon detector, then the central region would become clean. The boiling point of liquid xenon is higher than other rare gas, like argon or neon, so it is relatively easy to handle. We can use liquid nitrogen to liquefy xenon. Because of rare gas and liquid, various purify methods can be applied.

Xenon doesn't has a long life radioactive isotopes. A possible problem of liquid xenon for the low background experiments is that there is a contamination of ^{85}Kr during manufacture and refinement. Usual commercial "krypton-free" xenon still has ppb level krypton. It could become a serious background for the low background experiments.

We have built a distillation system to reduce the Kr contamination mainly, then processed 100kg of xenon. We have measured remaining krypton and radon levels after the distillation purification.

DISTILLATION PURIFICATION OF XENON FOR KRYPTON

Impurities, like CO_2 or H_2O, in xenon can be removed by adsorption method. A distillation method can be used to remove Kr, O_2, N_2, H_2, or He in xenon, because their boiling points are lower than xenon. We have designed a distillation system to remove krypton in xenon. Table 2 shows a summary of the requirements for the purification system. The total stages in the distillation tower became 13.

We have built it in Kamioka Underground Observatory, ICRR. Univ. of Tokyo, which is located 1000m underground (2700m.w.e.) in Kamioka Mine in Gifu Prefecture in Japan. Figure 1 shows a photograph of the distillation system in Kamioka. Packings are used in the tower in order to accommodate 13 stages in the distillation tower, because the cavity height is limited in the underground laboratory.

We have processed 100kg xenon by using the distillation purification system in March 2004. It took about 1 week to process it. The raw xenon had about 3ppb krypton. We have separated about 1kg of "off gas" xenon with 330 ± 100

CP 785, *Topical Workshop on Low Radioactivity Techniques,* edited by B. Cleveland, R. Ford, and M. Chen

TABLE 2. Requirements for the distillation system of Xe for Kr.

item	Design value
Process speed	0.6 kg/hour
Yield (Kr-less gas / raw gas)	99%
Reduction factor for Kr in Xe	1/1000
Operation conditions	178 ~ 180K, 2 atm.

FIGURE 1. The distillation system in Kamioka Observatory. The height of the distillation tower is about 3m.

ppb krypton from the raw xenon. The amount of the purified xenon was about 99% of the raw xenon. The krypton contamination in the purified xenon was assessed by a special device made with an Atmospheric Pressure Ionization Mass Spectrometer (API-MS) system and a Gas Chromatography (GC) system.

KRYPTON ASSAY WITH GC + API-MS SYSTEM

The API-MS + GC system for krypton assay in xenon was originally developed at Taiyo Toyo Sanso Co. Ltd. (SAAN). We have measured the krypton contamination in the purified xenon by using their API-MS + GC system.

In an API-MS, primary ionization is done by usual corona discharge. However, it is done in atmospheric pressure. Therefore, if the carrier gas (C) has a higher ionization potential than the target molecules (X), there is the secondary ionization by ion-molecule reaction of

$$C^{+} + X \rightarrow C + X^{+}. \tag{1}$$

In this reaction, most of the target molecules are ionized by charge exchange, then the statistical sensitivity of the Mass Spectrometer part becomes high. Table 3 shows the ionization potential of related gases. In the API-MS detector, He or Ar is used as a carrier gas. They have higher ionization potentials than Kr, so it satisfies the condition. Xenon, however, has a lower ionization potentials than Kr. Therefore, xenon have to be removed before API-MS detector so as not to ionize xenon atoms instead of krypton atoms. So, a GC system is used to separate krypton from xenon.

TABLE 3. Ionization potential.

	He	Ar	Kr	Xe
eV	24.6	15.8	14.0	12.3

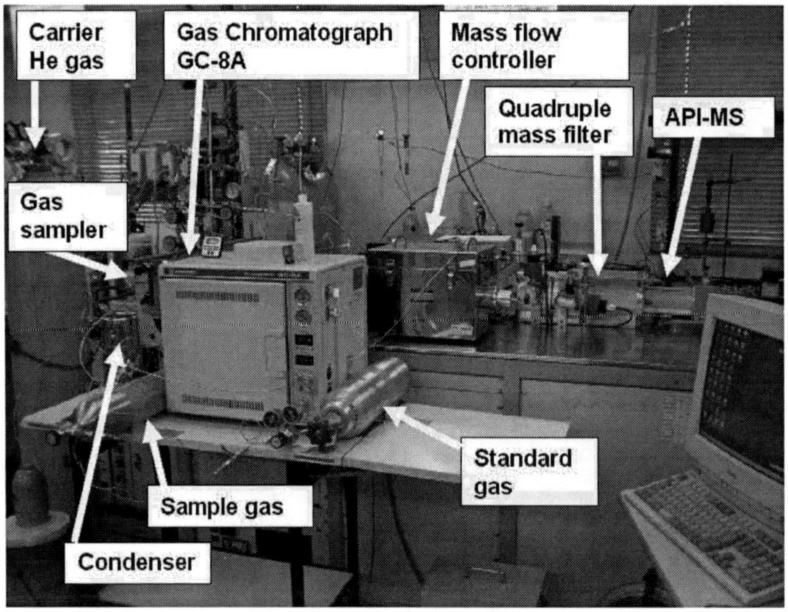

FIGURE 2. The Gas Chromatography and Atmospheric Pressure Ionization Mass Spectrometer system at SAAN.

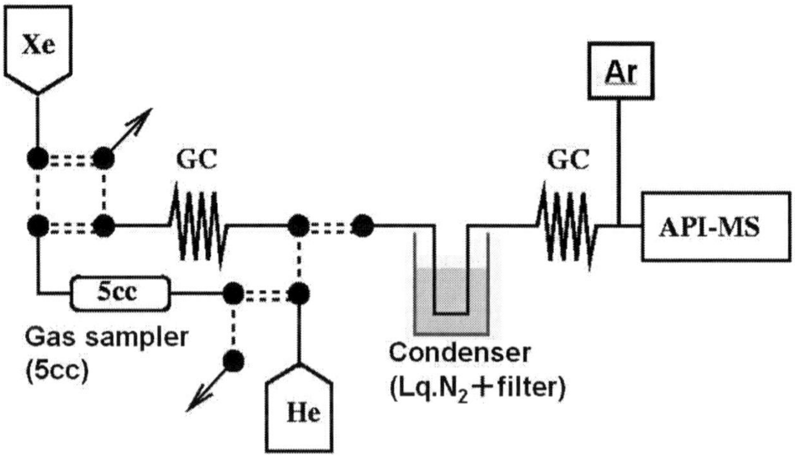

FIGURE 3. A schematic of the flow of the GC + API-MS system.

Figure 2 shows a photograph of the CG + API-MS system at SAAN. The condenser consists from liquid nitrogen and a pipe with a metal mesh filter to trap krypton. Figure 3 shows a schematic view of the GC + API-MS system.

We have measured the krypton contamination in our purified xenon by using their API-MS + GC system in September 2004. In the first step of the measurements, xenon gas was fed into the gas sampler, then 5cc of the gas was sampled. Then, the sample xenon gas was fed into the GC with He carrier gas. Only at the krypton timing, the sample gas was fed into the condenser, then krypton was trapped in the condenser. To increase statistical significance, we repeated these steps by 100 times. Finally, the condenser was heat up, the trapped gas was fed into the API-MS system with He and Ar carrier gases through the GC system. In the final step, only krypton timing was also selected. Actually, the API-MS counts ^{84}Kr which is most dominant krypton isotope. The abundance of ^{85}Kr is small. The ^{85}Kr / Kr ratio is about 10^{-11}.

Figure 4 shows the measured krypton signals in API-MS as a function of time. At the beginning in each measurement, the same 1ppm krypton standard gas was measured several times in order to correct environmental drift of the system. In the plot, relative correction factors are written. Then, the actual measurement was done at the end in each

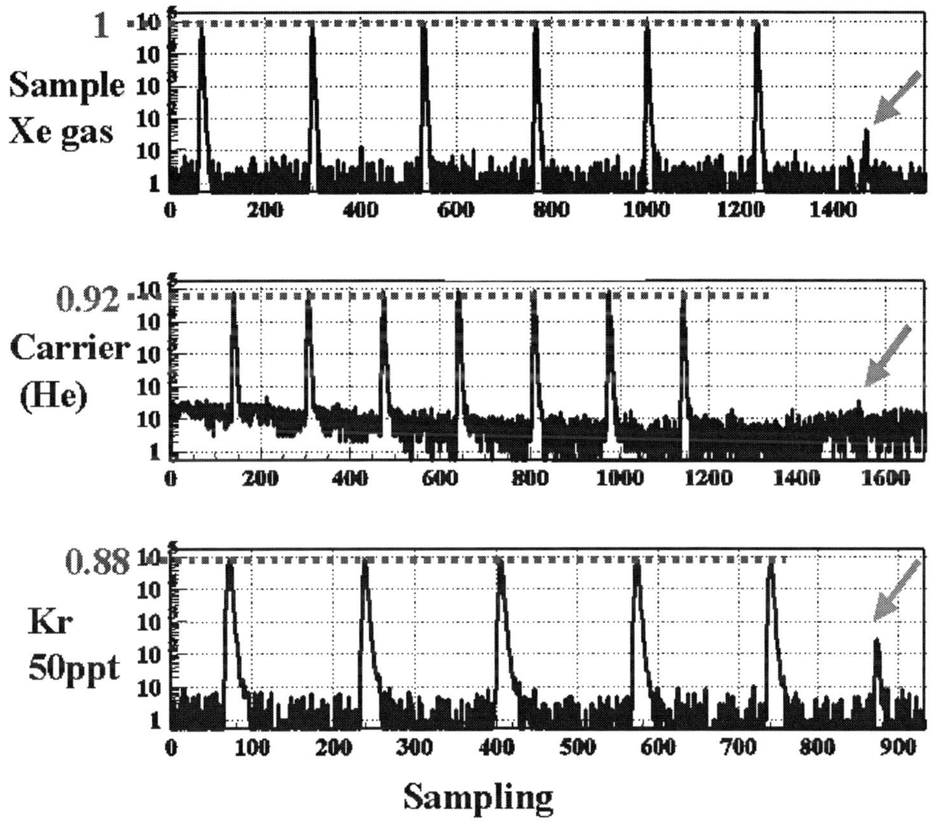

FIGURE 4. Krypton assay results from GC + API-MS system. The vertical and horizontal axes are APIMS count for ^{84}Kr and time. Top, middle, and bottom plots correspond to the sample xenon gas, carrier He gas, and krypton 50ppt calibration gas, respectively. The horizontal dashed lines show relative calibration factors among 1ppm krypton standard gas measurements. The arrows show actual measurements of the samples.

plot. The measured ^{84}Kr counts are the following.

Sample xenon gas	184.0 ± 15.8 count
Carrier He gas	$(88.5 \pm 23.9)/0.92$ count
Krypton 50 ppt standard gas	$(1176.9 \pm 35.7)/0.88$ count

Then, the krypton concentration in the sample xenon gas was obtained as $3.3 \pm 1.1(stat.)$ ppt. This corresponds to about 1/1000 factor reduction of krypton, because the initial krypton contamination in the raw xenon gas was about 3ppb.

RADON ASSAY WITH XMASS PROTOTYPE DETECTOR

XMASS is an underground experiment aimed at searching for rare phenomena under an ultra low background environment by using ultra pure liquid xenon in Kamioka mine, Japan. The main physics targets of XMASS are cold dark matter, neutrinoless double beta decays, and low-energy solar neutrinos. So far, we have developed a prototype detector and done 2 series of test experiments with it. By using the prototype detector, we have assessed the remaining radon in liquid xenon.

Figure 5 shows a schematic view of the prototype detector. The prototype detector consists of 100kg of liquid xenon and 54 low-background Photo-Multiplier Tubes (Hamamatsu R8778) developed for this experiments. It is installed in a heavy gamma ray shield in a clean room in Kamioka Observatory. The charge information from the PMTs are read by 54 channels of charge sensitive ADCs and a common Flash ADC for summed signals. The threshold of each PMT

FIGURE 5. XMASS prototype detector. 100kg liquid xenon and 54 low-background PMTs are used. PMTs are attached to the 30L OFC chamber through 5mm thickness MgF_2 windows. They are put into a heavy shield in a clean room. The XMASS prototype detector is located in Kamioka Observatory, ICRR, Univ of Tokyo, in Kamioka mine.

is about 0.4 photo electron. The data acquisition trigger is applied when there are 4 multiple hits within 100ns time window. The typical trigger rate for normal runs is about 1.5Hz.

The radon assay was done through the following coincidence reactions.

For ^{222}Rn: $^{214}\mathrm{Bi}\ (\beta,\ E_{max}\text{=3.3MeV}) \rightarrow {}^{214}\mathrm{Po}\ (\alpha,\ 7.7\mathrm{MeV},\ \tau = 164\mu\mathrm{sec}) \rightarrow {}^{210}\mathrm{Pb}$
For ^{220}Rn: $^{212}\mathrm{Bi}\ (\beta,\ E_{max}\text{=2.3MeV}) \rightarrow {}^{212}\mathrm{Po}\ (\alpha,\ 8.8\mathrm{MeV},\ \tau = 299\mathrm{nsec}) \rightarrow {}^{208}\mathrm{Pb}$

In the test run in August 2004, we found 67 candidate events of the ^{222}Rn during 1.8 days observation. Figure 6 shows the result of the ^{222}Rn assay. Fig. 6(a) shows the daughter energy distribution of all the time-coincidence events within 1msec. In the low-energy region, there are accidental coincidence events. Therefore, we have applied a cut at 3.5MeV in Fig. 6(a). Then, the remaining events became 67 events. Fig. 6(b), (c), and (d) show the distribution of these 67 candidate events. They agree with expectations.

This measurement was done on 4th and 10th August, 2004. The live times are 0.8 day and 1.0 day, respectively. If we assume radiative equilibrium, the remaining ^{238}U becomes like below.

4th: $^{238}\mathrm{U} = (72 \pm 11) \times 10^{-14}$ g/g
10th: $^{238}\mathrm{U} = (33 \pm 7) \times 10^{-14}$ g/g

The decrease of the remaining ^{238}U is consistent of the expected ^{222}Rn decay with 3.8 day half life. So, we think most of the remaining ^{222}Rn might come from outside the detector. We are planning to install a special device for radon reduction in future.

About ^{220}Rn, we didn't find any candidate events. Then we have set a limit as $^{232}\mathrm{Th} < 23 \times 10^{-14}$ g/g.

SUMMARY

We have built a distillation system of xenon for krypton removal in Kamioka Observatory. The measured reduction power was 1/1000 for krypton in xenon. The remaining krypton was measured by a GC + API-MS system. The radon contamination in liquid xenon was also assessed by the XMASS prototype detector. The current remaining impurities in purified xenon for XMASS experiment are the following.

$^{238}\mathrm{U} = (33 \pm 7) \times 10^{-14}$ g/g
$^{232}\mathrm{Th} < 23 \times 10^{-14}$ g/g
$\mathrm{Kr} = 3.3 \pm 1.1$ ppt

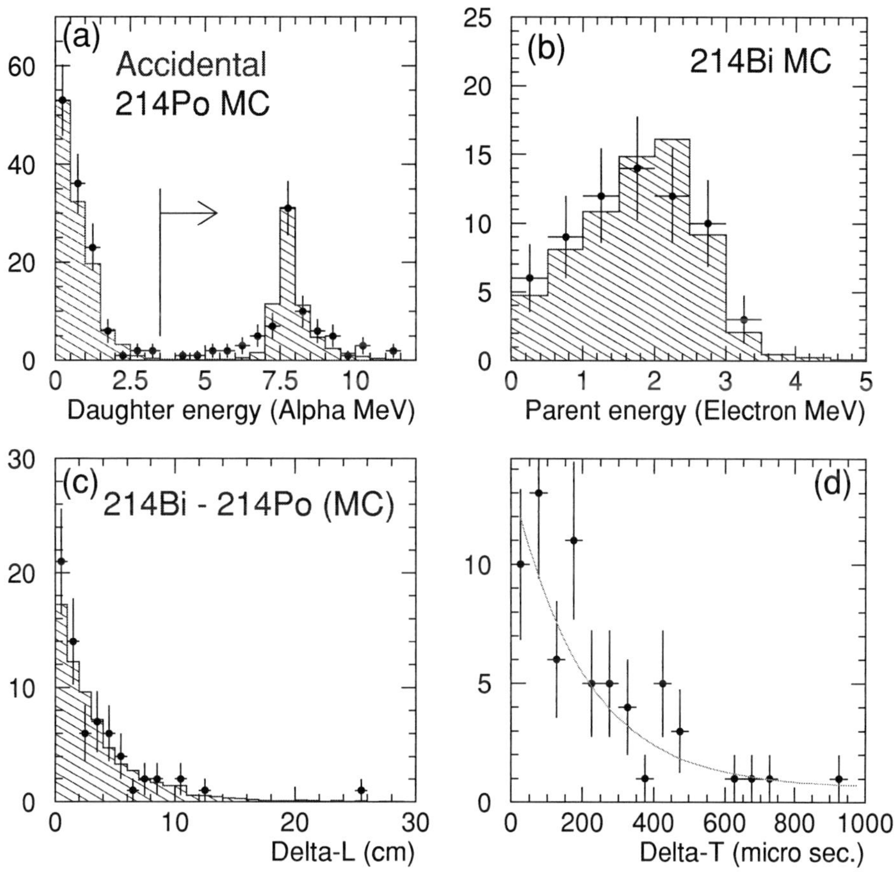

FIGURE 6. ^{222}Rn measurement in the test run in August 2004. (a) Energy distribution of the daughter event of the coincidence events within 1msec. (b) Energy distribution of the parent events. (c) Distance between the reconstructed vertex positions of parent and daughter events. (d) Time difference between the parent and daughter events. The fitted half life is $141 \pm 51 \mu$sec.

ACKNOWLEDGMENTS

The author would like to thank the cooperation of the Kamioka Mining and Smelting Company. This work is partially supported by Grant-in-Aid for Scientific Research on Priority Areas (A) of the Japanese Ministry of Education, Science and Culture.

Radioactive Background Evaluation by Atom Counting

Chad Orzel* and Daniel McKinsey†

*Department of Physics and Astronomy, Union College, Schenectady, NY 12308
†Yale University, New Haven, CT 06511

Abstract. We propose a new method of measuring ^{85}Kr background levels by direct counting of impurity atoms. The beta-decay of ^{85}Kr is a significant radioactive background for experiments that use liquified noble gases to search for dark matter and measure the low-energy solar neutrino flux. While there are several proposed methods for reducing Kr levels in these experiments, an independent technique is needed for measuring very low Kr levels. By selectively exciting Kr atoms to a metastable state, capturing them in a magneto-optical trap (MOT), and detecting fluorescence from the trapped atoms, individual Kr atoms can be counted with a high signal-to-noise ratio. This approach offers both higher sensitivity and shorter measurement times than more conventional techniques, with an estimated sensitivity of 3×10^{-14} in only 3 hours of integration.

Keywords: krypton-85, atom trap trace analysis, radioactive background, particle detectors
PACS: 32.80.Pj, 29.40.-n, 29.90.+r

INTRODUCTION

The characterization and removal of radioactive impurities from various materials is a critical part of many low-radioactivity experiments. Screening and purifying materials to the levels demanded by the next generation of astrophysical particle detectors is an extremely difficult problem, as demonstrated by many of the talks at this workshop.

In particular, the removal of ^{85}Kr from samples of other rare gases is of critical importance to the proposed detectors that use liquid rare gases as a scintillation medium (XENON[1], XMASS[2], and CLEAN[3]). ^{85}Kr is present in natural Kr at an abundance of 2.5×10^{-11}[4], has a half-life of 10.76 years, and beta-decays with an energy up to 687 keV, which is in the region of interest to solar neutrino and dark-matter search experiments. It is estimated that the XENON dark matter search experiment using liquid Xe can tolerate at most 150 ppt of natural Kr contamination[5], while the limits for the CLEAN (using liquid Ne) and XMASS (liquid Xe) neutrino detectors are much more stringent, on the order of a few parts in 10^{15}[6, 7, 8, 9].

Like all rare gases, Kr is chemically inert, and thus is very difficult to separate from Xe or Ne (Xe gas is commercially available with a guaranteed purity of only 20 ppb). It is also difficult to measure Kr/Xe or Kr/Ne concentrations at the necessary levels. At this workshop, Dr. Takeuchi of the XMASS group described a technique for measuring Kr/Xe levels down to 3 ppt using a combination of gas chromatography and accelerator mass spectroscopy[10]. This is a complicated endeavor, requiring days or weeks of processing before the measurement is completed.

The purpose of this paper is to propose a new method of measuring the Kr content of other gases, using the technique of Atom Trap Trace Analysis (ATTA), developed by Z.-T. Lu and colleagues at Argonne National Laboratory for radioisotope dating and environmental monitoring[11]. ATTA uses laser cooling techniques to trap and detect single atoms of a specific element or isotope, and determines abundances by directly counting atoms. This technique offers exceptionally high sensitivity to Kr impurities in a very short time relative to existing methods, reaching an estimated Kr/Xe sensitivity of 3×10^{-14} in 3 hours of integration[12].

In this paper, we will provide a brief description of laser cooling, and describe how it is used to detect single atoms for ATTA. We will describe how to adapt the ATTA technique to measureme Kr backgrounds in other gases, and discuss some of the advantages this technique offers over other measurement techniques. Finally, we will discuss some future prospects and other applications of the ATTA system.

LASER COOLING

The techniques of laser cooling and atom trapping were developed in the 1980's[13] and have revolutionized the field of atomic physics. Samples of tens or hundreds of millions of atoms at microkelvin temperatures (energies of

CP 785, *Topical Workshop on Low Radioactivity Techniques,* edited by B. Cleveland, R. Ford, and M. Chen

$\sim 0.1\,\mathrm{neV}$) can readily be prepared, and are the first step toward experiments in Bose-Einstein condensation and Fermi degenerate gases in dilute atomic vapors.

Laser cooling is based on the resonant scattering of photons. When an atom makes a transition between two electronic states by absorbing a photon of frequency ω (where the energy difference between states is $E = \hbar\omega$), the photon momentum $\hbar k$ (where $k = 2\pi/\lambda$, and λ is the photon wavelength) is also transferred to the atom. This atomic recoil results in a very small velocity change for a single photon scattering ($\sim 6\,\mathrm{mm/s}$ for ^{84}Kr, with a wavelength of 811 nm), but the cumulative effect of many photon scatterings can produce a large acceleration in the direction of the laser.

This light scattering force can be used to selectively cool a sample of atoms by exploiting the Doppler shift that results from atomic motion. If we tune the laser to a frequency slightly below the resonant frequency of the atoms ($\omega_L < \omega_o$), atoms moving toward the red-detuned laser will see the light Doppler shifted to a higher frequency, closer to resonance, and will absorb photons from the beam at a higher rate, reducing their velocity. Atoms at rest, or moving in the same direction as the laser, will absorb at a lower rate if at all, and will experience little or no force.

A single red-detuned laser beam can be used to bring a room-temperature ($v \sim 300\,\mathrm{m/s}$) beam of atoms to a stop over a distance of less than a meter. During this deceleration, the Doppler shift changes by more than 2 GHz, which must be compensated by the addition of a spatially varying magnetic field. This slowing magnet provides a changing Zeeman shift to compensate the changing Doppler shift as the atoms decelerate, keeping the moving atoms in resonance with the laser[14].

A counter-propagating pair of red-detuned lasers will produce a velocity-dependent damping force that opposes the motion of atoms in either direction[15]. This "optical molasses" can be used to slow samples of atoms to velocities of a few cm/s, corresponding to temperatures on the order of $100\,\mu\mathrm{K}$ or lower. The laser detuning required for optical molasses is on the order of the resonance width of the atomic transition, typically $1-10\,\mathrm{MHz}$ (about 1 ppb of the laser frequency). This degree of control over the laser frequency is not difficult to achieve using standard laser techniques.

Spatial confinement of the atoms can be achieved by adding a quadrupole magnetic field to optical molasses, to make a magneto-optical trap (MOT)[16]. The confining force in the MOT is provided by light scattering (typical MOT field gradients are 10 G/cm or less, far too small to magnetically confine the atoms), as the atoms moving out from the center of the trap are Zeeman shifted into resonance with the lasers by the increasing magnetic field. A MOT can be used to collect samples of up to 10^9 atoms at temperatures of $\sim 100\,\mu\mathrm{K}$, with each atom in the MOT scattering photons at a rate of $10^7 - 10^8\,\mathrm{s}^{-1}$.

Metastable Krypton

Laser cooling has been demonstrated for many different atomic species, including all of the alkali metals, most of the alkaline earths, and all of the stable rare gases, including Kr. The rare gases are particularly interesting, as the transition wavelengths required to cool these atoms in the ground state are prohibitively short given current laser technology. However, the lowest-lying excited states of these atoms are metastable states, with lifetimes in excess of 20 s, and these states can serve as effective ground states for laser cooling.

In Kr, the metastable state is the $5s[3/2]_2$ state (also called the $^3P_2$ state), with a lifetime of 28 s[17]. Laser cooling is accomplished by exciting the atoms to the $5p[5/2]_3$ ($^3D_3$) state, with a transition wavelength of $\lambda = 811.3\,\mathrm{nm}$. Fig. 1 shows some important levels in Kr.

In order to cool and trap Kr*, we must produce $5s[3/2]_2$ metastables in the atomic beam source. Ground-state atoms and atoms in other excited states will be unaffected by the cooling and trapping lasers. The excitation can be accomplished either by electron impact, or two-photon resonant optical excitation.

In the electron impact method (Fig. 1a), a plasma discharge is created and maintained using either high-voltage DC current[18] or radio-frequency electromagnetic fields[19]. Gas atoms passing through the source are excited by collisions with electrons in the discharge region. Discharge sources are robust and will work for any atomic species, but they have low efficiency, with metastable atom fractions of $10^{-4} - 10^{-3}$.

The two-photon resonant optical excitation method (Fig. 1b) uses a Kr lamp to excite atoms to the $5s[3/2]_1$ state using light at $\lambda = 124\,\mathrm{nm}$, followed by laser excitation at $\lambda = 819\,\mathrm{nm}$ to the $5p[3/2]_2$ state, which decays to the $5s[3/2]_2$ metastable level with a 77% probability. The remainder of the excited atoms return to the ground state, and can be re-excited.

Because this is a resonant two-photon process, it can, in principle, be more efficient than electron impact. For the present work, it offers the additional advantage of atomic selectivity: only Kr atoms will be excited, while background

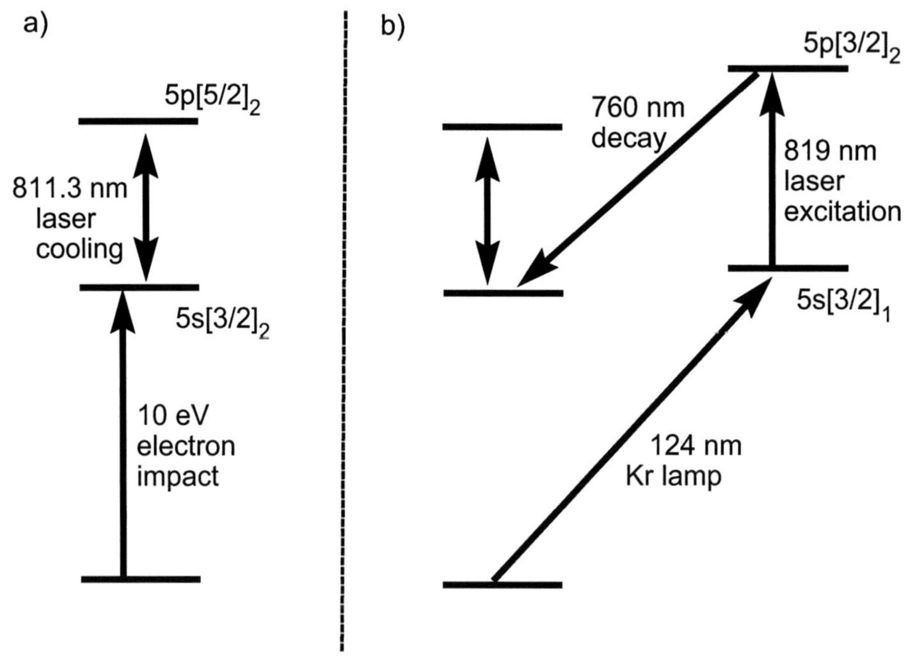

FIGURE 1. The atomic levels of Kr. Figure a) shows the energy levels relevant to atomic trapping of Kr*, and figure b) shows the energy levels relevant to optical excitation of Kr*.

gas atoms will be left in the ground state. Optical production of Kr* has been demonstrated in a gas cell, and is projected to give a metastable fraction of order 10^{-2} in an atomic beam source[20].

Either of these excitation methods, applied to pure samples of Kr, will produce an atomic beam with a Kr* flux of $10^{13}\,s^{-1}$ or better, which is more than sufficient for loading a MOT from a slowed atomic beam. These methods will also work to excite Kr* atoms from impurities in Ne or Xe samples.

ATOM TRAP TRACE ANALYSIS

Typical MOT loading rates for an abundant isotope are on the order of 10^8 atoms per second, with trap lifetimes of 1 s or longer, allowing large numbers of atoms to be accumulated. For an isotope like ^{85}Kr, however, with a natural abundance of 10^{-11}, the loading rate will be on the order of $10^{-3}\,s{-}1$, meaning that it will be rare for even a single atom to be trapped.

The ATTA technique works by detecting the fluorescence of single trapped atoms of rare isotopes[11]. As described above, an atom trapped in a MOT scatters photons isotropically at a rate of $10^7 - 10^8\,s^{-1}$. Some of this scattered light can be collected with a lens and focussed onto an avalanche photodiode detector (APD) capable of detecting single photons. A single trapped atom can give a signal of 10^3 counts/s above background, allowing the unambiguous detection of single trapped atoms.

The apparatus for ATTA is shown schematically in Figure 2. Krypton atoms are excited into the $5s[3/2]_2$ metastable state to produce a metastable atomic beam at the room-temperature thermal velocity of ~ 300 m/s. The beam is decelerated by the combination of a slowing laser and a spatially varying magnetic field, and slow ($v \sim 1$ m/s) atoms from the beam are loaded into a MOT, where they are further cooled and trapped. A lens outside the chamber collects approximately 1% of the scattered light from the trap, and focusses it onto an APD to provide the fluorescence signal.

ATTA has been used to measure the abundance of both ^{85}Kr and ^{81}Kr (^{81}Kr/Kr $\sim 10^{-13}$) by comparing the MOT loading rate for the rare isotope to that of a stable isotope of known abundance (^{83}Kr/Kr~ 0.11)[11]. This measurement is precise enough to allow radioisotope dating of ground water samples from oases in the Sahara, using dissolved ^{81}Kr ($\tau_{1/2} \sim 200{,}000$ years)[21].

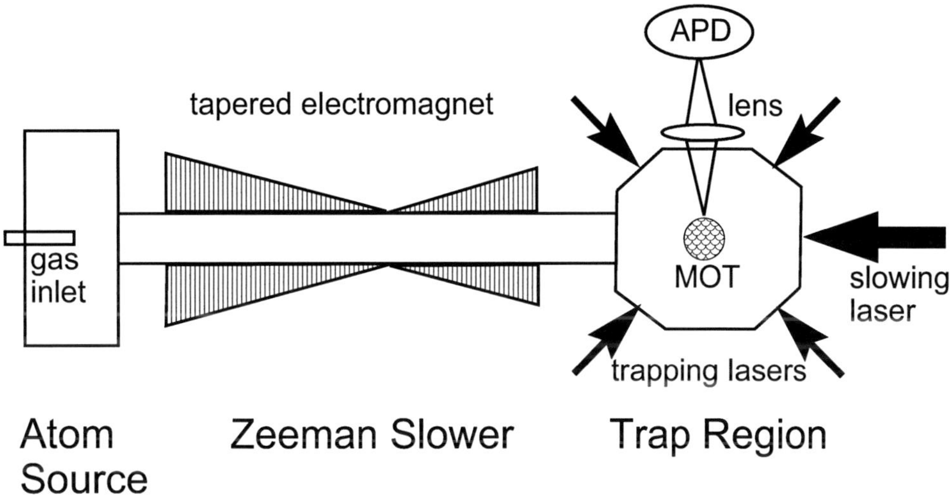

FIGURE 2. Schematic of the proposed experimental apparatus for measuring krypton impurities. Kr atoms are excited into the metastable state in the source region, decelerated by light pressure forces in the Zeeman slowing region, and loaded into a magneto-optical trap (MOT), where they are confined by light pressure and magnetic fields. A small fraction of the light scattered from the MOT beams is collected by a lens, and focussed onto an avalanche photodiode (APD) to produce the fluorescence signal used to detect trapped atoms.

ATTA FOR RADIOACTIVE BACKGROUNDS

The ATTA technique is easily adapted to the measurement of radioactive contamination, by loading the system with a sample of ultra-pure Ne or Xe gas, and trapping and detecting ^{84}Kr, the most abundant stable isotope (^{84}Kr/Kr $= 0.57$). The fractional ^{84}Kr abundance can be determined by comparing the trap loading rate for an ultra-pure sample to that for a sample prepared with a known Kr abundance (initial reference samples can be prepared at the ppb level using conventional measurement techniques, and lower limits can be reached using samples measured by ATTA). This will place a limit on the ^{84}Kr content of the sample of interest, which implies a proportionally smaller limit on the ^{85}Kr content.

The sensitivity of ATTA measurements can be estimated from typical values for Kr* atomic beam flux and trap loading rates. An atomic beam source of the type described in Ref [11] consumes 7×10^{16} atoms per second, and beam-loaded MOT systems have overall capture efficiencies of about 1×10^{-7} (including both the metastable excitation efficiency and capture from the atomic beam). Making the simple assumption that the excitation and capture efficiency for Kr* would be the same in a different gas, the detection of a single atom in three hours of integration implies a limit of 3×10^{-14} for natural Kr contamination[12].

This estimate assumes a metastable production efficiency of 10^{-4}, typical for discharge sources. It may be possible to improve this efficiency by an order of magnitude or more using an optical excitation source, with a corresponding increase in sensitivity. Smaller increases may also be attainable by improving the capture efficiency of the MOT, and by increasing the integration time. A first-generation system should easily be able to reach the Kr/Xe levels of interest to the XENON project, and it is not unreasonable to expect that a future system will be able to reach the 10^{-15} level required for CLEAN and XMASS.

Background Contamination

Due to the resonant nature of the laser cooling and trapping process, and the narrow resonance widths of atomic transitions, ATTA offers exceptional isotopic selectivity. Slowing, trapping, and detecting requires the absorption and emission of well over 10^5 photons, meaning that there is essentially no off-resonant background signal. Lu *et al.* are able to trap and detect ^{81}Kr and ^{85}Kr atoms, with no background from the abundant ^{83}Kr whose trapping laser frequency is within 150 MHz[22]. Transition frequencies in other atomic species will be several orders of magnitude farther away, and can safely be ignored.

The selectivity and sensitivity of the ATTA technique also mean that no further distillation or isotopic enrichment is necessary for the measurement of Kr levels. ATTA measurements can be made on purified gas as it is produced, with no additional sample preparation.

The metastable Kr system offers an additional advantage in the area of background contamination. Kr atoms may be cooled and trapped only in the $5s[3/2]_2$ metastable state, so contamination outside the metastable beam source will not contribute to the signal. Ground-state Kr atoms entering other parts of the system (either from atmospheric Kr, or outgassing within the chamber) are unlikely to pick up the 10 eV of internal energy needed to excite them to the $5s[3/2]_2$ state, and thus will not be trapped or detected.

The most likely sources of contamination in the source region are from sample handling, Kr outgassing from the vacuum chamber, and cross-contamination when switching between test and reference samples. With regard to the first, standard high-vacuum techniques should be sufficient to prevent contamination by atmospheric Kr. As for the second, we estimate the background level due to outgassing in the source chamber to be on the order of 10^{-16}[12]. Neither of these is expected to be a major obstacle.

The most significant source of background contamination for ATTA will be cross-sample contamination due to the operation of the source. The normal operation of a plasma discharge source will create Kr^+ ions and accelerate them into the chamber walls. Once embedded in the walls, these atoms can emerge later, either through normal outgassing, or by being driven out by later ion impacts. Du *et al.*[23] report this as an important source of cross-sample contamination, accounting for $\sim 1\%$ of their Kr signal.

This "memory effect" is potentially a serious problem for an ATTA system using a discharge source. Assuming that the magnitude of the effect is not changed by the presence of other gases, a calibration sample with a measured Kr abundance at the commercial limit of 20 ppb would contribute a background of 2×10^{-10}, on the same order of the Kr/Xe level of interest for XENON. Cross-contamination can be minimized by using calibration samples with a lower Kr abundance, and by aligning and testing the system with Ar (the laser cooling transitions in Ar and Kr differ by less than 1 nm), but this effect will likely limit the sensitivity of discharge source systems.

Sample contamination through this "memory effect" can be eliminated by switching to an optical excitation source, which will excite atoms directly to the $5s[3/2]_2$ without producing ions or embedding atoms in the walls of the source chamber. A discharge source system should be sufficient for preliminary measurements, but optical excitation will be essential in order to reach the sensitivity demanded by CLEAN and XMASS.

FUTURE PROSPECTS

The ATTA technique can also be used to measure other radioactive contaminants of interest to low-level detectors. A particularly attractive application would be ^{39}Ar, which has a half-life of 269 years and a beta decay endpoint of 565 keV. ^{39}Ar is not as significant a problem as ^{85}Kr (CLEAN only requires Ar/Ne $\leq 10^{-10}$[9]), but an independent measurement of Ar content will be useful. The same apparatus used for Kr can easily be adapted to Ar, detecting one of the stable isotopes (^{40}Ar (99.6%), ^{38}Ar (0.06%), or ^{36}Ar (0.34%)). The laser cooling transitions for Kr* and Ar* differ by less than 1 nm, so the same lasers and optics can be used for both elements.

In principle, the ATTA technique can be used to measure trace quantities of any element that can be laser cooled and magneto-optically trapped. In practice, there are two factors that may limit its application to systems other than rare-gas metastables.

First, ATTA is probably not useful for detecting radioactive impurities directly, as the 10^{-14} fractional abundance detectable with ATTA will correspond to a level of radioactivity far beyond what is acceptable for detector applications. ATTA works for ^{85}Kr background evaluation because we can use the stable isotope ^{84}Kr as a marker for ^{85}Kr at a much lower level. ^{84}Kr at the 10^{-14} level implies ^{85}Kr content of 10^{-25}, which is well below the sensitivity one can hope to measure directly with ATTA. ATTA should prove useful for measuring the levels of any radioactive contaminant with a stable isotope that is not itself a problem (such as ^{84}Kr or ^{40}Ar), but the technique is not sensitive enough for elements that lack stable isotopes (such as Rn).

Metastable atom systems also provide the advantage of insensitivity to background contamination. Kr contamination in the detection region is not a problem in this system, because only atoms in the $5s[3/2]_2$ state can be trapped and detected. This will not be true for species that can be laser cooled in the ground state, and contamination from other parts of the system may limit the sensitivity that can be obtained. This is not an absolute bar to using ATTA with ground-state atoms (for example, using ^{39}K as a marker for ^{40}K or ^{42}K), but the sample handling requirements will be much more stringent for those systems.

For the metastable rare gas systems where it works well, ATTA offers some intriguing possibilities. The relatively short integration time required (3 hrs for 3×10^{-14}) makes it one of the fastest measurement techniques available, and opens the possibility of using it for nearly continuous monitoring. Particularly for applications with less stringent purity demands (such as the ~ 100 ppt for XENON), one could imagine using an ATTA system to continuously monitor the Kr impurity level, by bleeding off a small amount of the gas (the source consumption rate is on the order of 3μL/s STP) during the purification process. A similar process might also be used to check the purity during the detector operation, to help detect any outside contamination of the detector medium.

CONCLUSION

The technique of Atom Trap Trace Analysis (ATTA) potentially offers a fast and accurate technique for measuring Kr contamination in other rare gases, to determine the radioactive background expected from ^{85}Kr in liquid rare-gas scintillators. The estimated sensitivity is 3×10^{-14} in only three hours of integration, compared to the days or weeks required to reach the same sensitivity with more conventional measurement techniques.

It is important to note that ATTA is a detection method only, and not a technique for purifying bulk gases, as the throughput of the source is much too low ($\sim \mu$L/s) and the system cannot easily be scaled up. The technique is independent of the production method, and can be used in conjunction with any of the current gas purification schemes to provide an independent measurement of Kr levels with different systematic errors. ATTA should be viewed as a complement to existing schemes for the purification of Xe and Ne.

The apparatus required for ATTA is relatively inexpensive compared to accelerator mass spectrometry and the liquid rare gas scintillation detectors themselves. Together with its high sensitivity to Kr contamination and fast measurement times, this means that ATTA has the potential to make an important contribution to the next generation of particle detectors.

ACKNOWLEDGMENTS

We thank Dr. Zheng-Tian Lu, Dr. Peter Mueller, and Dr. Herman Beijerinck for valuable discussions. C.O. is supported by a grant from the Research Corporation.

REFERENCES

1. E. Aprile *et al.*, astro-ph/0207670.
2. Y. Suzuki, hep-ph/0008296.
3. D. N. McKinsey and K. J. Coakley, astro-ph/0402007.
4. H. H. Loosli, B. E. Lehmann, and W. M. J. Smethie. Noble gas radioisotopes: 37Ar, 85Kr, 39Ar, 81Kr. Pages in P. Cook and A. L. Herczeg, editors. Environmental Tracers in Subsurface Hydrology. Kluwer Academic Publishers. 379-397, 2000.
5. T. Shutt, personal communication.
6. D. N. McKinsey and J. M. Doyle, J. Low Temp. Phys. **118**, 153 (2000).
7. R. A. Michniak et. al., Nucl. Inst. and Meth. A. **482**, 394 (2002).
8. C. J. Horowitz, K. J. Coakley, and D. N. McKinsey, Phys. Rev. D **68**, 23005 (2003).
9. D. N. McKinsey and K. J. Coakley, Astroparticle Physics **22**, 355 (2005).
10. Y. Takeuchi, "Rn Assay and Distillation Purification of Liquid Xe for Kr," these proceedings (2005).
11. C.Y. Chen *et al.*, Science **286**, 1139 (1999).
12. D. N. McKinsey and C. Orzel, *Nucl. Inst. and Meth. A* (in press, 2005) (see also astro-ph/0406526 (2004)).
13. H. Metcalf and P. van der Straten, *Physics Reports* **244**, 204 (1994).
14. W. Phillips and H. Metcalf, Phys. Rev. Lett. **48**, 596 (1982).
15. P. D. Lett *et al.*, J. Opt. Soc. Am B **6**, 2084 (1989).
16. E. L. Raab *et al.*, Phys. Rev. Lett. **59**, 2631 (1987).
17. J. Lefers *et al.*, Phys. Rev. A **66**, 012507 (2002).
18. W. Lu *et al.*, Rev. Sci. Instrum. **72**, 2558 (2001).
19. C. Y. Chen *et al.*, Rev. Sci. Instrum. **72**, 271 (2001).
20. L. Young *et al.*, J. Phys. B: At. Mol. Opt. Phys. **35**, 2985 (2002).
21. N. C. Sturchio *et al.*, *Geophys. Res. Lett.* **31**, L05503 (2004).
22. Z.-T. Lu *et al.*, Atomic Physic 17, edited by E. Arimondo, P. DeNatale and M. Inguscio (AIP, 2001)
23. X. Du *et al.*, Geophys. Res. Lett. **30**, 2068 (2003).

Highly Sensitive Measurements of ^{222}Rn Diffusion and Emanation

Grzegorz Zuzel

Max Planck Institute for Nuclear Physics
Saupfercheckweg 1, 69117 Heidelberg, Germany

Abstract. Highly sensitive techniques for determination of the ^{222}Rn emanation from solids and diffusion through different membranes are presented. ^{222}Rn and its daughters are measured via the alpha decays in special proportional counters at the absolute sensitivity of ~30 µBq. Radon diffusion can be measured at the level of ~10^{-13} cm^2/s. Several samples were examined, e.g. stainless steel, teflon, various gaskets (emanation and diffusion measurements) and tanks. A combination of measurements of the ^{222}Rn diffusion and emanation of thin nylon foils (used in the Borexino experiment) allowed the determination of ^{226}Ra in the materials of interest at the level of ~10^{-12} g/g ^{238}U-equivalent.

Keywords: Radium, radon, radon emanation, radon diffusion, polymers, proportional counters
PACS: 23.60.+e , 29.40.Cs, 51.20.+d

INTRODUCTION

^{222}Rn and its short-living daughters form an important background component in low-level radioactivity measurements. A direct precursor of ^{222}Rn is ^{226}Ra, which belongs to the ^{238}U chain and therefore is omnipresent in all materials. ^{222}Rn as an inert noble gas can very easy penetrate and diffuse through several materials before it decays ($T_{1/2}$ = 3.82 days). Additionally a decay of ^{222}Rn is followed by a chain of progeny until stable ^{206}Pb is reached (see Fig. 1). Since the half-life of ^{210}Pb is significantly longer than that of ^{222}Rn it breaks the equilibrium in the chain.

^{222}Rn can contribute to the background in many experiments, especially those sensitive to low energies, like solar neutrino experiments, dark matter detectors or neutrino-less double beta decay experiments. This is because decays of radon and its daughters are followed by alphas, betas and gammas of very different energies covering a region of interest for different experiments.

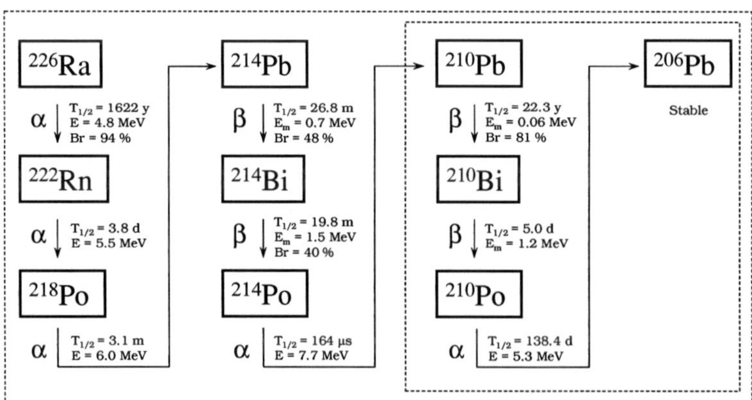

FIGURE 1. ^{238}U chain starting from ^{226}Ra with the most important decay modes of the following isotopes. Dashed line shows the sub-chain of ^{210}Pb.

CP 785, *Topical Workshop on Low Radioactivity Techniques,* edited by B. Cleveland, R. Ford, and M. Chen

Also in the very sensitive detectors e.g. germanium spectrometers used for material screening ^{222}Rn poses a serious problem. Usually after the detector setup and its shield against external radiation has been completed (taking into account intrinsic radioactivity of the materials and their radiation stopping power) the ^{222}Rn and progenies often dominate the residual background. Highly sensitive detectors equipped with an anticoincidence shield or located deep in underground laboratories generally need ^{222}Rn suppression systems.

It is clear that in the modern low background experiments ^{222}Rn monitoring is very important. All materials used in the experiments/detectors have to carefully tested for ^{222}Rn emanation and diffusion. This requires techniques, which allow for the detection of radon at very low activities.

RADON EMANATION MEASUREMENTS

In general emanation of ^{222}Rn is measured by placing a sample into a hermetically sealed chamber. Next, enclosed with the sample air is replaced with radon-free helium or nitrogen and after reasonable time (usually a few days) ^{226}Ra supported ^{222}Rn is extracted and measured in a suitable detector.

Radon Assay Systems

The base emanation assay system consists of two stainless steel chambers. a vacuum line, a ^{222}Rn-free He supply line and two traps: a water trap and a radon transport trap. Both chambers are electro-polished in order to reduce ^{222}Rn emanation from their inner surfaces. All fittings and connections are metal sealed, including the chambers top flanges for sample insertion (sealed with copper o-rings). The volumes of the chambers are 20 l and 80 l. Detail description of the system can be found in Ref. [1].

After placing the sample in one of the vessels it is sealed, evacuated and filled with pure He (at the slight overpressure). After a chosen time accumulated ^{222}Rn is pumped out together with helium over the transport trap containing activated carbon and cooled in a liquid nitrogen bath. If necessary the water trap (filled with silica gel) can be used to remove water vapors. The trap with collected radon is then connected to the counter filling line, where the sample is purified and filled into the counter.

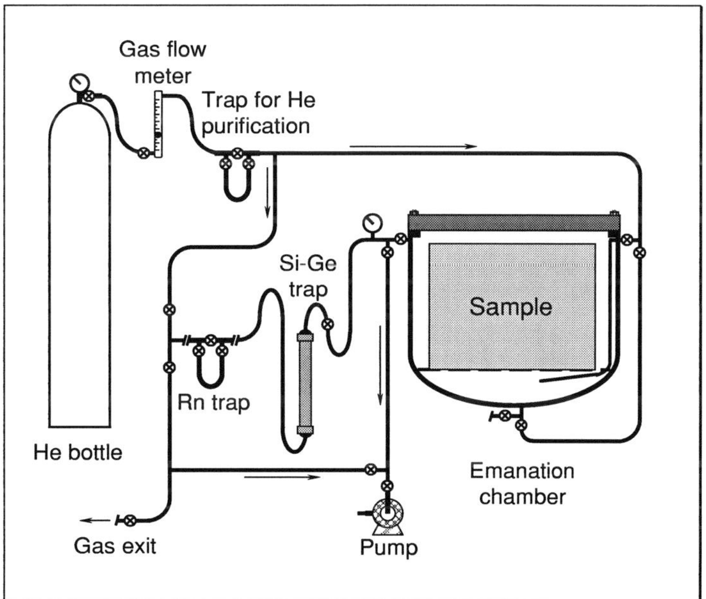

FIGURE 2. A scheme of the emanation chamber with the vacuum and He supply line. The ^{222}Rn trap containing a sample can be disconnected and hooked up to the filling line.

The blank value for the small chamber was measured to be (64 ± 15) μBq and for the big one (153 ± 25) μBq. Taking into account the sensitivity of the proportional counting about 70 μBq for the small and about 100 μBq for the big chamber can be measured as an absolute minimum emanation activity. Limit for specific activities will depend on the size/surface of the samples and can be very low, especially for the big vessel. Since each chamber is equipped with a distribution tube (sparger) for a gas it is possible to do measurements of the emanation not only under the He atmosphere, but also in water or other liquids. Here also ^{226}Ra content in liquids can be investigated at the level of 1 mBq/m^3 (using the big chamber).

In case of small samples (up to 2 liters) it is possible to use glass vials, which are hooked up directly to the filling line. Usually blank activities for these small containers are below the detection limit. Fig. 2 shows the scheme of one chamber (80 l) with auxiliary systems.

Tests of "stand alone" units (tanks, columns, pipes etc.) are also possible. In this case the investigated object is first purged with pure gas (He or N_2) until its volume is exchanged at least 5 times. Then, depending on the tested volume or required sensitivity, ^{222}Rn is extracted completely using MoREx [2] or partly, taking a known volume of the gas and collecting the radon in a small transport trap. In the first case the system is evacuated or if it is not possible, helium/nitrogen is flushed so that the accumulated radon is removed completely.

Detection of ^{222}Rn

For detection of ^{222}Rn miniaturized proportional counters are used. They were originally developed for the GALLEX solar neutrino experiment [3]. The active volume of the counter is ~1 cm^3. As a counting gas P10 (90 % argon, 10 % methane) is used. Only α decays of ^{222}Rn and its daughters are recorded. The counting efficiency measured for the equilibrium between ^{222}Rn and the short-lived daughters and for the energies above 50 keV (threshold) is (148 ± 2) % [1]. This means that about 1.5 counts are registered for one radon decay (since in the radon chain there are two more α emitters the efficiency is ~50 % for each α decay). The background depends on the individual counter and varies from 0.1 to 2 cpd (counts per day). The schematic view and the picture of the counter are shown in Fig. 3.

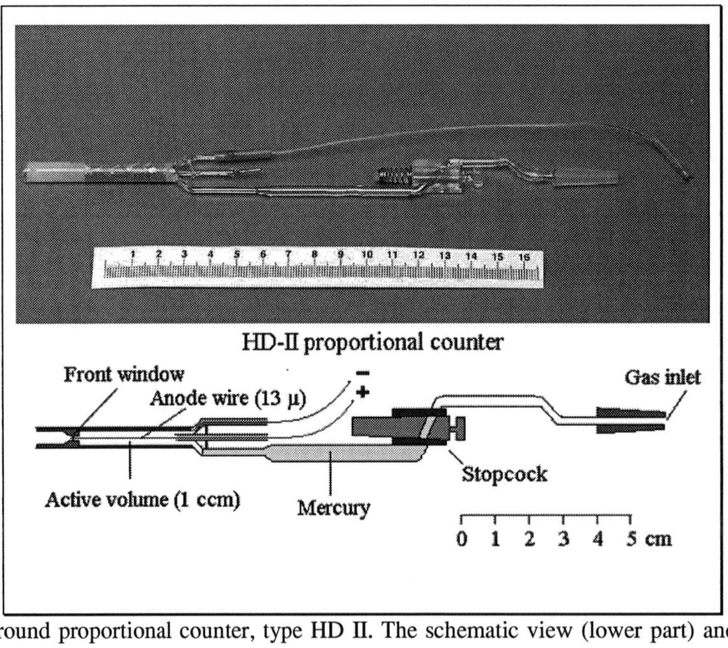

FIGURE 3. Low-background proportional counter, type HD II. The schematic view (lower part) and the picture (upper part) showing real dimensions.

Radon from the sample is purified, mixed with the counting gas and filled into the counter on a dedicated glass vacuum line. ^{222}Rn is flushed out from the transport trap (hooked up to the line) with pure He and collected in a molecular sieve trap (kept at the LN_2 temperature). Next it is transferred into a small chromosorb trap (also cooled by LN_2) where it is mixed with P10 and finally pushed into the counter by using a Töpler pump. Gaseous impurities present in the sample are collected on a NaOH trap (mainly CO_2), a silica-gel trap (H_2O vapors) or absorbed by a ZrAl getter pump (mostly reactive gases like nitrogen or oxygen). Taking into account the background value of the line it is possible to give an absolute detection limit for ^{222}Rn at the level of 30 μBq. It means using proportional counters ~15 atoms can be seen.

Results

Several samples have been investigated, mainly materials that were planned to be used in the Borexino detector [4] or its smaller version CTF [5]. Some units in the auxiliary systems were examined as well. Obtained results are collected in Table 1.

TABLE 1. ^{222}Rn emanation rates for different samples

Sample Description	^{222}Rn Emanation Rate	Remarks
Stainless Steel	(10.2 ± 0.8) μBq/m^2	Foil, 71 m^2
Stainless Steel After Cleaning	(4.6 ± 0.9) μBq/m^2	Foil, 71 m^2, rinsed with water
Electro-polished Stainless Steel	< 3 μBq/m^2	Steel package, 25 m^2
Teflon Foil	< 9 μBq/m^2	29 m^2
Borexino PMT	< 40 μBq/m^2	Complete PMT
Activated Carbon CarboAct	(0.3 ± 0.1) mBq/kg (0.6 ± 0.3) mBq/kg	Used in the LTA I*) Used in the TLA II**)
Polyurethane	(<0.3 – 40) mBq/m^2	Different gaskets
Silicon rubber	(196 ± 4) mBq/m^2	Gaskets
Nitrogen Stripping Column	(2.3 ± 0.2) mBq	200 l, 280 m^2 of steel inside
Borexino Pseudocumene Storage Tank	(25 ± 3) mBq	TK4, 114 m^3, 140 m^2

*) LTA – Low Temperature Adsorber of the Borexino nitrogen plant [2]
**) In 2002 second purification column was built.

Stainless steel foil was measured in the big emanation chamber before and after cleaning. It was "normal" steel, not electro-polished. After rinsing with water the ^{222}Rn emanation rate dropped by a factor two. Emanation from electro-polished steel was measured to be below 3 μBq/m^2 (90 % C.L.). This value was found for the steel package (25 m^2) planed to be used in the nitrogen stripping column of the Borexino liquid handling system. It can be compared directly with the emanation rate from the complete column containing 26 similar packages of 10.6 m^2 each (~280 m^2 in total) which was ~8 μBq/m^2. The comparison shows that the system in its final state produces more radon than expected. The reason is that to the measured signal contribute also all gaskets, joints, welds, valves and residual contaminations. In case of stainless steel especially welds may be critical because of their high porosity and possible high content of uranium and thorium (rods material).

A rather high ^{222}Rn emanation rate was measured for the Borexino Pseudocumene (PC) storage tanks. E.g. for the tank No. 4 we found ~180 μBq/m^2 (see Table 1). For three others the results were very similar. The inner surface of each tank was electro-polished and cleaned with detergent (containing phosphoric and citric acid) after construction [6].

Teflon foil seems to emanate a little bit more radon than steel, however here we measure not only the surface contamination, but also partly ^{222}Rn emanated from a thin (~0.4 mm) sub-surface layer. This is due to the rather high radon diffusion coefficient in teflon of ~1×10^{-9} cm^2/s [7].

Investigations of different types of gaskets show that the best material from the point of view ^{222}Rn emanation seems to be polyurethane. However, to be sure about its quality each batch of material has to be tested individually.

Activated carbon CarboAct was used as a radon adsorber in the nitrogen purification columns (LTAs) of the Borexino nitrogen plant. Samples from two batches were measured showing emanation at the same level of ~0.5 mBq/kg, which is at least two orders of magnitude lower rate than any other activated carbon [2].

MEASUREMENTS OF RADON DIFFUSION

Presented here method enables the study of diffusion and solubility of radon in thin membranes and thus leads to the fundamental constants – the diffusion and permeability coefficients (which are generally not available in the literature). Material of interest can be examined under different conditions, e.g. may be exposed to liquid or a gas of different humidity.

The Apparatus

An apparatus used to investigate radon diffusion through membranes placed in the nitrogen atmosphere of different relative humidity (RH) is shown in Fig. 4. Two-liter glass flasks, partly filled with saturated aqueous solutions of various salts, were used as generators of gas humidity standards. By using selected chemical compounds, it was possible to obtain various humidity values in the gas phase above the salt (in equilibrium with the solution) over the whole range of RH (0 – 100 %).

The ^{222}Rn source supplies the system with a constant ^{222}Rn activity of about 18.5 kBq. With the help of the pump ^{222}Rn is circulated together with nitrogen between the source container (4.6 l), the humidity standard and the ^{222}Rn chamber. The light generated in the scintillator (ZnS(Ag)) by alpha particles emitted by radon and its daughters is detected by photomultipliers. ZnS(Ag) forms a thin layer on the quartz windows placed between the photomultipliers and the active volumes of the detectors. The first detector monitors the ^{222}Rn concentration (meant to be constant) in its active volume (and in the ^{222}Rn chamber). This volume is separated from the chamber by a membrane filter, which is very permeable for radon but stops alpha particles emitted by the ^{222}Rn and Po isotopes decaying in the chamber. The active volume of the second detector is separated from the ^{222}Rn chamber by a filter, identical to that in the first one (they are used to set the same counting efficiencies for both detectors), and the membrane to be investigated. The latter is carefully sealed by a special indium gasket. The count rate of the second detector (background-corrected) is proportional to the total amount of radon diffused through the membrane, and it starts to increase after a certain time-period directly related to the film thickness and its ^{222}Rn diffusion coefficient. The background counting rate of the detectors allows measurements of the diffusion coefficient down to 10^{-13} cm^2/s. The detailed description of the system can be found in Ref. [8].

To find the diffusion and solubility coefficients from the registered count rates it is necessary to construct a mathematical model of diffusion with these two parameters. They can be next found by fitting the model curve to the experimental data using the least-square method. Detailed description of the applied model can be found in Ref. [8] and Ref. [9]. In Ref. [8] there is also described a system, which allows the diffusion and solubility constants for membranes surrounded by liquids to be measured

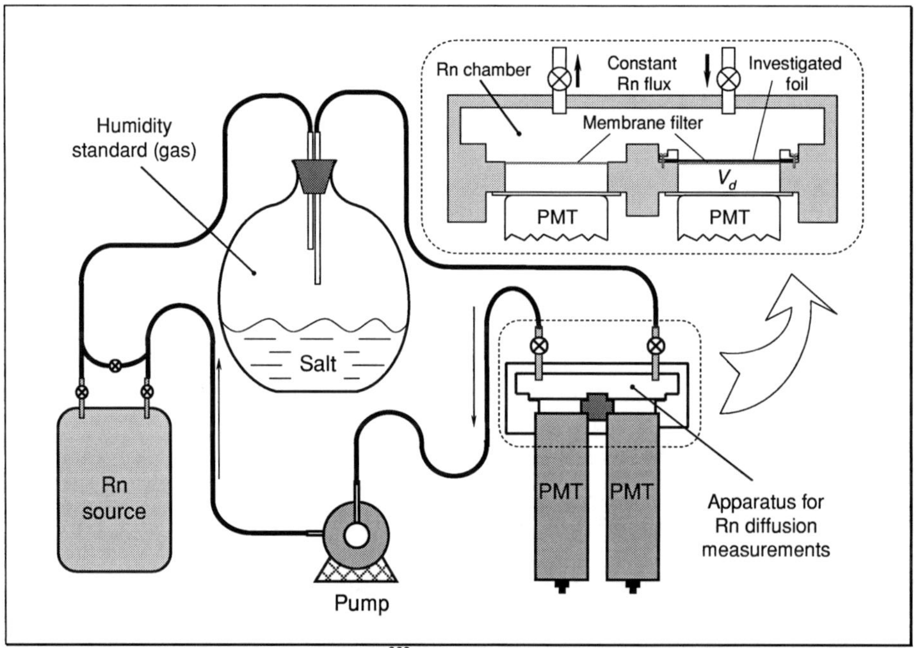

FIGURE 4. Schematic view of the apparatus measuring ^{222}Rn diffusion and solubility coefficients for thin membranes in the gas (nitrogen) atmosphere. Radon from the source is pumped with nitrogen through a glass flask filled partly with salt (humidity standard) to the chamber of the apparatus used for measurements.

Results

Several samples have been tested, mainly nylon foils used in the Borexino experiment. The dependence of the radon diffusion on the humidity of the inner vessel film is of great importance from the background point of view [8]. Obtained results for nylon C38F (its physical and chemical parameters are very close to those of Sniamid [10] – the film used to construct the Borexino inner vessel) show a strong relation between these two parameters. For RH lower than 20 %, the diffusion coefficient is small and within the measurement uncertainty remains equal to its value for dry nylon (~2×10^{-12} cm^2/s). Above RH = 30 %, it begins to grow slowly along with ambient humidity. Above 80

%, the diffusion coefficient increases rapidly with RH of the circulating gas. For RH = 100 %, it is roughly three orders of magnitude larger than in the dry membrane.

The results show that radon diffusion does not change significantly (with respect to the dry film) with relative humidity increasing from 0 to 60 %, what results in a low expected ^{222}Rn emanation/permeation rate for the material. This is also the range of the inner vessel humidity environment optimal from the point of view of the mechanical parameters of nylon (50 % < RH < 60 %) [11]. It is therefore important to keep the scintillator vessel in such conditions in Borexino in order to avoid damage and lower as much as possible the ^{222}Rn-induced background of the detector.

Table 2 gathers complete set of diffusion coefficients obtained for different humidities. All the measurements were performed for the same sample of nylon. In the last column radon permeability is given. It is a product of the diffusion and solubility coefficients.

TABLE 2. ^{222}Rn diffusion, solubility and permeability coefficients for nylon C38F as a function of the relative humidity in the surrounding nitrogen atmosphere

RH standard Salt	RH in gas phase [%]	Diffusion coefficient D [cm^2/s]	Solubility S	Permeability P [cm^2/s]
$Mg(ClO_4)_2$	~ 0	(2.1±0.4)×10^{-12}	4.5 ± 0.7	(9.5±2.3)×10^{-12}
$H_3PO_4 \cdot ½H_2O$	9 ± 1	(2.3±0.3)×10^{-12}	2.5 ± 0.3	(5.8±1.1)×10^{-12}
$LiCl_2 \cdot H_2O$	12 ± 1	(2.2±0.3)×10^{-12}	2.2 ± 0.3	(4.8±0.9)×10^{-12}
$CaCl_2 \cdot 6H_2O$	32 ± 2	(4.3±0.5)×10^{-12}	1.8 ± 0.2	(7.7±1.2)×10^{-12}
$Na_2Cr_2O_7 \cdot 2H_2O$	52 ± 2	(1.9±0.3)×10^{-11}	1.4 ± 0.2	(2.7±0.6)×10^{-11}
$Na_2S_2O_3 \cdot 5H_2O$	76 ± 2	(6.5±0.9)×10^{-11}	1.5 ± 0.2	(9.8±1.9)×10^{-11}
K_2CrO_4	88 ± 3	(1.3±0.2)×10^{-10}	1.5 ± 0.2	(2.0±0.4)×10^{-10}
$Na_2SO_4 \cdot 10H_2O$	93 ± 3	(3.3±0.4)×10^{-10}	1.0 ± 0.1	(3.3±0.6)×10^{-10}
H_2O vapors	100 ± 3	(1.3±0.2)×10^{-9}	0.7 ± 0.1	(9.1±2.0)×10^{-10}

MEASUREMENTS OF RADIUM CONTENT IN THIN POLYMERS

A combination of highly sensitive measurements of radon diffusion and emanation makes it possible to measure the surface and bulk activities of ^{226}Ra in thin nylon foils at unprecedented concentration levels, as low as 0.5 μBq/m^2 and 10 μBq/kg, respectively [12]. As described in the previous paragraph, when the foil is fully saturated with water (RH = 100 %), the permeability can increase by almost 2 orders of magnitude as compared to completely dry film. This opens up the possibility to investigate surface and bulk ^{226}Ra contamination of the sample by measuring ^{222}Rn emanation when it is dry (^{222}Rn signal predominantly from surface contamination, ^{222}Rn diffusion length[1] in nylon is in the range of 7 μm) and when it is wet (predominately bulk contamination, ^{222}Rn diffusion length in nylon is in the range of 300 μm). A mathematical diffusion model is used to distinguish the surface and bulk activity of ^{226}Ra from the ^{222}Rn emanation rates of dry and wet film. One foil could be identified with ^{226}Ra contamination low enough (~10^{-12} g/g ^{238}U-equivalent) for application in the solar Borexino detector.

Sampling procedure

For the nylon foil investigations, a foil tent equipped with a clean-room air filter was erected over the emanation chamber to protect the samples during insertion against particulate contamination from the environmental dust containing ^{226}Ra. In addition, filtered air was continuously flushed at the rate of about 1 m^3/min from the bottom tube through the chamber during all operations with open chamber. Four stages of filters were applied: for particulates down to 3 μm and down to 0.01 μm, charcoal against humidity, and an additional filter for (charcoal) particulates down to 0.003 μm.

In the first step of the preparation procedure, a film sample of the mass of about 10 kg (in the form of a roll of 31 cm high and about 30 cm in diameter) was placed on the bottom plate of the chamber. This plate had many holes so that the sparging gas entering from a distribution tube below was uniformly dispersed among the foil layers. The original tightly rolled film was then arranged in such a way that the neighboring layers no longer adhered to each

[1] Diffusion length is defined as $d = \sqrt{D / \lambda_{Rn}}$, where λ_{Rn} is the radon decay constant

other and ^{222}Rn could freely emanate from the whole surface (the average distance between the layers was about 0.1 mm). Next, the chamber was sealed, evacuated, and filled with He.

After a sufficiently long waiting time, when the air-born ^{222}Rn (absorbed in the nylon or introduced during sample insertion) has decayed and the ^{226}Ra-supported ^{222}Rn has grown in for several half-lives (usually about three weeks after the chamber was closed), radon was extracted. Three to four runs with variable ^{222}Rn grow-in times were performed for each foil sample (dry and wet). In this way the reproducibility of the obtained emanation rates could be verified.

When the runs for a dry test were completed, about 6 l of high purity water (^{226}Ra concentration below 5 mBq/m^3) was filled into the chamber through the bottom tube. The water level stayed below the foil sample, so that after about one week the nylon was fully saturated with water by vapor only. This was verified by the increased weight of the sample at the end of the experiment (foil fully saturated with water increased its weight by ~10 %). Next, after an additional reasonable grow-in time, the series of wet extraction measurement runs was started. Each single extraction was done in the same way as for dry foil, and lasted at least about 10 hours. This time was long enough for quantitative ^{222}Rn removal from the gaps between the foil layers, but also for ^{222}Rn diffusion from the whole volume of the wet nylon (for wet foil the ^{222}Rn diffusion length is higher than the film thickness). This allowed the collection of almost all ^{222}Rn, which has been produced due to ^{226}Ra decay.

Finally, the foil was removed and the blank activity of the chamber (still filled with water) was measured. In this way it was verified that no ^{226}Ra activity was introduced into the chamber during sample preparation and water addition. The total time needed for a complete investigation of one nylon sample extended to almost three months.

Results

Five nylon samples, differing in origin and type of material, were investigated by applying the procedure described above. According to the measured ^{222}Rn emanation rates from dry and wet foils the bulk and surface ^{226}Ra concentrations were calculated (using a mathematical model of ^{222}Rn emanation from membranes [12]) and given in Table 3. For the fifth sample only a clear signal was observed with the wet foil, under dry conditions the measured ^{222}Rn activities were below the detection limit and we could only give limits for the bulk and surface ^{226}Ra contaminations. The total ^{226}Ra concentration is given according to the signal measured for the wet foil; therefore, in contrast to the bulk and surface concentrations, it is not an upper limit and corresponds to 1.3×10^{-12} g/g ^{238}U-equivalent.

TABLE 3. Deduced bulk and surface ^{226}Ra contaminations for different nylon foils

Sample No.	Nylon type, thickness, mass surface area (both sides)	Measured ^{226}Ra equiv.* [mBq] dry foil	Measured ^{226}Ra equiv.* [mBq] wet foil	Bulk ^{226}Ra [mBq/kg]	Surface ^{226}Ra [µBq/m^2]	Total ^{226}Ra [mBq/kg]
1	Capron DF400 0.1 mm, 9.7 kg, 171 m^2	—	5.7 ± 0.2	0.59 ± 0.05	—	0.59 ± 0.05
2	Sniamid/Capron** 0.1 mm, 9.5 kg, 168 m^2	0.65 ± 0.15	1.3 ± 0.2	0.10 ± 0.02	3.0 ± 0.9	0.14 ± 0.03
3	Capron B73ZP (1st batch) 0.1 mm, 11.8 kg, 208 m^2	0.63 ± 0.06	2.6 ± 0.1	0.21 ± 0.03	1.1 ± 0.3	0.22 ± 0.04
4	Capron B73ZP (2nd batch) 0.125 mm, 9.5 kg, 134 m^2	1.45 ± 0.11	5.0 ± 0.2	0.46 ± 0.07	6.4 ± 1.6	0.55 ± 0.08
5	Sniamid (blend) 0.125 mm, 9.2 kg, 130 m^2	< 0.11	0.15 ± 0.04	< 0.021	< 0.8	0.016 ± 0.004

*) Ra equivalent – mean value of measured ^{222}Rn activities (extrapolated to saturation) under given conditions
**) Transition roll, partly Capron B73ZP (20%), partly Sniamid ADS40T (80%)

ACKNOWLEDGMENTS

Many people were involved in the presented work, mainly members of the Heidelberg and Kraków group of the Borexino collaboration: C. Buck, B. Freudiger, G. Heusser, W. Rau, H. Simgen and M. Wójcik.

REFERENCES

1. Rau, W., and Heusser, G., Appl. Rad. Isot. 53, 371-375 (2009).
2. Heusser, G., et. al., Appl. Rad. Isot. 52, 691-695 (2009).
3. Wink, R., et al., Nucl. Instr. and Meth. A 329, 541-550 (1993).
4. Alimonti, G., et al., Astrop. Phys. 16, 205-234 (2001).
5. Back, H. O., et al., Phys. Lett. B 563, 23-34 (2003).
6. Perraso, L., „Report on the Borexino Storage Vessels“, Borexino internal report, March 2000.
7. Müller, W., Thesis, Universitaet Heidelberg, 1978.
8. Wójcik, M., and Zuzel, G., Nucl. Instr. and Meth. A 524, 355-365 (2004).
9. Wójcik, M., et al., Nucl. Instr. and Meth. A 499, 158-171 (2000).
10. Cadonati, L., "The Borexino Solar Neutrino Experiment and its Scintillator Containment Vessel", PhD Thesis, Princeton University, Princeton 2001.
11 McCarty, K., et al., "Glass transition, moisture content and mechanical characteristics of nylon", Borexino internal report, Princeton University, March 2001.
12. Zuzel, G., et al., Nucl. Instr. and Meth. A 498, 240-255 (2003).

VI – CLEANLINESS STUDIES AND CLEAN FABRICATION TECHNIQUES

Low background techniques for the Borexino nylon vessels

Andrea Pocar

Physics Department, Stanford University, Stanford, CA 94305 USA

Abstract. Borexino is an organic liquid scintillator underground detector for low energy solar neutrinos. The experiment has to satisfy extremely stringent low background requirements. The thin nylon spherical scintillator containment vessel has to meet cleanliness and low radioactivity levels second only, within the detector, to the scintillator itself. Overall, the background from the vessel in the fiducial volume of the detector must be kept at the level of one event per day or better. The requirements, design choices, results from laboratory tests, and fabrication techniques that have been adopted to meet this goal are presented. Details of the precautions taken during the installation of the vessels inside the Borexino detector are also discussed.

Keywords: Borexino, solar neutrinos, low-background, radon, nylon, adsorption, charcoal
PACS: 29.40.Mc, 26.65.+t, 81.05.Lg, 81.65.Cf, 68.48.-h

THE BOREXINO DETECTOR

Borexino is a solar neutrino experiment approaching completion at the underground Gran Sasso national laboratories in central Italy [1] . It is designed to detect low energy solar neutrinos (^{7}Be, pep) as well as antinuetrinos emitted in beta decays in the earth's crust [2] and neutrinos from possible supernovæ in our galaxy [3].

Borexino will detect neutrinos that elastic scatter off the electrons of its 300 ton organic scintillator target. The inner 100 tons are designed to have minimal background in the enrgy observation window for solar neutrinos (between 250 keV and 1.5 MeV). The scintillator is contained inside an 8.5 meter diameter spherical balloon-like nylon vessel (Inner Vessel, IV). The scintillation light flash produced by recoiling electrons will be detected by more than 2200 photomultiplier tubes (PMTs) aimed at the scintillator volume and mounted on a stainless steel sphere (SSS), concentric with the IV and 13.7 meters in diameter. A second nylon vessel (Outer Vessel, OV), concentric and essentially identical to the IV but with a diameter of 11 meters is placed between the IV and the PMTs; its role is to prevent radon emanated by the PMTs and the supporting steel sphere from diffusing into the core of the detector. The volume inside the SSS in then divided into three regions: these are, from the inside out, the Inner Vessel, the Inner Buffer (IB), and the Outer Buffer (OB). Each vessel has two sets of low activity ropes that carry the buoyant loads due to density differences between the volumes and temperature gradients.

The scintillator is a solution of 1.5 g/l dyphenil oxazole (PPO) in 1,2,4-trimethylbenzene (pseudocumene). The volume between the SSS and the IV is filled with a solution of dimethylphthalate (DMP) in pseudocumene. This volume of fluid acts as a buffer for background radiation from the SSS and PMTs and the DMP quenches the light such radiation would generate at high rate in the detector. The choice of scintillator and buffer solutions with quasi zero buoyancy allowed the vessels to be extremely thin thus limiting their total mass. As explained in the following, this was one of many necessary steps in order to minimize the radioactive contamination close to the active region of the detector. Ultra pure water contained in a tank around the SSS provides shielding for external neutrons from the experimental hall and is instrumented with PMTs to detect Čerenkov light from muons that cross it. A detailed sketch of the Borexino detector is shown in fig. 1

The nylon vessels

The Borexino nylon vessels are two concentric spherical balloons, with diameters of 8.5 and 11 meters respectively. They are made out of a thin nylon membrane, just 125 μm thick. Nylon was chosen because it is transparent and optically clear, chemically inert in pseudocumene, radioactively very pure, and presents low permeation for gases (especially the radioactive noble gasses ^{39}Ar, ^{85}Kr, and ^{222}Rn [4, 5, 6]). The choice of a thin film was driven by the effort of minimizing the amount of material close to the scintillator. An extensive treatment of the properties of the nylon membrane can be found in [4]. At the poles, the vessels are connected to a set of pipes that carry the scintillator

CP 785, *Topical Workshop on Low Radioactivity Techniques,* edited by B. Cleveland, R. Ford, and M. Chen

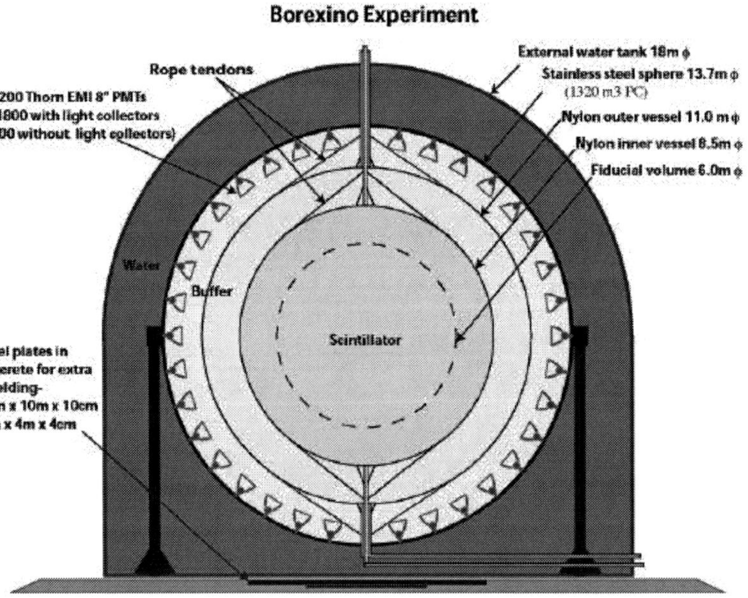

FIGURE 1. The Borexino detector. The active scintillator target is contained in a transparent, spherical nylon vessel, 125 μm thick. The low background requirements for the scintillator and the nylon vessel are extremely stringent, having to contribute on the order of one event per day in the energy window for neutrino detection below 1 MeV.

and buffer fluids into the IV, IB, and OB.

The following sections describe the low radioactivity requirements for the nylon vessels and the details of their design and fabrication explicitly meant to enhance radio-purity and cleanliness. Low radioactivity techniques employed for all the Borexino components are reported in [7].

LOW RADIOACTIVITY ASPECTS OF THE BOREXINO NYLON VESSELS

Providing an ultra pure container for the 300 ton scintillator target is arguably the hardest challenge for the Borexino nylon vessel. The first natural step to take is to minimize their mass. As mentioned above, the quasi-zero buoyancy design between scintillator and buffer fluids made it possible to use 125 μm thick film, yielding a mass of the IV envelope of less than 50 kg. A strict material selection and screening was nonetheless necessary. Gamma-rays emitted at the IV by isotopes in the ^{238}U and ^{232}Th natural radioactive chains (especially ^{214}Bi and ^{208}Tl) and by ^{40}K (naturally present in potassium at the 0.01% level) are a serious potential background (the attenuation length for 1 MeV γ rays in pseudocumene is $\sim$25 cm). Last but not least, contamination from particulate and radon exposure during fabrication and installation inside the detector had to be kept at an absolute minimum. Indeed, whatever deposits on the inner surface of the IV can end up directly into the scintillator. In particular, ^{210}Pb (32 year mean life) accumulation following radon exposure yields ^{210}Bi (1.17 MeV β decay) and ^{210}Po (5.3 MeV α decay), both extremely dangerous backgrounds for Borexino. For particulate contamination, assuming its ^{238}U and ^{232}Th concentration to be 1 ppt, no more than 3 mg were allowed on the entire inner surface of the IV.

The thin film design made it possible to build the vessels as folded stacks on tables inside a clean room. Contamination from radon daughters was a main concern: a straightforward estimate made with somewhat pessimistic assumptions on the dynamics of radon daughters says that a normal fabrication procedure (in a clean room with a typical radon activity of 30 Bq/m^3) could lead to backgrounds three to four orders of magnitude too high inside the detector from this source [8, 9]. Several aspects of the fabrication technique, illustrated below, were driven by the need to minimize the exposure of the vessel surfaces. Strict cleanliness standards and clean room policies were adopted. In addition, the clean room was specifically constructed to accommodate a radon filter for its supply air, which further reduced the radon contamination of the vessels. The filter and laboratory-scale studies of radon contamination of nylon surfaces are specifically discussed below.

FIGURE 2. Gluing one of the joints of the IV. The large size of the clean room can be appreciated, as well as the effort made to cover all surfaces of the vessel not being bonded and to individually bag the clamps (used to apply pressure to the bonded panels during glue curing) to prevent particulate contamination of the vessel.

The clean room. The Princeton clean room for nylon vessel fabrication is approximately $22 \times 6.5 \times 4.5$ meters in size (fig. 2). It is certified to class 100, and was measured to be around class 10 when unperturbed (the class is the number of particles with diameter $\geq 5\,\mu$m per cubic foot of air). Recirculation time of the air through HEPA filters is about 30 seconds. The water for humidification ($\sim$50% RH) was aged for about 100 days for radon reduction. The clean room was designed to be fed with low-radon air supplied by a radon filter. This requires the clean room system to be leak-tight in order to prevent back-diffusion of radon and to reduce the amount of radon-free air needed to give the room the needed over-pressure. The radon filter delivered $\sim$85 m^3/h of radon-scrubbed air during vessel construction and is described more in detail below. More details on the clean room are found in [8], while the problem of radon contamination of surfaces is addressed in [8, 9].

Film selection and production. Two raw nylons ("Capron" by Allied Signal/Honeywell and "Sniamid" by Nyltech) were selected among many due to their high purity in the pellet state. The measured concentrations of ^{238}U and ^{232}Th were $\sim$1 ppt for both, with Capron being slightly less contaminated and thus the first choice. For both, the concentration of ^{nat}K was < 10 ppb. For details on these measurements and for an explanation on the differences between the two nylons, refer to [4] and references therein.

Two extrusions campaigns were carried out: one at Honeywell with Capron and the other at MF-Folien with Sniamid. For both, great care was put in keeping the area around the extruding machines clean. After extrusion, the film was precision cleaned at CleanFilm with a non-contact technique where particles were first ultrasonically disrupted from the surface and then removed by negative pressure. Its surface was certified to level 25 Mil. Std. 1246C (less than one particle 25 μm in diameter per square foot [10]).

Only after the OV, made out of Capron, was completed, radon emanation results became available for both films. This was the ultimate test of the final extruded film after it underwent the cleaning process. It also measured the concentration of ^{226}Ra rather than ^{238}U higher in the decay chain, automatically including the possibility that the two isotopes are not in equilibrium. Capron turned out to be about an order of magnitude more contaminated than Sniamid, with the latter having a ^{226}Ra activity of 16 μBq/kg [11], the lowest ever measured in any material (this translates to 1.3 ppt ^{238}U assuming secular equilibrium, or approximately 1000 ^{226}Ra atoms per gram of nylon!). The technique adopted for the radon emanation measurement from the film was able to separate bulk and surface contaminations by using the humidity dependance of the diffusion coefficient of radon through nylon. A ^{226}Ra surface contamination <0.8 μBq/m^2 could be set, thus proving the effectiveness of the surface cleaning procedure [11] .

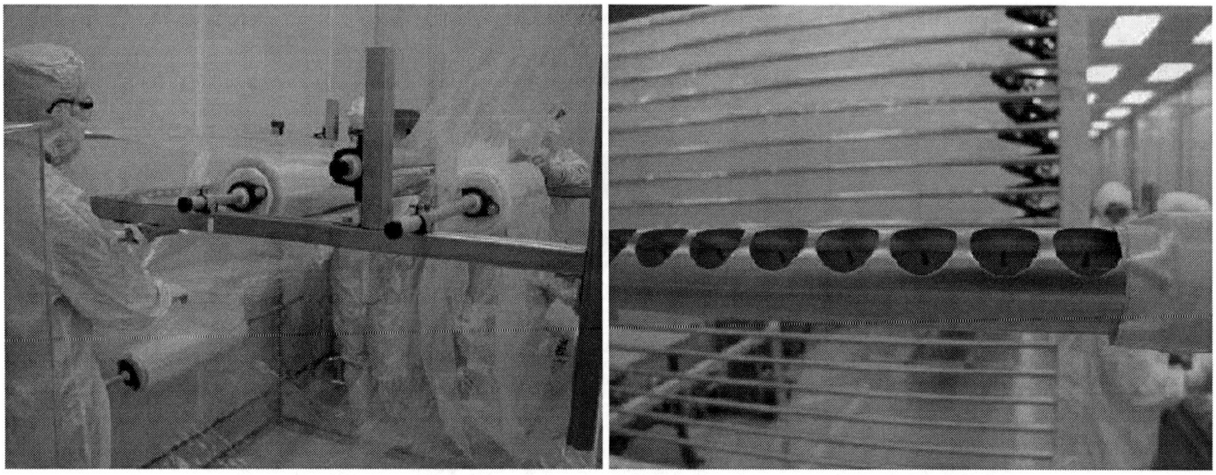

FIGURE 3. The nylon film was immediately sandwiched between sheets of clean nylon film to protected it from radon daughter deposition and residual particulate (left). To minimize radon daughter contamination during this operation, the nylon membrane was discharged from the accumulated static charge due to friction by two high voltage de-ionizing bars (right).

Other components. The material used to build the polar "end caps" and the pipes for the scintillator were also carefully screened. The section of the IV pipe closest ($\sim$1 m) to the vessel is made of nylon, while stainless steel was used for the sections at larger radii. Nylon is in fact less dense and much less radioactive. In addition, extruded nylon parts were found to be more radio-pure than cast ones. The mass of the end caps was kept to a minimum (about 15 kg), and the activity of the material presented $\sim$50 ppt concentrations of ^{238}U and ^{232}Th and $\sim$500 ppb of ^{nat}K. Nylon and stainless steel sections were precision cleaned at Astropak and their surfaces also certified to level 25 Mil. Std. 1246C [10]. Most of the ropes considered for the Borexino vessels were found to have a high potassium content. Those selected for Borexino are made out of ultra high density polyethylene (UHDPE) and have a ^{nat}K contamination of approximately 1 ppm.

Vessel fabrication. Fabrication of the nylon vessels started in the summer of 2001 and lasted for a year, until August 2002. Film was first unrolled and let acclimatize to the humidity of the clean room to their equilibrium elongation (few days). Then almond-shaped panels were cut out (36 for the IV, 40 for the OV): glued together, these would define a sphere when inflated. During these steps, a big effort was put into keeping the nylon film clean by protecting it from residual particulate and deposition of radon daughters. The nylon coming off the spool was immediately sandwiched between two layers of extremely clean (level 25 Mil. Std. surface cleanliness [10]), 25 μm-thick nylon sheets (fig. 3, left). In order to prevent deposition of radon daughter ions during this step where the dry film typically accumulates static charge on its surface, two high voltage de-ionizing bars were employed (fig. 3, right).

The panels were glued together on tables in a flat accordion-shaped stack (fig. 4). Each panel is bonded to the one below along a folded edge. Glue is applied on both sides of the joint and pressure is applied for several hours to obtain strong adhesion. During this step, all panels are kept covered on both sides and only the area being glue is directly exposed to the clean room atmosphere for the time needed. More details on this procedure are found in [8]. The stack geometry has the advantage of being efficiently self covering, with the top panel protecting all the underlying ones.

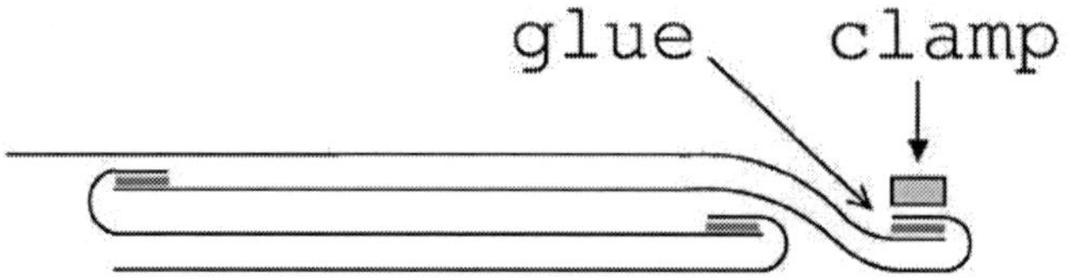

FIGURE 4. The nylon vessels were built in accordion-shaped stacks which have the advantage of being self covering for the panels below the top one. This geometry also suppresses radon diffusion by reducing the possible air gaps in contact with the nylon surface.

FIGURE 5. Left: the IV during the nesting operation into the OV. The presence of a very leak-tight shroud in a single package assembled in a radon-suppressed clean room environment adds a layer of protection for maintaining the cleanliness of the IV surface. Right: the completed nylon vessel system (IV nested inside the OV) is wrapped with several layers of clean film of various thicknesses and sealed in a thick aluminized foil to minimize radon diffusion.

The air gaps between panels are extremely small so that radon diffusion into the stack is highly suppressed. The whole stack was also kept covered with additional layers of clean film and a sheet of thick, aluminized film during down time to reduce radon exposure.

The IV was then sealed with the polar end regions and leak checked (see [8] for details). The vessel assembly sequence foresaw nesting the IV into the OV into a single package. This operation took place in the clean room (fig. 5, left). The presence of a very leak tight envelope around the IV, assembled in a clean room environment added a layer of protection for keeping the surface of the IV clean. Once the vessels were nested into each other, the OV was sealed along the last open seam and the assembly was packaged and folded for shipping. Several layers of clean film of various thicknesses were used. The package was finally sealed in a thick layer of aluminized film against mechanical damage and radon diffusion (fig. 5, right).

THE RADON FILTER

A radon filter of novel design was built to supply radon-suppressed air to the clean room for nylon vessel assembly. The filter is based on room temperature vacuum-swing adsorption on activated charcoal. It operated continuously during the whole year of vessel assembly, radon-scrubbing the entire flow of make up air required by the clean room. Such flow averaged around 85 m^3/h ($\sim$ 25 liters/minute). The device is shown in fig. 6 and is described in detail in [8]. The main results from the radon filter and the prototypes on which it was founded are illustrated in the next few paragraphs.

Principle of operation. Vacuum-swing adsorption (VSA) is conceptually identical to pressure-swing adsorption (PSA), a widely used process for gas separation [12, 13]. Adsorption on a surface is a bond between an atom or molecule and molecular sites on the surface. It is, in its simplest form, a dipole-dipole (Van der Waals) electromagnetic interaction whose strength falls as the distance to the sixth power. It's strength depends on both the species to be adsorbed and the surface material used. In the case of activated charcoal, the adsorption binding energy is stronger for a radon atom than for the other components of air (oxygen, nitrogen, argon, ...). Thus, when traveling through a charcoal bed (column), radon is effectively slower than other species in the flow, and the air emerging from the output is radon-depleted.

A pressure (vacuum)-swing system is composed of at least two identical charcoal columns which undergo a cycle of feed and purge (fig. 7, left). One of the columns filters the full air flow, while a small fraction of the clean output is expanded at lower pressure and fed in counter flow through the other column. This high velocity purge flow collects the impurities (radon in this case) accumulated during the previous cycle when said column was used as the filter, and discharges them. At the end of this stage, the purged column is clean and ready to work as a filter again, while the

FIGURE 6. The radon filtered that supplied the entire make up air flow to the vessel assembly clean room. The device operated continuously for the whole year of vessel fabrication. The main components are labelled: a,b) vacuum pump (roots blower + backing pump, respectively), c) charcoal columns, d) hepa filter at the output, mounted in a frame together with the valve manifold, e) duct to the clean room, f) column repressurizarion line.

other is loaded with impurities and ready to undergo a purge cycle. The time after which the two columns switch roles depends on the application and degree of purity required. A vacuum swing system uses feed at atmospheric and purge at sub-atmospheric pressure. Swing systems thus provide continuos supply of filtered gas stream by using multiple columns and continuous inline regeneration.

The regeneration step relies on the pressure-dependent properties of adsorption. In particular, the adsorptive capacity per unit area of a surface increases with pressure. Typically, for small concentrations, a linear relation holds between such two quantities. At higher loads, the adsorbent approaches saturation and linearity is lost. To first order, the radon

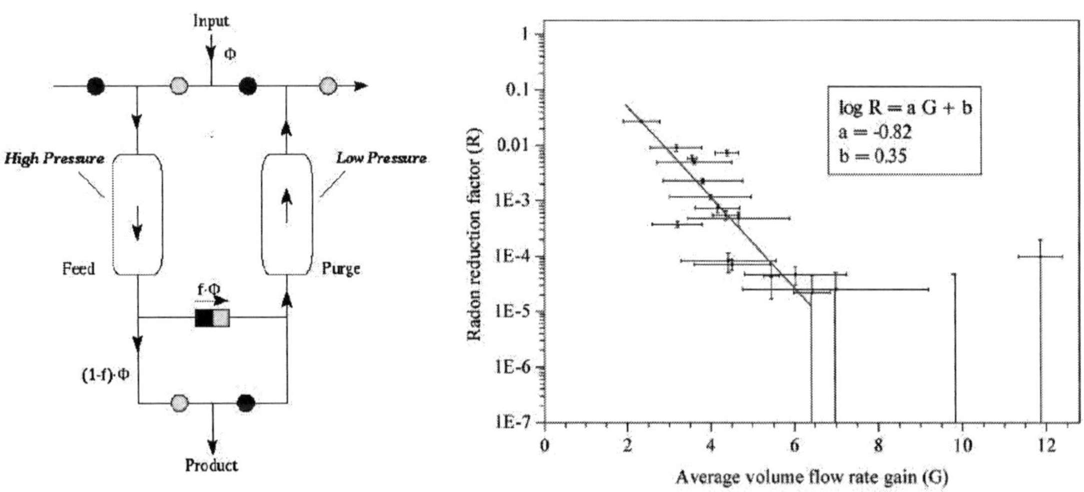

FIGURE 7. Schematic representation of a pressure-swing system. A fraction f of the clean product Φ from the feed column is expanded and sent through the other column in counterflow. Light an dark dots represent open and closed valves. **Right**: radon reductions (R) obtained with a small-scale VSA system on activated charcoal. Data are obtained contaminating the input air stream with a radon source. The line represents the best fit to the data (that show a non-negligible spread). The gain G is expressed as an average to account for pressure drops that are more significant across the purged column.

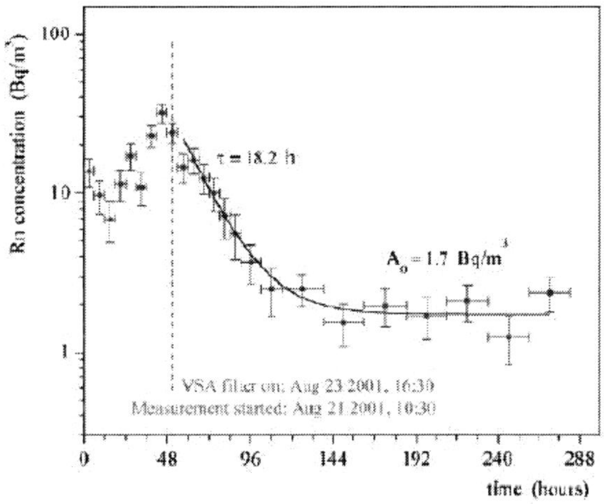

FIGURE 8. Radon concentration inside the Borexino clean room before and after the radon filter was turned on. The radon activity rapidly decreases at first, settling at a value ~1.5 Bq/m^3. This value is determined by radon emanation inside the clean room, since the radon activity at the output of the filter was measured to be at least one order of magnitude smaller.

front moves along the column with a velocity proportional to the *volume* flow rate. Ideally then, if the purge *mass* flow rate is a fraction f of the feed *mass* flow rate, purging must be carried out at a pressure $P_{\text{purge}} < fP_{\text{feed}}$ in order for the purge volume flow rate to be greater than the feed volume flow rate. Practically, radon doesn't move as a step-like front along charcoal columns but displays very pronounced tails in the concentration profile [8]; the purge pressure must then be set significantly lower than the ideal value. The performance of a filter can be described as a function of a quantity G (gain) defined as

$$G = \Phi_{\text{purge}}/\Phi_{\text{feed}},$$

where Φ_i (i = purge, feed) represents the total volume flow rate through the column (proportional to the inverse of the pressure for a given mass flow rate). The bigger this ratio, the stronger the purge. It is clear how a VSA system can easily achieve very big values of G with small purge fractions f by purging at low pressure.

Preliminary results. Initially, a prototype PSA charcoal filter, operated between 1 and 8 bar, was run with a radon source at the inlet; it was able to significantly reduce the radon concentration at the output only with very high purge fractions (>70%), proving extremely inefficient. It was found that nitrogen (the carrier gas and hence present in an essentially infinite amount in the system) would saturate the adsorption sites on the charcoal. This behaviour, far from linear between 1 and 8 bar, would make radon adsorption in this pressure range highly non-linear and explain the inefficiency of the device [8]. Vacuum-swing adsorption was then investigated to overcome this problem.

Studies performed on a prototype VSA device (~1.5 m^3/h of inlet air, 1.35 kg of charcoal per column) using a radon source at the input have, on the other hand, demonstrated radon reduction factors in excess of 10^4 with limited purge fractions (<10%). The device was operated at various gains and operating pressures. Results for radon reduction are shown in fig. 7 (right): the data, characterized by a non-negligible spread due to uncertainties in flow rates and pressure drops, show a clear trend in the radon reduction capability as a function of gain as:

$$\log R = -0.82G + 0.35$$

where R is the ratio of the radon concentration at the output to that at the input of the filter. The plot shows that, for example, a gain of ~4 is needed for a 1000-fold radon reduction. Details of the device and extensive experimental results are found in [8].

Performance of the full-scale filter. The full-scale VSA radon filter (fig. 6) was operated continuously during the year of vessel production (Aug. 2001 - Jul. 2002). It consisted of two charcoal columns with ~250 kg of activated charcoal each. It employed a 500 m^3/h vacuum for the purging stage, composed of a roots blower and a rotary vane roughing pump. The inlet flow rate (dried before it reached the charcoal columns) was ~100 m^3/h and the typical purge flow rate and pressure ~12 m^3 and ~20 mbar respectively. The gain G was about 4, yielding an the expected

FIGURE 9. The Borexino nylon vessels installed and inflated inside the SSS. Ultra-low radon synthetic air carried via electroplated stainless steel pipes was used. The particulate level was better than class 1 and the radon activity was continuously monitored during inflation and was below the sensitivity limit of the detector employed of 1 mBq/m^3 [14].

radon reduction $\sim$1000 when scaling up the prototype results. These figures strongly depend on details of the filter such as evacuation and re-pressurization times of the columns, pressure drops across them, and residual water in the input flow, all of which would make the filter less efficient. On the other hand, the amount of charcoal is proportionally more than in the small-scale prototype, thus compensating for some of these unavoidable inefficiencies. Details of the filter design and performance are found in [8].

The initial exponential radon reduction inside the Borexino clean room is shown in fig. 8. The figure also shows that the radon concentration reaches a plateau at $\sim$1.5 Bq/m^3. The radon concentration during the year was consistently 1-2 Bq/m^3; on the other hand, the radon concentration measured at the output of the VSA filter before the clean room was at least one order of magnitude smaller, at $\sim$0.1 Bq/m^3. This is only an upper limit since it is the sensitivity limit of the radon detector used for the measurement (Durridge RAD7 flow-through electrostatic alpha counter) and is hence compatible with the expected radon reduction stated above. The origin of the radon inside the clean room that limits the filter's performance is still unclear, and seems to be a recurring problem in radon-suppressed clean rooms. Small leaks ($\sim$ 5-10 m^3) into the clean room from the outside, maybe in places of the room with the lowest over-pressure (such as the air handling regions with high velocity flow) could explain the results. Another possible explanation is radon emanation from the HEPA filter material: this could be due to intrinsic uranium contamination of the material (that displays a very large surface area) or to dust and particulate trapped inside the filters following startup and long-term operation. In any case, it is certainly an issue worth looking into with care for future experiments; measuring the intrinsic radioactivity (specifically ^{238}U) of HEPA filter materials could give useful insight for future applications.

VESSEL INSTALLATION AND INFLATION

The Borexino nylon vessels were shipped from Princeton in Aug. 2002. Their installation, greatly delayed by a partial shut down of operations at the Gran Sasso laboratory following a scintillator spill, began in January 2004. During this time, the vessels were kept wrapped, as shown in fig. 5 (right). The tight self-covering folds together with the presence of several gallons of water to keep the film moist have certainly provided a barrier against radon contamination (radon solubility in water is rather small). The extent of such protection is hard to further quantify accurately.

The Borexino SSS was operated as a class 10000 clean room until it was finally sealed (after vessel installation). The access to it was possible through a class 100 clean room, where the vessel system was partially unwrapped, hung to a horizontal rail, and carried into the SSS. It was then stretched inside the SSS by lifting it from the top polar tube assemblies. The excess film (the length of the folded vessels is $\pi/2$ times the radius of the OV) was supported on a platform mounted on a 5 m high scaffolding tower built with extremely clean material. This lower section of the vessel system was kept wrapped to prevent it from opening up and drag on the bottom of the SSS.

The two vessels were then inflated into their final spherical shape with ultra-low radon synthetic air (air instead of nitrogen was used for safety of the personnel working inside the SSS in the event of a leak from the vessels). The air was made by mixing ultra-low radon boil-off nitrogen and oxygen stored in stainless steel bottles and aged for many months underground. The synthetic air was humidified to 60% RH with a membrane inline humidifier and fed into the vessels from the top through electroplated stainless steel pipes, part of the final Borexino fluid handling manifold. The lines were thoroughly purged with clean high pressure nitrogen until the particle count of the output flow was better than class 1. During vessel inflation the radon concentration of the synthetic air was continuously monitored using a large flow-through electrostatic chamber equipped with a solid state α-counter [14]. The readings were always below the sensitivity limit of the detector of 1 mBq/m^3. SF_6 gas was added to the inlet air stream into the IV at $\sim$1% concentration for subsequent vessel leak checking. Once inflated, though, the most sensitive leak test for both IV and OV turned out to be given by direct measurement of the differential pressures between IV, OV, and OB, recorded with great accuracy by the pressure gauges installed in the Borexino filling lines. Gas leak rate limits of $\sim$0.5 cc/s for the IV and a leak rate $\sim$ 10 cc/s were achieved. These limits are better than the IV and OV design specification of 1 cc/s and 100 cc/s gas leak rates at 1 mbar over-pressure (equivalent to $\sim$0.01 and 1 cc/s liquid leak rate when the liquid to gas viscosity ratios are taken into account). A picture of the Borexino vessels installed and inflated inside the SSS, taken with one of the 7 fixed cameras of the experiment, is shown in fig. 9.

WATER FILLING THE DETECTOR PRIOR TO SCINTILLATOR FILLING

It is foreseen to fill the Borexino detector with ultra-pure water prior to scintillator filling in order to test the Borexino filling plants. This will also provide a powerful rinsing step for all the volumes that will be in contact with the scintillator. In particular, water has been proven to be very effective at removing ^{210}Pb from nylon surfaces. This is very important in light of possible residual radon exposure the vessels, especially the IV, could have experienced (e.g. during the long storage at Gran Sasso).

Square nylon film samples (side of 4 cm) were exposed to radon-loaded air from a flow-through source for a period of $\sim$ 2 months. The samples were placed inside a stainless steel chamber (volume $\sim$10 litres) and laid out on 6 perforated steel shelves. The air was filtered by a clean room grade filter to mimic the particulate contamination during vessel fabrication. The nylon samples were cut out from the left over material used for the IV (Sniamid, see previous sections). The 46 keV γ-ray activity from ^{210}Pb, as well as the 5.3 MeV α activity from ^{210}Po were measured with an intrinsic germanium and a silicon surface barrier detector respectively. The activities of both isotopes were measured just before and just after the samples were immersed for a given time (10 minutes to 10 days) in DI water (1 litre, $\rho >$ 16 MΩ·cm). One sample was spray-washed for $\sim$15 minutes with running DI water. The ^{210}Po activity, that had not reached equilibrium yet with ^{210}Pb due to its long 200-day lifetime, was measured a few times at different times prior to water soaking to account for the buildup rate. The main results from these measurements were:

a. essentially all ^{210}Pb is removed after 1 day of soaking. A 10-minute soak washes off $\sim$75% of the lead. There is some indication that 5-10% of the lead is not removed even after 10 days, but this result has a large error due to the very low gamma activity of the samples;
b. ^{210}Po is released much more slowly than lead. Even after 10 days, approximately 20% of the activity is still on the surface (see fig. 10, right).
c. there was a significant asymmetry of activity between the two sides of the samples. The side directly facing the input flow inside the exposure chamber had a much higher (10-100 times) polonium activity than the other, facing the perforated metal shelves. It was not possible to directly measure this asymmetry in the lead activity since the gamma rays would go through the 125 μm sample.

These results strongly support water filling the Borexino detector. Soaking the vessels in ultra-pure DI water for a period of the order of one month would dramatically reduce the residual ^{210}Pb and ^{210}Po contamination on their surface. The real need for such procedure depends on the actual radon exposure of the vessels during fabrication and installation and on the dynamics of lead and polonium release from nylon surfaces into the organic scintillator. In any case, an aggressive spray-wash of the IV, which was considered as an option, is proven unnecessary. Other interesting information can be extracted from the data. The striking asymmetry in ^{210}Po activity is presumably related to the lead deposition mechanism. There could be a correlation between the observed residual lead activity and the polonium

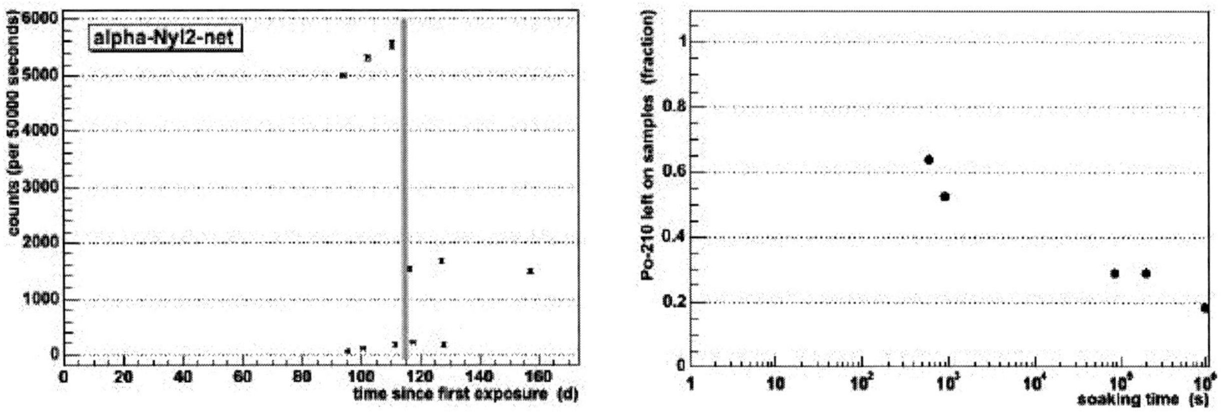

FIGURE 10. ^{210}Po removal from nylon surfaces following water soaking. Left: ^{210}Po surface activity for both surfaces of one of the samples. The vertical band is the time the sample has been immersed in DI water. The large exposure asymmetry is clearly visible in the difference in activity. It is interesting to notice that the polonium on the low activity side is essentially unchanged after immersion. Right: fraction of polonium remaining on the surface as a function of soaking time in DI water. The point around 1000 seconds corresponds to the 15-minute DI water spray-wash, similar in effect to a soak in static DI water of about twice the duration. The ^{210}Po desorption depends on the logarithm of time; the same behaviour, with a different time constant, has also been observed in the release of ^{210}Pb from stainless steel in water [9].

activity on the low activity side of the samples, unchanged by soaking in DI water, that could be due to different levels of isotope implantation in the material. Further studies of this sort could help better understand radon daughter deposition on surfaces in clean room environments that could be very valuable for many low background experiments.

ACKNOWLEDGMENTS

The results reported in this paper were possible thanks to many years of R&D carried out within the Princeton/Borexino group. The installation and inflation phase at Gran Sasso saw the participation of many other members of the Borexino collaboration who contributed to make the commissioning of the nylon vessels successful. Some crucial low radioactivity measurement of the film were performed by the Heidelberg/Borexino group, also responsible for the supply of ultra-low radon synthetic air and nitrogen at LNGS.

REFERENCES

1. Borexino Collaboration, G. Alimonti et al., *Astropart. Phys*, **16**, 205–234 (2002).
2. C. G. Rothschild, F. P. Calaprice, and M. C. Chen, *Geophys. Res. Lett.*, **25**, 1083–1086 (1998).
3. L. Cadonati, F. P. Calaprice, and M. C. Chen, *Astropart. Phys.*, **16**, 361–372 (2002).
4. L. Cadonati, *The Borexino Solar Neutrino Experiment and its Scintillator Containment Vessel*, Ph.D. thesis, Princeton University, Princeton, NJ 08544 (2001).
5. M. Wójcik et al., *Nucl. Instr. Meth. A*, **449**, 158–171 (2000).
6. M. Wójcik et al., *Nucl. Instr. Meth. A*, **524**, 355–365 (2004).
7. The Borexino Collaboration, C. Arpesella et al., *Astropart. Phys*, **18**, 1–25 (2002).
8. A. Pocar, *Low Background Techniques and Experimental Challenges for Borexino and its Nylon Vessels*, Ph.D. thesis, Princeton University, Princeton, NJ 08544 (2003).
9. M. Leung, "Surface Contamination from Radon Progeny," in *Low Radioactivity Techniques 2004 Conference*, December 2004, Sudbury, ON, Canada, these proceedings, 2005.
10. Military Standard 1246C. Product Cleanliness Levels and Contamination Control Program, Tech. rep., Department of Defense (1994).
11. G. Zuzel et al., *Nucl. Instr. Meth. A*, **498**, 240–255 (2003).
12. D. M. Ruthven, S. Farooq, and K. S. Knabel, *Pressure Swing Adsorption*, VCH Publishers, 1994.
13. R. T. Yang, *Gas Separation by Adsorption Processes*, Butterworths Series in Chemical Engineering, 1987.
14. J. Kiko, *Nucl. Instr. Meth. A*, **460**, 272–277 (2001).

Clean Fabrication and Cleanliness Monitoring in SNO

E. D. Hallman

Department of Physics & Astronomy, Laurentian University, Sudbury, ON Canada P3E 2C6
for the SNO Collaboration

Abstract. Cleanliness during the construction and operations of the Sudbury Neutrino Observatory has been an essential part of SNO's successful achievement of the required low backgrounds for its solar neutrino measurements. The setting up of the laboratory as a clean room, some aspects of clean construction and the monitoring systems used are described.

Keywords: clean room, surface dust levels, dust deposition, x-ray fluorescence.
PACS: 26.65.+t, 82.80.Ej

1. INTRODUCTION

In order to achieve a measurement threshold of ≤ 5 MeV for solar neutrino detection in the Sudbury Neutrino Observatory and a precision measurement of SNO's neutral current signal, targets for material purity, radon reduction and surface cleanliness during construction were set[1]. An extensive program was set up to measure the radioisotope impurities in candidate detector materials, and to monitor the actual impurities in detector materials and components. Techniques including direct γ-ray counting, neutron activation analysis, radon emanation and radium leaching measurements were used. Many components were tested and assembled in clean room facilities at SNO institution laboratories. During the construction of the laboratory, the assembly of the phototube sphere and acrylic vessel for SNO were complex tasks requiring many months of work. To ensure that targets for cleanliness were met, the whole underground laboratory was made a clean room facility. Clean materials and equipment were delivered in sealed containers or cleaned in the carwash at the laboratory entry. Personnel were required to take showers and change to clean room clothing as they entered the laboratory. Characteristics of the laboratory clean rooms, procedures used and cleanliness monitoring implemented are described below.

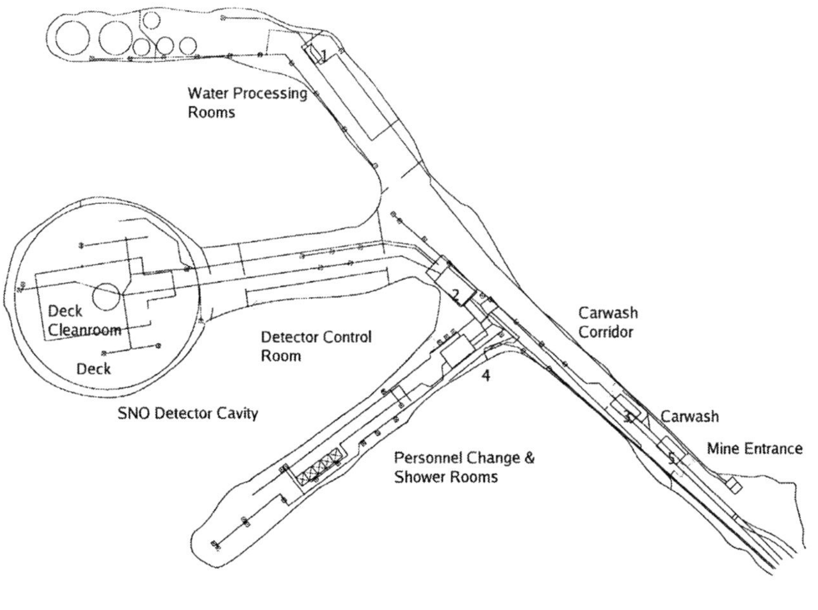

FIGURE 1. The Sudbury Neutrino Observatory room layouts showing laboratory areas and the air handlers (#1 to 5) and ventilation ducts for clean room operations.

CP 785, *Topical Workshop on Low Radioactivity Techniques,* edited by B. Cleveland, R. Ford, and M. Chen

2. Establishing and Maintaining Clean Room Conditions

The SNO laboratory (Figure 1) was designed to be isolated from the normal mining environment of the host Inco Creighton Mine through a specially-designed carwash room where delivery containers can be isolated and cleaned before the clean material or equipment inside is delivered to the laboratory. Larger pieces of equipment or materials delivered underground can be cleaned in the carwash before they are moved inside. Personnel entering the laboratory wash footwear at the outer entry door, then remove footwear and hardhats inside. Mine (dirty) and laboratory (clean) change rooms and showers for men (main floor) and women (mezzanine level) in the personnel area are used by all personnel on entry. A lunch/meeting room in the clean section of this area is also a mine refuge station for emergencies or fire drills; it can be isolated from the mine, is equipped with a compressed fresh air supply, water and communications. Fresh air from the mine ventilation system is supplied through an entry duct to an air handler (AHU 05) which provides filtration and cooling.

More than 10 years of successful clean room operations in the SNO laboratory are a result of a program which includes:

- making cleanliness a very high priority for personnel during entry, housekeeping and laboratory operations.
- wherever possible, delivering clean materials in sealed "blue box" containers to the lab and monitoring the cleanliness before use in the laboratory or detector.
- maintaining cleanliness during the detector operations phase by constant monitoring and regular air systems maintenance. Materials and equipment to be placed in the detector are checked for surface dust levels.
- providing extra isolation for critical areas – for example the *deck clean room* provides restricted personnel entry and extra air recirculation to lower air particulate levels and dust deposition near the neck entry for the acrylic vessel and the heavy water core.

Air circulation, cooling and filtration is provided by a system of five air handler units (AHU) and exhaust fans, as summarized in Table 1. Doors between laboratory areas (deck clean room, deck above the detector, control room, junction area, water processing room, entry corridor, personnel area, car wash, laboratory entry corridor and mine passageway) control air flow and air pressure differentials ($\Delta P \sim$ several inches of water) are set up so that any leakage flow is directed from the cleanest areas (deck clean room) to the less clean and entry areas. Cooling is provided by a dual chiller unit located in the mine entry corridor 300 m from the laboratory; it supplies chilled water to cool air (to 20° C) and the light water of the detector (to 10° C) to counter heat flow from the surrounding rock which has a core temperature of 41° C at this depth in the mine.

TABLE 1. Air circulation and filtration in the SNO Laboratory

Air Handling Unit	Air flow (cfm)	Filter(s) installed	Room Volume (cu.ft.)	Air changes/hr (ACH)
AHU 01 – water processing room	13,300	90% ASRAE	66,000	12.0
AHU 02 – deck, control room junction area (during operations)	16,000 (with 25 HP fan)	Activated charcoal 90% ASRAE 99.7% HEPA	100,000	9.6
AHU 02 – deck, control room junction area, cavity (during construction)	40,000 (with 125 HP fan) 16,000 (portable fan/filter units)	Activated charcoal 90% ASRAE 99.7% HEPA 99.7% HEPA	350,000	9.6
AHU 03 – personnel room	7,200	90% ASRAE	44,000	9.8
AHU 04 – Carwash, entry area	4,600	90% ASRAE	20,000	13.8
AHU 05 – Lab air intake	6,700 operations	20, 30% preliminary 60, 90% ASRAE 99.7% HEPA	Fresh air supply from mine	~ 1 hour for complete air change
Exhaust fan 1	1,640 operations	From personnel area		
Exhaust fan 4	1,290 operations 500 through door	From car wash		
Exhaust fan 5	1,740	From water processing room		
TOTAL / AVERAGE			230,000	10.7

The clean room air quality and dust deposition levels maintained in the separate parts of the SNO laboratory are outlined in Table 2. Accumulated dust deposition is measured by a set of witness foils placed in separate sections of the laboratory; experience has shown that reading out the accumulated deposition monthly gives a satisfactory measure of lab cleanliness. In the deck clean room, a continuous measurement of air particle concentrations is carried out. These levels change with the personnel activity and specific operations and extra monitoring is performed during critical or unusual operations periods.

TABLE 2. Air quality parameters in the SNO Laboratory

Location	Air Particulate Concentration (CLASS)	Dust Deposition Rate ($\mu g/cm^2/month$)
Mine passageway	~ 1,000,000	-
Lab entry & carwash	5,000	≤ 1.0
Corridor junction area	2,000	≤ 0.1
Deck clean room	200	≤ 0.05

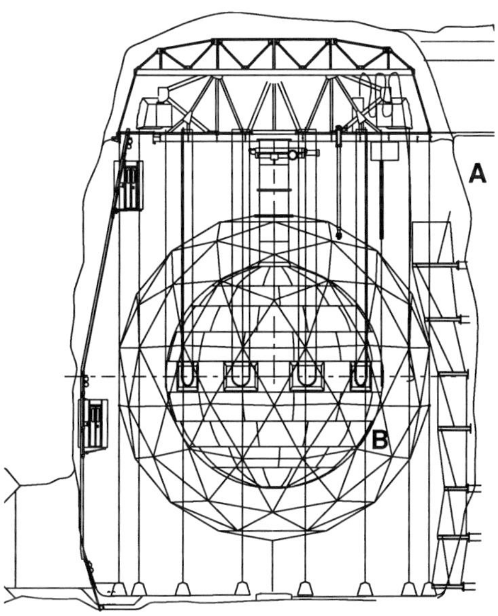

FIGURE 2. The Sudbury Neutrino Observatory water Cerenkov detector is located in a 22 m diameter by 30 m high cavity, 2000 m underground at the Inco Creighton Mine. The central acrylic sphere holds 1000 tonnes of heavy water (D_2O). The elevator (at left) and ladderway (at right) were removed at the end of the assembly phase.

The SNO Laboratory clean room systems were installed in 1994 after the detector cavity had been coated with a 8 mm thick polyurethane liner (sprayed on the smoothed concrete layer covering the rock walls - see Figure 2 A) and laboratory surfaces had been coated with epoxy paint (over the *shotcrete* sprayed concrete surfaces on walls and poured concrete floors). Ducts and air handlers were delivered clean to the laboratory, installed and then cleaned *in situ* where necessary, after an entry air filtration unit was put in operation to provide cleaner fresh air. Initially, only the coarser filter units were operated, with the high efficiency HEPA units installed last. A cleaning program for floors and surfaces was set up (using *swipe* testing, described in section 4 below, as a fast measure of satisfactory cleaning) and personnel first changed into clean safety shoes, later into clean room coveralls, and finally, the entry protocol with showers and clothing change was implemented by the fall of 1994. Materials were delivered in a clean state to the laboratory in large sealed containers mounted on mine rail system flatcars, which fit, with minimal clearance, in the mine elevator.

3. Detector Assembly Schedule and Critical Operations

Major components of the SNO heavy water detector were assembled under clean room conditions so that minimal mine (or concrete) dust would be present in the final assembly. A target of ≤ 50 g of *equivalent mine dust* (defined below) was set for SNO cavity and ≤ 2.2 g for the acrylic sphere and chimney surfaces (see Figure 2 B);

this target was set by evaluating the background resulting from this amount of dust (using the known trace concentrations of uranium and thorium for average dust in the lab entry region), and comparing it to other estimated background contributions[1]. The assembly of the photomultiplier tube support sphere (PSUP) and panels, and the bonding of the 120 curved acrylic panels for the heavy water vessel took place over many months (see the schedule outline in Table 3). Clean room conditions were maintained and monitored in the cavity, and, wherever possible, dust covers were used to protect clean components and surfaces as the assembly proceeded.

TABLE 3. SNO Detector Assembly Schedule

Assembly Operation	Start Date	Completion Date
Clean assembly and operations	November 1994	-
Upper section – phototube (PSUP) sphere	November 1994	August 1995
Acrylic vessel and chimney	August 1995	November 1997
Lower section – phototube (PSUP) sphere	September 1997	January 1998
Phototube cabling, ladder and elevator removal, ramp entry bulkhead, liner completion and final cleaning	January 1998	April 1998
Water filling (H_2O and D_2O) & detector commissioning	April 1998	May 1999
Start of neutrino measurements	May 1999	

Critical operations during this assembly period were closely monitored and special cleaning was carried out as needed. These included the mounting of the photomultiplier tube panels (previously assembled in a SNO clean assembly facility not far from the mine), the bonding of the acrylic vessel panels (more than 1 km total bond length), the smoothing and polishing of vessel joints, the installation of the acrylic vessel chimney and fittings and the installation of a co-axial cable for each of the 9600 photomultiplier tubes. The production or selection of ultra pure materials (sometimes including special high purity production runs) for detector components and their pre-assembly in other clean room facilities was a major effort of the SNO collaborating institutions critical to the success of the detector, but this extensive work cannot be summarized in this review.

4. SNO's Unique Dust Monitoring Program

Given the special requirements for cleanliness during the several year assembly period for the detector, a monitoring program to assess surface cleanliness and deposition rates for dust in the laboratory air was required. The very low levels of dust which had to be measured precluded the use of standard gravimetric methods. An x-ray fluorescence (XRF) method was developed by Robert Stokstad and collaborators at the Lawrence Berkeley National Laboratory. The method is based on the observation that finely divided material in the mine passages near the laboratory contains about 7 % iron by weight. The entry of this dust into the laboratory is a critical background concern (it contains, on average, about 1.2 x 10^{-6} g/g ^{238}U and about 5.5 x 10^{-6} g/g ^{232}Th). The presence of an iron peak in a dust deposition x-ray spectrum can be taken as an indication that mine dust is possibly present, and further cleaning can be carried out until this signal is not seen or minimized. The *equivalent mine dust* level is simply the iron concentration measured multiplied by 1/0.07.

A small XRF spectrometer, with a molybdenum target x-ray tube and silicon detector, based on a design used for environmental air particulate monitoring, was built for this work and installed in the SNO laboratory. For a 20 minute count, iron levels on plastic witness foils could be detected down to about 0.007 μg/cm^2 corresponding to mine dust deposition on the foil of 0.1 μg/cm^2. To test surface cleanliness, thin adhesive tape was used to lift any dust from the surface. By using one tape to lift five different dust samples, levels of mine dust down to 0.015 μg/cm^2 can be detected. Other elements are sometimes seen in air deposition and tape lift spectra. For example, during construction, zinc from galvanized scaffolding in the cavity was usually seen and for a tape lift from a stainless steel surface, Ni and Cr as well as Fe can be detected.

During the detector assembly, a set of witness foils in the cavity were regularly monitored. During periods of extensive assembly activity, dust deposition rates of about 1-2 μg/cm^2/month were measured, while for inactive periods, deposition rates dropped to about 0.1 μg/cm²/month. Laser particle counter measurements showed air particulate levels up to CLASS 10,000 (particles of diameter ≥ 0.5 μm per cu.ft. (28 L)), with average levels of CLASS 2500 ± 500 during light construction activities. The deposition measurements in the SNO cavity were used to estimate possible dust accumulation on acrylic panel edges (and in the bonds) as the panels were aligned and bonded. The tape lift program was used to ensure that the assembled detector components had acceptable final surface cleanliness.

For less critical surfaces in the laboratory (ducts, floors, walls), surface cleanliness was assessed by means of a *swipe test*. A white (or black) cloth strip was mounted over the edge of a rubber eraser and wiped along a 15 cm section of the surface. A line on the cloth from any dust present could be compared to standard swipes and the presence of dust above a level of 0.5 μg/cm² could be quite reliably determined. The test was used frequently during construction and is used periodically during the operations phase to assess cleaning operations in the laboratory.

5. Examples of Cleanliness Monitoring During Detector Assembly and Operations

(a) Accumulated Dust Deposition In The Laboratory

Witness foils located at standard sites in the different parts of the laboratory are read out periodically to give an integrated measure of total dust deposition over time. Figure 3 shows the results of XRF measurements on two foils located at the same site in the car wash corridor, during the detector operations phase (no heavy construction). All materials and equipment are moved from the carwash down this corridor into the laboratory, so this site monitors the lab entry air cleanliness over time. Figure 3(a) shows mine dust (in μg/cm²) as a function of time (in months). Both foils indicate an average accumulation of about 0.10 ± 0.03 μg/cm²/month in this area. Figure 3(b) shows zinc accumulation on the same foils (0.06 ± 0.02 μg/cm²/month). Traces of other metals including copper can also be seen.

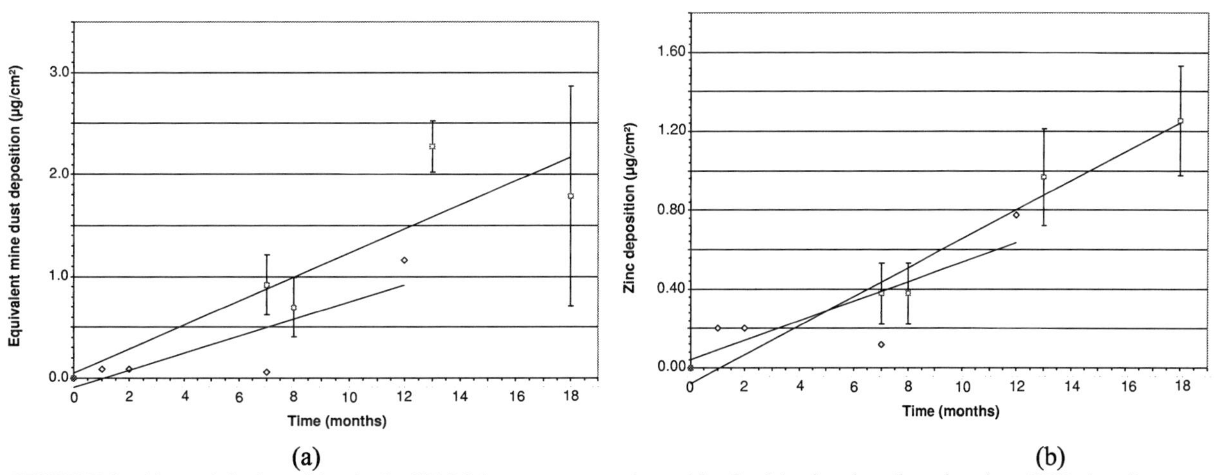

FIGURE 3. Air particle deposition in the SNO laboratory carwash corridor for (a) mine dust (based on iron XRF signal) and (b) zinc.

(b) Final Cleanliness Of Detector Components After Assembly

After the 12 meter diameter acrylic vessel had been assembled and cleaned, a tape lift survey was carried out to ensure that the acrylic surfaces met a cleanliness criterion of ≤ 0.10 μg/cm² mine dust. For 34 tape lift samples from the vessel interior, mine dust concentrations averaged 0.007 ± 0.039 μg/cm², while for 35 samples from the vessel exterior, concentrations averaged 0.030 ± 0.034 μg/cm². For the total acrylic vessel surface area of approximately 1000 m², an upper limit of 0.55 g of surface mine dust was obtained, well below the target.

(c) Examples Of Routine Monitoring During The SNO Detector Operations Phase

Detector calibration devices are routinely inserted into the D_2O through the neck in the acrylic vessel, every few weeks, to assess the detector performance. The cleanliness of the equipment inserted is critical, and a tape lift program has been established to provide a final cleanliness test prior to deployment.

Umbilical cables provide electrical power and/or fiber optic connections to calibration devices. Tape lifts are used when a new cord is installed or major changes are made to an existing cord. Further cleaning is performed if tests show a mine dust level ≥ 0.50 μg/cm^2. In one test, initial tape lift tests showed a level of 0.95 μg/cm^2 for an umbilical cord which was to be deployed. Following a wash and rinse of the cord, final tape lifts showed a surface dust level of 0.25 μg/cm^2, and the cord was cleared for use.

Stainless steel (SS) components are also tested prior to use in the D_2O. Typically some SS particles are lifted by the tape from machined surfaces; the x-ray spectrum shows typically 60% Fe, 30% Cr and 10% Ni signals. A Fe signal in excess of this composition is attributed to mine dust.

In the third phase of SNO's solar neutrino measurements (begun in late 2004), vapor-deposited high purity nickel *Neutral Current (neutron) Detector* tubes were deployed in a vertical array in the D_2O. Tape lifts of these microscopically-textured surfaces show a nickel signal at the 0.2 to 0.8 μg/cm^2 level. Possible mine dust can still be detected (using the Fe signal) down to levels of ≤ 0.1 μg/cm^2.

6. Clean Room Facilities in SNOLAB

The expansion of the SNO laboratory currently underway, to accommodate new international experiments in neutrino physics, dark matter detection and particle astrophysics incorporates SNO's successful approach to cleanliness.

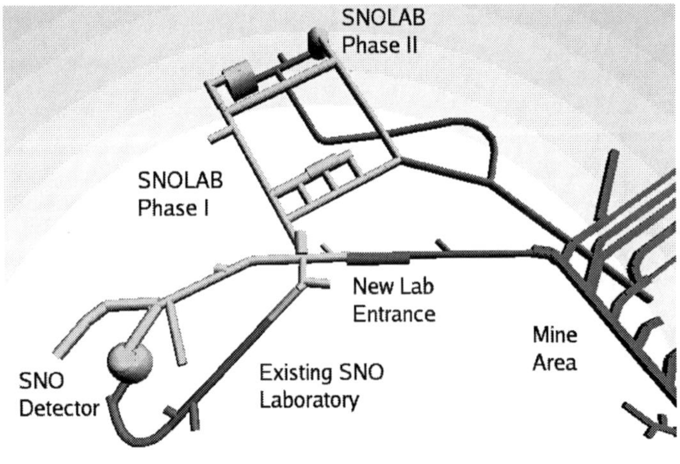

FIGURE 4. The SNOLAB facility, currently under development adjacent to the existing SNO Laboratory. The expanded clean room areas are shown in light gray.

Figure 4 shows the sections being added, with a new laboratory entrance including new showers, change rooms and meeting area for personnel, and the new experimental halls and connecting passageways. Clean room wall and floor coatings will be installed throughout, and a second air handling system will provide cooled and filtered air to each section of the new facility. As in the SNO laboratory, a system of doors and pressure differentials will establish air flows from the high cleanliness experiment areas into the corridors and towards the laboratory entrance. Air quality targets of $\leq$ CLASS 2,500 have been set for the facility as a whole (as achieved in the SNO Laboratory), with the option of additional recirculating filter installation to give higher cleanliness in local areas as necessary.

A new SNOLAB surface building (Figure 5) to be completed in July 2005, will provide administrative offices, meeting and change rooms, a group of clean room development laboratories, a clean room chemistry laboratory and a clean assembly area, in support of the underground experiments. The development and chemistry laboratories will have air change rates of 10/hour, and a target air quality of ~ CLASS 10,000, while the clean assembly area will have 30 air changes per hour and a target air quality of CLASS 2000-3000. Additional recirculating filter units can be used to enhance these levels as needed.

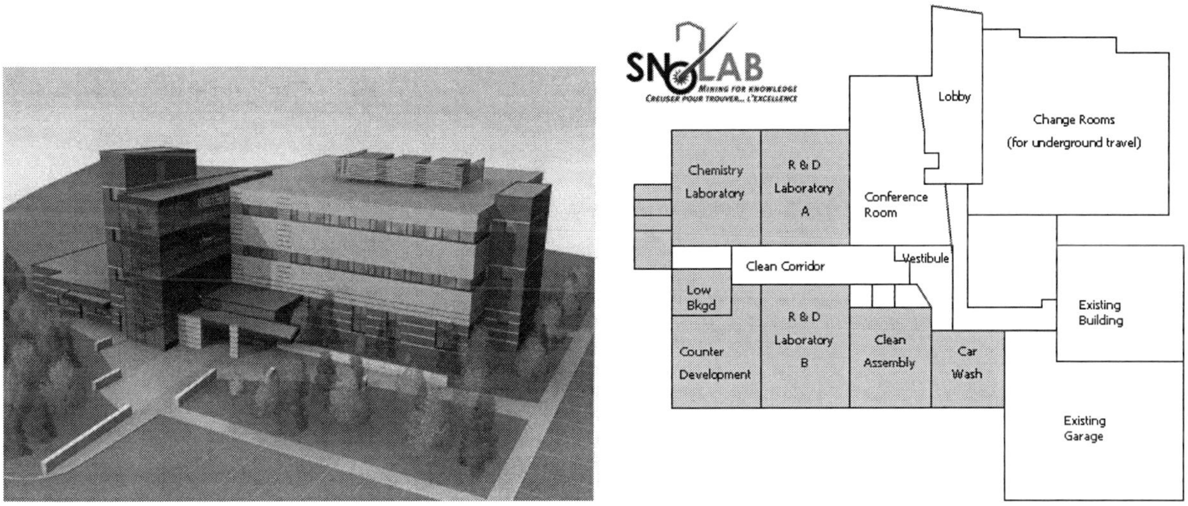

FIGURE 5. The SNOLAB Surface Building and a layout view of the experiment development and assembly clean room areas.

7. Conclusions and Future Outlook

The high priority given to the establishment of clean room conditions, the monitoring of surface and airborne dusts during construction and operations, and the protocols developed for personnel entry and activity in the SNO laboratory have led to unprecedented low backgrounds and a vital contribution to the success of SNO's solar neutrino measurements. The development of unique monitoring and operations systems has many applications in further experiments in which cleanliness will be even more critical. It is clear from SNO's experience that options for even higher cleanliness can include, for example, special clean room areas within the laboratory (e.g. the SNO deck clean room) in which dust deposition can be reduced by an order of magnitude, and further isolation from underground environments achieved. Undoubtedly, these specialty clean rooms will become more necessary and common in the future.

Where a suitable element tracer is available, the XRF dust monitoring technique developed at SNO can be extended to provide *in situ* measurements of surface dusts down to possibly 10 ng/cm^2 levels.

The increasing interest in measuring and controlling the deposition of radionuclides such as ^{210}Pb will necessitate more extensive cleaning and wrapping/protection steps for detector components and more extensive use of radon-reduced environments.

I acknowledge the work of a dedicated team of SNO scientists and laboratory employees in embracing a strong cleanliness philosophy and priority, in implementing and maintaining protocols for this over more than a decade of clean room operations, and in developing novel methods of monitoring and maintaining clean conditions. In particular, Robert Stokstad developed much of the cleanliness protocol as well as the XRF monitoring tool, and Hugh Evans, Peter Skensved, David Schroeder and Carol Woodliffe were key contributors to the monitoring and cleanliness maintenance program.

REFERENCES

1. SNO Collaboration, "Sudbury Neutrino Observatory Proposal", Report SNO-87-12, October 1987
2. J. Boger *et al*, "The Sudbury Neutrino Observatory", Nuclear Instruments and Methods A 449, 2000, pp 172-207.

Ultra-Low-Background Copper Production and Clean Fabrication

C.E. Aalseth, R.L. Brodzinski, O.T. Farmer, E.W. Hoppe, T.W. Hossbach, H.S. Miley

Pacific Northwest National Laboratory
902 Battelle Boulevard, Richland, WA 99354, USA

J.H. Reeves

Reeves & Sons, LLC
216 Abert, Richland, WA 99352, USA

Abstract. Commercial high-purity copper is an attractive material for constructing ultra-low-background radiation measurement devices. When even higher purity is desired, additional electrolytic and chemical purification can be combined with the final fabrication step. This process results in "electroformed" copper parts of extreme purity. Copper electroforming can be done underground, providing a way to eliminate cosmogenic activation products seen in copper that has had above-ground exposure. A brief summary of the history, cosmogenics, process chemistry, cleaning, and passivation of this material is given. Examples of finished parts illustrate the method. The required infrastructure is summarized.

Keywords: Material purity cosmogenic electroforming clean spectrometer underground physics.
PACS: R23.40.-s, 14.60.Pq

HISTORY AND MOTIVATION

Commercial high-purity copper is an attractive material for constructing ultra-low-background radiation detectors. Intrinsically low in primordial radioisotopes (U, Th) and having no long-lived radioisotopes itself, copper benefits from the "electrowinning" process used during refinement. This processing step electroplates the material from a sulfate solution onto cathodes. Most impurities do not follow this electrochemical step, and the resulting material, known as "electrolytic tough pitch" copper, is rather pure. Further processing, including smelting and hot rolling, results in copper of excellent radiopurity and with the attractive mechanical, electrical, and thermal properties for which copper is known.

When even higher purity is required, additional electrolytic and chemical purification can be combined with the final fabrication step, resulting in "electroformed" copper parts of extreme purity. This electroforming process can be done underground, providing a potential way to eliminate cosmogenic activation products seen in copper with above-ground exposure, e.g. ^{60}Co.

The low-background spectrometer history leading to the development of electroformed copper for this application is illustrated in Figure 1. The graph shows downward evolution of background over fifteen years of radioassay and double-beta decay research at PNNL. By the end of this period, cosmogenic backgrounds were recognized as being limiting factors. Electroformed copper had emerged as a material-of-choice for purity, as it allowed copper cosmogenic isotopes to be removed and its production technique allowed components of reduced mass. Early data showing spectroscopic identification of cosmogenic radioisotopes is shown in Figure 2.

CP 785, *Topical Workshop on Low Radioactivity Techniques,* edited by B. Cleveland, R. Ford, and M. Chen

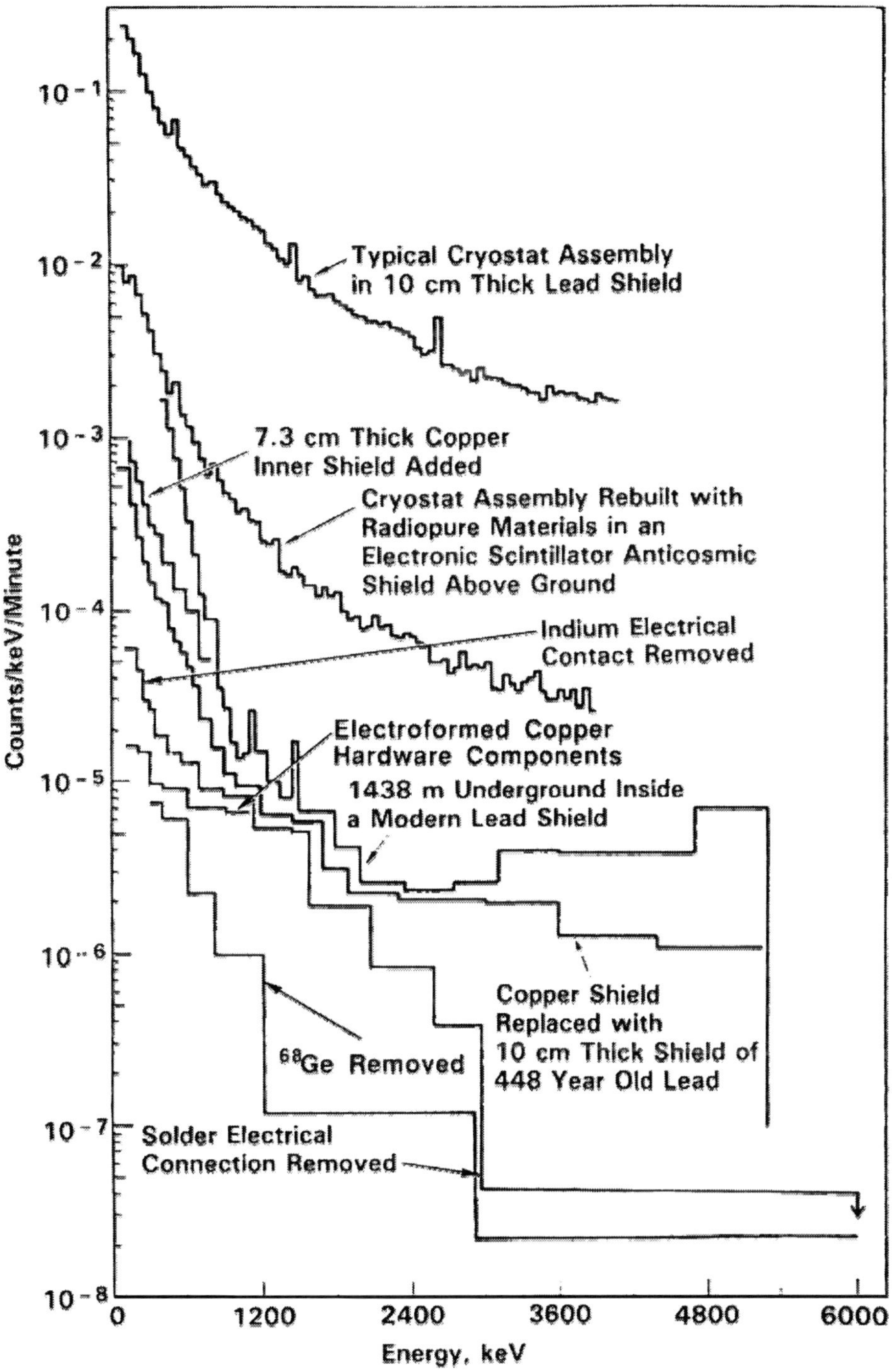

FIGURE 1. Successively lower background spectra obtained over fifteen years of spectrometer refinement. From reference [1].

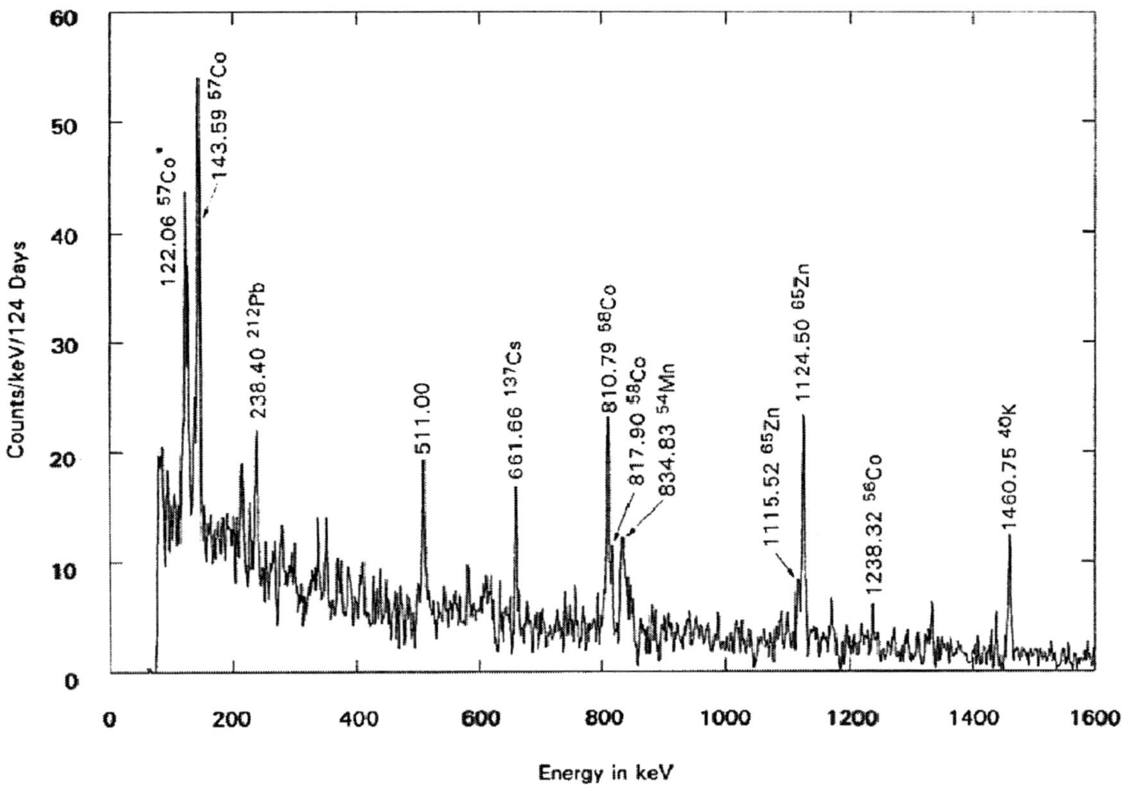

FIGURE 2. Early measurement of cosmogenic radioisotopes in high-purity germanium spectrometer with copper cryostat. Isotopes ^{56}Co, ^{57}Co, ^{58}Co, and ^{54}Mn are examples of production from copper spallation. An example of germanium spallation production is ^{65}Zn. From reference [2].

CHEMISTRY AND PURITY

After electroformed copper was first produced for low background spectrometers, chemistry improvements continued. Uranium and thorium progeny were reduced substantially over early work. A factor of 100 reduction in uranium progeny and a factor of 10 reduction for thorium progeny was achieved [1].

These radiochemistry gains were driven by three main improvements. First, the purity of the H_2SO_4 was improved through sub-boiling distillation and processed and stored without contact with glassware. Second, the $CuSO_4$ used was subjected to recrystallization, enhancing its purity. Third, a small amount of $BaSO_4$ was introduced into the electroforming chemistry, providing a source of Ba+ ions that exchange with the radium (Ra+ ions) present in the bath. The results of these advances are discussed in [1] and are similar to those referenced for the best newly-produced commercial cryogenic-grade (NOSV) high-purity copper [3].

ELECTROFORMING OVERVIEW

To give a broad view of the process of electroforming, the plating bath process parameters, key electroforming technology elements, and final cleaning and passivation steps are described. Figure 3 shows a schematic diagram of an electroplating setup.

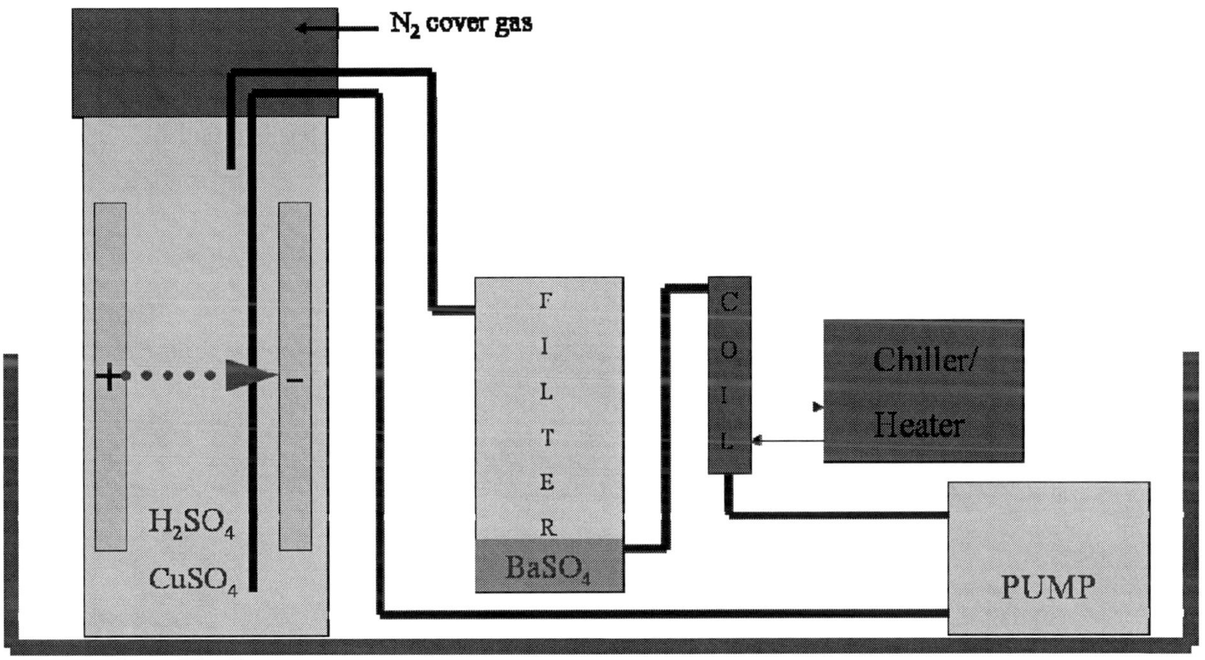

FIGURE 3. Schematic diagram of electroforming.

Plating Bath Process Parameters

Plating is done onto polished, cleaned, stainless steel mandrels in the shape of the desired parts. The current density used is ~40 mA/cm^2, corresponding to a plating rate of ~0.05 mm/h. A small amount of $BaSO_4$ is added to the recirculating bath. This collects in the micro-filtration stage and acts as a radium scavenge. Additionally, $CoSO_4$ was added as a holdback carrier for the cosmogenic $^{56,57,58,60}Co$ present in the starting copper. Small amounts of HCl and thiourea affect copper crystal nucleation and grain size. This process chemistry is summarized in Table 1.

Key Technology Elements

There are several key elements in the plating process required to produce extremely pure copper. To summarize:

1. Use of semiconductor-grade acids
2. Glassware-free processing and handling
3. Recrystallized copper sulfate for increased purity
4. Continuous microfiltration of the bath to remove oxides and precipitates
5. Continuous barium scavenge to remove radium.
6. Inert cover gas in plating tanks to reduce oxide formation.
7. Periodic surface machining during production to minimize dendritic growth.

Cleaning and Passivation

Surface cleaning and preparation is critical to ensuring high quality and purity copper parts. The final machining process often leaves unwanted copper fragments that make assembly difficult. Additionally, copper oxides will easily form on the surface during production handling. Finally, radon daughter isotopes will deposit on exposed surfaces and must be removed to achieve the best radiopurity for finished parts. Traditionally, cleaning was accomplished with a strong nitric acid etch. This was a very problematic process since nitric acid will attack the copper metal, thus altering critical mechanical tolerances. To alleviate this problem, a refined technique was developed that removes loose fragments and surface contaminants. This process relies upon ultrasonic cleaning for gross fragment removal and acidified peroxide etchant for fine fragment, surface contamination, and oxide removal.

A surface passivation step is required to prevent the reformation of oxides and control the IR emissivity of the copper. After testing ~30 passivation procedures, the preferred method for passivation was found to be room-temperature immersion in a 1% solution of citric acid. Cleaned copper parts are immediately immersed in the citric acid solution for several minutes. Excellent results have been achieved with this passivation step. Parts passivated will remain free of obvious oxide formation for many weeks after cleaning if maintained in a dry, controlled, atmosphere.

TABLE 1. Plating Bath Chemistry.

Constituent	Concentration
$CuSO_4$	188 g/l
H_2SO_4	75 g/l
HCl	30 mg/l
Thiourea	3 mg/l
$CoSO_4$	1 mg/l
$BaSO_4$	~1 mg/l

INFRASTRUCTURE REQUIREMENTS

Electroforming underground provides the opportunity to escape the cosmogenic radionuclides found in all copper with above-ground post-purification exposure. Because operations underground generally require more detailed infrastructure planning than typical laboratory operations, the infrastructure requirements of an electroforming facility are described. This prototypical facility would be adequate to produce low-background cryostat parts for a next-generation ^{76}Ge neutrinoless double-beta decay experiment that has on the order of 10^3 kg of spectrometer crystals. Figure 4 is a schematic layout of an electroforming facility. Co-location of these areas is important so that copper under process can be removed from a bath, rinsed, dried, moved to the clean shop and machined, then returned to the plating room, etched, and returned to a plating bath. After final machining in the clean shop, copper will be moved to the material storage area for radon- and/or oxygen-free storage until needed.

Plating Room

The plating area requires a 4 x 8 x 3 m room. In addition to the actual plating, this room also contains space for etching parts between plating steps and for acid storage. This space requires a spill containment lining to protect against leaks in the plating baths. For rinsing and bath formulation a Milli-Q water system with de-ionized supply water is required. A shared 10^{-6} Torr dry vacuum system will serve the plating room along with the material storage area. The room will require an acid-rated fume extractor ventilation system for etching and for the small amount of acid vapor evolved during the plating process. Flammable and hazardous gas sensors are also needed, as plating has the potential to electrolyze water and produce hydrogen gas if excessive voltages are used. To protect the radiopurity of the baths and the copper during the etching and rinsing steps, a HEPA-filtered air supply is needed. For the lowest-level work, radon-scrubbed air is desirable. As with typical cleanrooms, an air-lock entry and washable walls are needed. The power required is ~ 12 kW 120/240 VAC for the plating operation. Air conditioning to ~20°C is needed both for operator comfort while wearing acid protective gear and to allow efficient plating power supply operation. For some baths, additional cooling via liquid-liquid heat exchanger may be needed to maintain an optimal bath temperature.

Materials Storage Area

The materials storage area requires a 3 x 4 x 3 m room. Radon-proof storage lockers with purge gas and vacuum capability provide a way to store parts without either the deposition of radon daughters on the surface or surface oxidization. These storage lockers are also served by a shared 10^{-6} Torr dry vacuum system. Argon or nitrogen purge gas allows easier short-term storage of material without requiring evacuated chambers. This room also requires a HEPA-filtered air supply to protect the radiopurity of the copper during transfer into and out of storage. As with the plating room, an air-lock entry and washable walls are needed. The power required is ~ 2.4 kW 120 VAC. The room should be air conditioned to ~ 20°C as is the plating room.

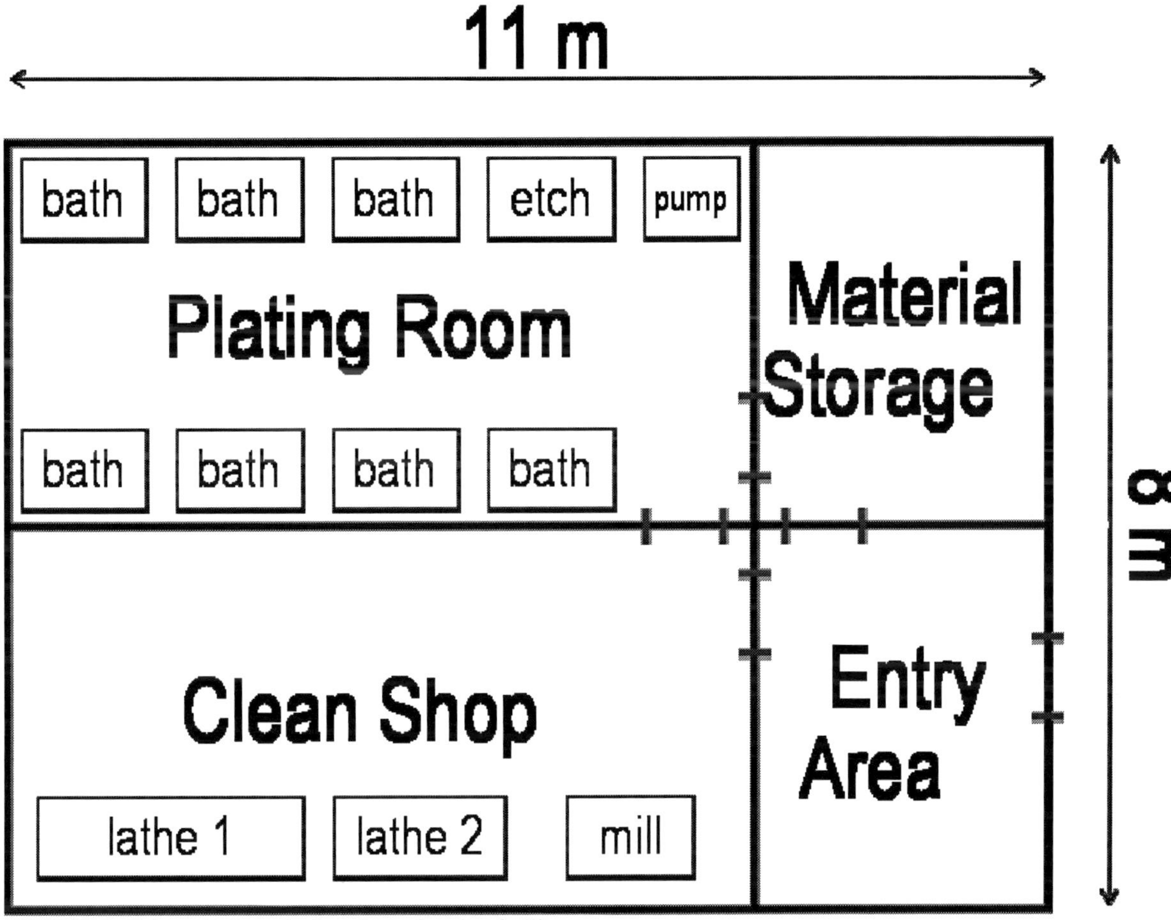

FIGURE 4. Schematic layout of an electroforming facility.

Clean Shop Area

The clean shop area requires a 4 x 8 x 3 m room. It requires a HEPA-filtered air supply, air-lock entry, and washable walls. Driven primarily by the machine tools, this area has a power requirement of ~ 24 kW 120/240 VAC. It should be air conditioned to ~ 20°C.

SUMMARY

Electroformed copper was a key material in the International Germanium Experiment (IGEX). It continues to play a key role in the current generation of research detectors, providing ever-greater purity levels and the required electrical, mechanical, thermal, and vacuum performance. Electroformed copper combines high purity with fabrication of low-mass parts, further reducing materials backgrounds. Underground manufacturing is feasible, allowing cosmogenic activation to be reduced or eliminated. The chemical process for electroformed copper has excellent potential for continuing purity improvements.

As one of the most pure materials available, it contributes centrally to the design of proposed next-generation neutrinoless double-beta decay experiments such as the Majorana Project.

ACKNOWLEDGMENTS

This work was performed at the Pacific Northwest National Laboratory with Government support under Contract Number DE-AC06-76RLO-1830 awarded by the United States Department of Energy. Valuable contributions were made by Reeves and Sons, LLC.

REFERENCES

1. Brodzinski, et al., Journal of Radioanalytical and Nuclear Chemistry, 193 (1) 1995 pp. 61-70.
2. Brown, M. P., and Austin, K., Appl. Phys. Letters 65, 2503-2504 (1994).
3. M. Laubenstein, et al., Applied Radiation and Isotopes, 61 (2004) 167-172.

Cleanliness, backgrounds and surface contamination in CUORE

S. Pirro*, S. Capelli*, O. Cremonesi*, S. Nisi†, E. Palmieri**, M. Pavan* and E. Previtali*

*Dipartimento di Fisica dell'Universita' di Milano Bicocca and INFN Sezione di Milano, Milano I-20126, Italy
†INFN - Laboratori Nazionali del Gran Sasso, I-67010, Assergi (L'Aquila), Italy
**INFN - Laboratori Nazionali di Legnaro, Via Romea 4 I-35020 Legnaro, Italy

Abstract.
CUORE is a proposed array of 988, 750 g, TeO_2 crystal bolometers. The experiment has been approved by the Scientific Committee of Gran Sasso Laboratories and the special dilution refrigerator, that is intended to house the detector has been funded. The Experiment will search for the 0ν-Double Beta Decay of ^{130}Te. As in all the proposed next generation Double Beta Decay Experiments, the main task is the reduction of the radioactive background. A peculiar property of thermal detectors is that they are active over the entire volume and therefore strongly subject to radioactive surface contaminations. Unlike radioactive bulk contaminations, that can be measured through High-Purity Ge Detectors, radioactive surface contaminations are not easily measurable at very low levels. Different techniques were developed in order to reach the required sensitivity. Present results already achieved and studies that are underway are here presented and discussed.

Keywords: Double Beta, Radioactivity, low trace levels
PACS: 23.40.Bw ; 91.65.ND

INTRODUCTION

The goal of CUORE is to achieve a background rate in the range $0.001 \div 0.01$ counts/keV/kg/year at the 0ν-Double Beta Decay transition energy of ^{130}Te (2528.8 keV). A low counting rate near threshold (that will be of the order of $5 \div 10$ keV) is also foreseen and will allow CUORE to produce results for Dark Matter. CUORE will be a tightly packed cylindrical array of 19 towers made of 13 planes of 4 5x5x5 cm^3 crystals each. One such tower has been successfully constructed and is being operated in the Gran Sasso Laboratory as a test Experiment, and also as an independent, first-step next generation double-beta decay experiment called CUORICINO [1].

The comprehension of CUORICINO background origin will therefore lead the choices in the final CUORE design and assembly. With this aim, a Monte Carlo code has been developed in order to simulate decay processes occurring in the bulk and on the surfaces of the various parts of the experimental setup, in order to disentangle the sources responsible of measured background by comparing the simulated data with the experimental ones. The information thus extracted are used to evaluate the CUORE background in the 0ν-DBD region.

BULK CONTAMINATIONS

The main contribution to background due to the bulk contaminants arises from the cryostat structure (radiation shields), from the heavy structures close to the detectors (the Copper mounting structure of the array, the Roman lead box and the two lead disks on the top and the bottom of the array) and from the detectors themselves (the TeO_2 crystals). The radioactivity levels used in the computations for these materials are given in Table 1. These levels are nearly equal to the best upper limits obtained for the radioactive content of these materials, measured with HPGE Detectors and with bolometers.

The results of the Monte Carlo simulations can be seen in Fig. 1. The assumed threshold is 10 keV and only the values obtained after requiring an anti-coincidence between detectors are indicated. These values have to be considered as upper limits on the possible contribution of bulk contaminations to the CUORE background; they prove that with the *already available materials*, background levels of $\sim 2 \cdot 10^{-3}$ counts/keV/kg/y in the 0ν-DBD decay region and $\sim 3 \cdot 10^{-2}$ counts/keV/kg/d in the low energy (10-50 keV) region are foreseen.

CP 785, *Topical Workshop on Low Radioactivity Techniques,* edited by B. Cleveland, R. Ford, and M. Chen

TABLE 1. Bulk contamination levels (in 10^{-12} g/g) used in the simulation for TeO_2 , Copper and lead.

Contaminant	^{232}Th	^{238}U	^{40}K	^{210}Pb	^{60}Co
TeO_2	0.5	0.1	1	10 μ Bq/kg	0.2μ Bq/kg
Copper	4	2	1	0	10μ Bq/kg
Roman Lead	2	1	1	4 mBq/kg	0
Low Act. Lead*	2	1	1	16 Bq/kg	0

* Low in ^{210}Pb

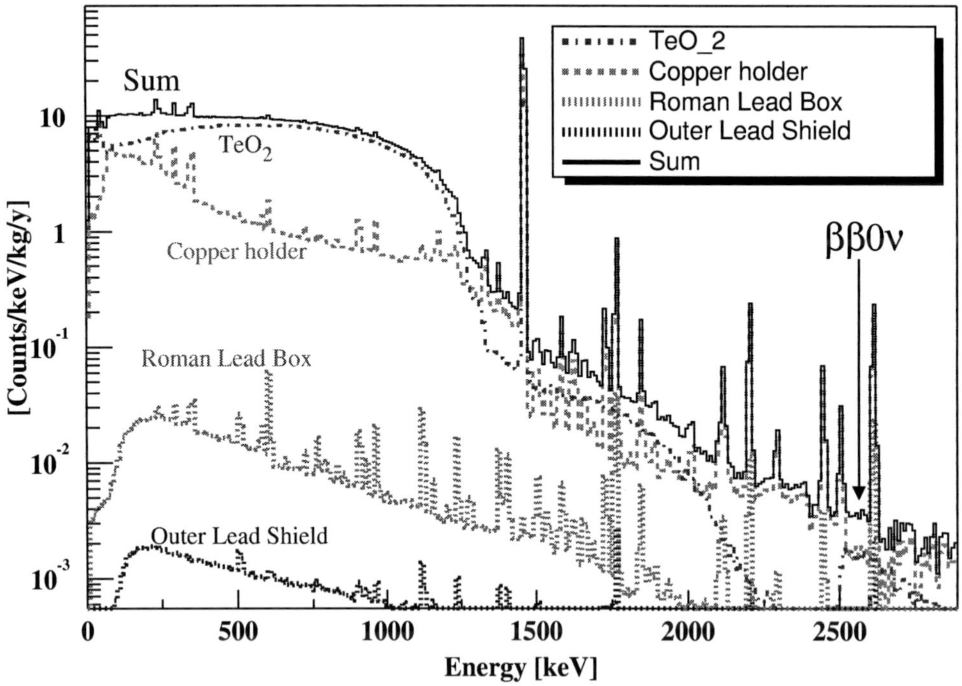

FIGURE 1. Simulated spectra for bulk contaminations of the TeO_2 crystals, the Copper structure, the Roman Lead shield and the outer Lead shield. Each spectrum is obtained by summing the simulated anticoincidence spectra of all the CUORE detectors.

SURFACE CONTAMINATIONS

Surface contaminations contribute to the background only when they are localized on the crystals or on any kind of materials faced to them; in the adopted setup, more than 90% of this surface is Copper. As learned in MiDBD 20 detector array Experiment [2] and CUORICINO, the presence of even a very low level of radioactive impurities on the surfaces of the detectors can produce a non negligible contribution to the $\beta\beta(0\nu)$ background level. Unfortunately this kind of contamination, that is important mainly for detectors without a surface dead layer (as is the case of bolometers), is poorly studied. Lacking any data coming from direct measurements of the typical impurity levels present on TeO_2 and Copper surfaces, we rely on the results obtained in MiDBD and CUORICINO.

These data indicate that both a surface contamination of the crystals and of the Copper surface of the mounting structure are present. In the case of TeO_2 , we know that the contamination is mainly ^{238}U and that its presence is strictly connected to the kind of surface treatment undergone by the crystals. The data of CUORICINO suggest that this contamination is responsible of $\sim$20% of the total background counting rate in the $\beta\beta(0\nu)$ region, corresponding to $(4\pm 3)\times 10^{-2}$ counts/keV/kg/y .

Concerning the Copper surface contamination, we still miss a clear indication of its origin and identity (i.e.

TABLE 2. Estimated upper contribution to the CUORE $\beta\beta(0\nu)$ region from surface contaminations obtained by using the surface contamination levels evaluated for CUORICINO and assuming an exponential density profile with λ=1 μm for TeO_2 crystals (^{238}U) and λ=5 μm for Copper (^{238}U and ^{232}Th).

Element	Contamination Bq/cm^2	Contribution to the$\beta\beta(0\nu)$ region counts/keV/kg/y
TeO_2	1.9E-08 ± 9.8E-09	1.6E-02 ± 8.4E-03
Copper	4.9E-08 ± 1.9E-08	5.8E-02 ± 6.6E-03
Total	3.4E-08 ± 1.3E-08	7.4E-02 ± 1.1E-02

we are not yet able to distinguish between contributions generated by different possible sources), but we have an evaluation of the $\beta\beta(0\nu)$ background counting that can be ascribed to them (whatever they are, ^{238}U or ^{232}Th): ~0.1± 0.04 counts/keV/kg/y.

According to a preliminary Monte Carlo simulation of the CUORE detector, using the actual contamination levels evaluated with CUORICINO, we obtain a background counting rate in the $\beta\beta(0\nu)$ region (after the anticoincidence cut) of about 1.6×10^{-2} counts/keV/kg/y and 5.8×10^{-2} counts/keV/kg/y for the TeO_2 crystals and the Copper structure respectively, as shown it Tab. 2

The goal of CUORE is to reduce the surface contribution by a factor at least 10 with respect to this evaluation, obtaining a background coming from surfaces of about ~7 $\times10^{-3}$ counts/keV/kg/y. Improvements of a factor of ~1.5–2 are expected simply by possible reductions of the Copper mounting structure dimensions (the Monte Carlo simulation we used so far refers to an identical mechanical structure for the CUORE and CUORICINO single detector modules). A reduction by a factor of at least 10 of the Copper surface contamination and a factor 4 for the crystals is therefore the first milestone of CUORE.

In order to understand the contribution of both internal and surface contamination to the $\beta\beta(0\nu)$ a MC code based on GEANT 4 was developed. By comparing MC simulation with measured data it has been deduced that the shapes and rates of the observed degraded alpha peaks are well reproduced assuming surface contaminations on the crystals and Copper holders. The simulation is not straightforward and several points have to be addressed; the most delicate are

- Roughness of the surfaces on the scale of μm.
- Density profile of the contaminations.
- Breaks in the Secular Equilibrium.

Also the low statistic collected plays a crucial role. Using an exponentially decaying density profile with characteristic lengths between 1 and 5 μm a good agreement between data and simulation was obtained in the alpha region of the CUORICINO spectra, as shown in Fig. 2.

EXPERIMENTAL TECHNIQUES

Using the results of the simulation we expect contaminations on the Copper surface of the order of $\sim1.6\times10^{-10}$ g/g for ^{238}U and $\sim7\times10^{-10}$ g/g for ^{232}Th within the first μm of Copper decreasing exponentially with depth. In order to perform a *direct measurement* on the emitted alpha particles one has to convert the contamination into activity; using the background value given above and considering the dimension and weight of each crystal we find an activity of ~0.003 c/cm^2/day/MeV. Commercial low-level Si-Barrier alpha detectors have a intrinsic background of the order of 0.1 c/cm^2/day/MeV, far away from our needs. We therefore developed different techniques, both direct and indirect, in order to evaluate the radioactive surface contamination of materials.

HR-ICPMS Measurements

The techniques using High Resolution Inductively Coupled Plasma Mass Spectrometry are *indirect* measurements. This kind of detection measures practically all kind of nuclei with a sensitivity of the order of $\sim10^{-14}$ g/g . This

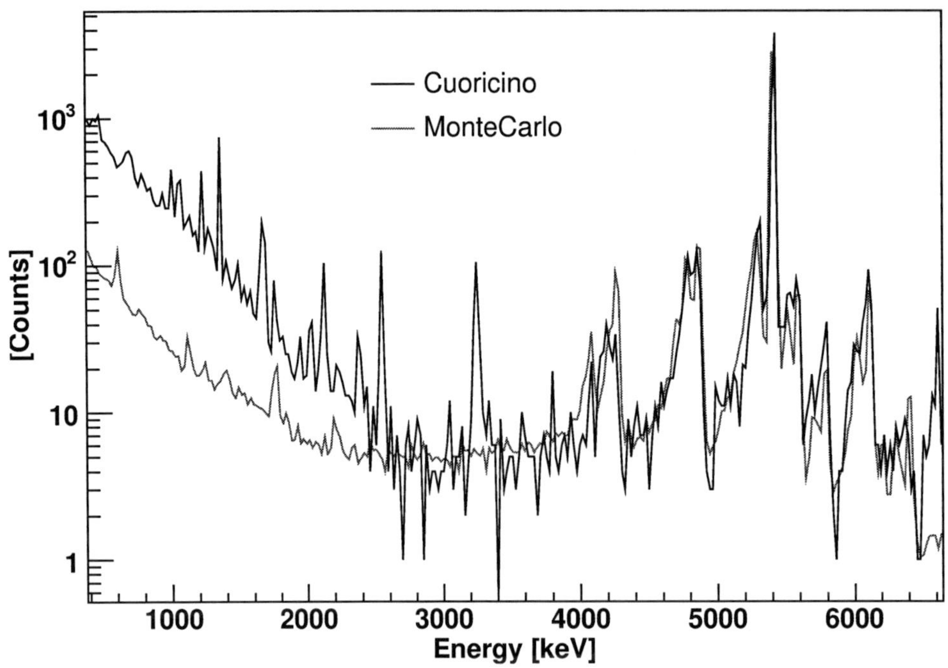

FIGURE 2. Comparison between CUORICNO spectrum and MC-spectrum obtained simulating the contaminations on Crystals and Copper surfaces.

means, practically, that we can measure (within our needs), only long living radioactive nuclei, i.e. *only* ^{238}U ^{232}Th , the parents of the two main radioactive chains. Unfortunately, in case of breaks in the secular equilibrium, this method can over/underestimate the actual radioactive activity in the sample. The procedure is, in principle, very simple. The aim is to measure the concentration of ^{238}U and ^{232}Th in a thin surface layer (1÷10 μm) of Copper. The idea is to remove by chemical attack the Copper from the surface and then to measure the concentration of the contaminants in the solution via Mass Spectrometry. The first step (carried out in the chemical lab of LNGS) was to establish the best condition (reagent type, concentration and temperature) in order to achieve a suitable corrosion rate. In the second step the concentration was measured through HR-ICPMS (VG-Axiom-Plus) in the Joint Centre laboratory of ISPRA (Italy). The chosen etching solution is 2.5 % HNO_3 solution at 45 Celsius. The corresponding etching rate strongly depends on the surface treatment of the Copper and the values range between 0.2 and 1 μm/min. Due to the solution-to-removed Copper weight ratio (1000÷5000), in order to have a sensitivity of the order of $\sim 10^{-11}$ g [^{238}U ,^{232}Th]/g[Cu] all the measurements have to be carried out with reagents/container with radiopurity of the order of $\sim 10^{-14}$ g/g. Moreover, in order to preserve the instrument, the maximum allowed reagents concentration in the final solution for HR-ICPMS is 0.1% of salts and 5 % of HNO_3. The Copper tests samples consists of a Copper slab of dimension 90x35x8 mm^3; the volume of nitric acid solution is $\sim$150 ml. The procedure (see Fig. 3) is as follows: the solution is placed in a PTFE container. A *blank* sample is drawn in order to measure the initial contamination of the solution. After that the Copper bar is placed in the container. At different times (i.e. different etched depths) a sample is drawn from the solution. Each of this samples is divided in 4 samples. One of this is used for the actual measure, while into the remaining three different calibrated amounts of ^{238}U and ^{232}Th are added in the sample in order to calibrate the instrument for *each* measure. This means that 24 measures are needed in order to sample a Copper slab at 5 different depths. The results obtained on a Copper slab treated with the same procedure adopted for CUORICINO are presented in Fig. 4; the results are in good agreement with the expected contaminations. Nevertheless some problems arise in trying to further pursue this technique. We experienced that there are still some open questions that need to be addressed. In particular, working at very low trace level, some (even small) accidental contamination are possible. Since the measure is an *integral measure*, in the sense that when the *i*-sample is taken it contains also the *i-1* etched layer of Copper, some problems can arise during the subtraction of the different values in order to evaluate the actual density profile. Furthermore the large percentage of Copper in the solution decreases the performances and the overall sensitivity of the ICPMS, resulting in a long time (few days) in order to purge it.

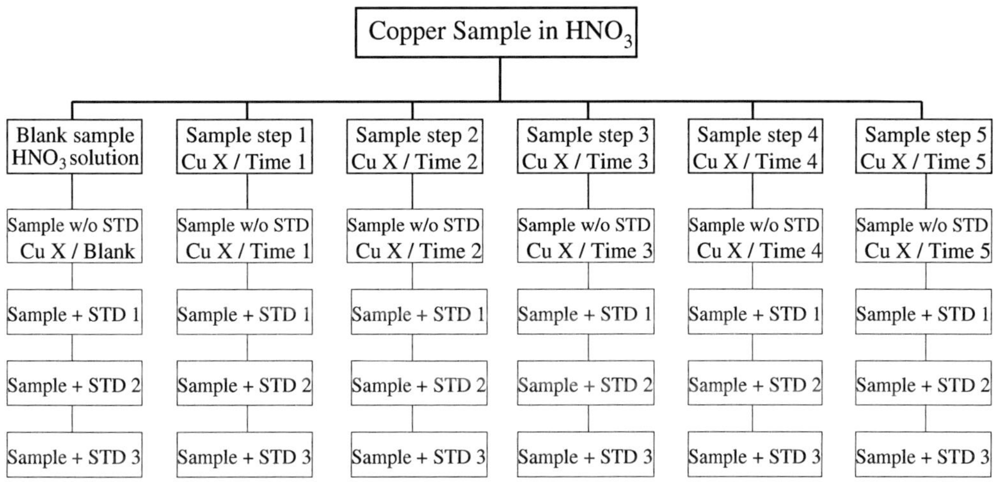

FIGURE 3. Scheme of the procedure used in the ICPMS method.

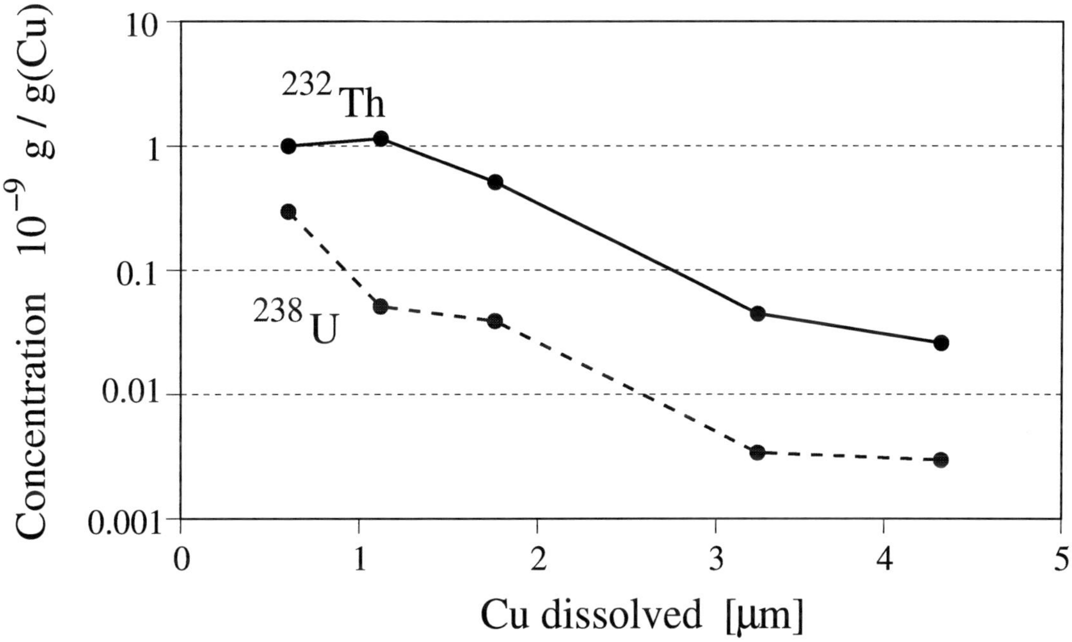

FIGURE 4. Contamination of ^{238}U and ^{232}Th in the Copper as a function of the depth of the etched layer

Direct Measurements

Due to the fact that the ICPMS is sensitive only on ^{238}U and ^{232}Th contaminations, a direct measurement of surface radioactivity is essential. We therefore constructed a "small" CUORICINO detector (Radioactive Array Detector, RAD), consisting of eight $5{\times}5{\times}5$ cm^3 TeO_2 crystals. The active area of the overall detector is therefore 1200 cm^2. Thanks to the information given by the CUORICINO data, we changed the procedure of the crystal polishing, in order to avoid surface contaminations on the crystals. The procedure consists in etching the crystals in an ultrapure 4.5 M nitric acid solution. A mean layer of $\sim$10 μm is removed. The aim was to remove even small inclusions of the previous used lapping powder. Due to the fact that the etched surface is not suitable to run the crystals as bolometer, the second

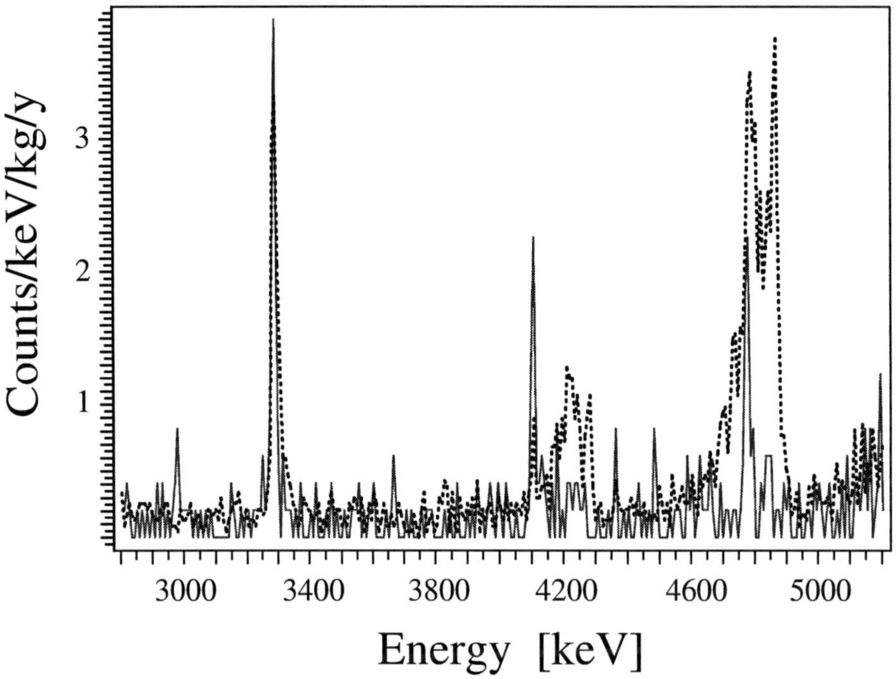

FIGURE 5. Comparison of the backgrounds spectra in the alpha energy region obtained with the RAD (bold line) with respect to the one of CUORICINO (dashed line).

step was to polish them with a new selected high-pure SiO_2 lapping powder.

The Copper structure was cleaned trough electro-erosion in a ultra-clean solution of citric acid; it acts as electrolyte as well as passivation. The first "blank" run, performed in the Hall C of LNGS gave encouraging result. As showed in Fig. 5, a clear decrease in the peaks due to the ^{238}U chain was obtained.

The data collected with this run enable us to put new limits on the internal activity of the TeO_2 crystals. We found a limit of 1×10^{-13} g/g for ^{238}U and a value of $\sim 2\times 10^{-13}$ g/g for ^{232}Th . The surface radioactivity on the TeO_2 crystals was decreased by a factor $\sim$4. Thus all the requirements for the crystal contamination (both internal and superficial) are fulfilled.

Unfortunately the background spectrum in the continuum region of the alpha spectrum was decreased only by a factor $\sim$1.8 with respect to CUORICINO, with a background activity of $\sim 1.7 \times 10^{-3}$ Counts/cm^2/day/MeV.

The RAD is now being used as a actual surface alpha detector. We screened all the materials used in CUORICINO that faces directly on the detectors, namely silicon heaters, gold bonding wires and PTFE Teflon. In this way we finally proved that these materials contribute for no more than 10 % of the observed continuum, restricting our surface contamination problem only to the Copper.

Further Developments

As mentioned before, the RAD is a powerful detector in order to measure surface contaminations. Once we find the correct procedure to clean the Copper, the RAD will be able to prove the CUORE feasibility. The drawback is the time needed in order to collect enough statistic. To measure the background-goal of CUORE, with an accuracy of 20 % at 1σ level, we need three months of live time background run. Thus the need to have new, different and *quick* screening procedures becomes crucial. We therefore are developing two different new techniques. The first is the use of the Laser-Ablation ICPMS. With respect to the standard ICPMS there is the great advantage that there is no use of solvents and other materials that can *accidentally* contaminate the samples; furthermore there is, obviously, no dilution factor. The main problem of this technique lies in the measurement of the removed Copper. The problem arises from the fact that it is not possible to measure at the same time (i.e. during each laser pulse) the Copper removed (big quantity) and the low traces.

The removed Copper depends crucially on the local properties of the surface and can change, within the same sample by more than one order of magnitude. The new approach, developed at the CNR of Padova consists in a statistical approach based on averaging over a matrix (large number) of spots (repeated in the same position). The statistical calibration of the removed Copper *vs* the number of the repeated spots can be evaluated by investigating the profile of the craters created on the Copper surface using an Electron microscope. We got the first encouraging result both in terms of sensitivity and reproducibility. Furthermore we are investigating the possibility to *directly* calibrate the amount of the ablated Copper through the measurement of low contamination traces commonly present in Copper (for example Ag , $0.1 \div 10$ per 10^{-6} g/g).

A second technique is being developed in Milano. The goal is the construction of an ionization chamber to do direct alpha spectrometry: the idea is to build up a large volume ionization chamber with low intrinsic background, to be able to measure the surface contaminations of different materials. The difference in the pulse development of the surface alpha events with respect to the gamma ones enable to discriminate between the two backgrounds. This detector could therefore be used to measure, among the others, surface contamination of Copper in a time scale of less than a week. The advantage of this method is that it allows spectroscopic measurements, i.e. it will be possible to recognize the various contaminants by the energy of the alpha detected by the instrument. Another important advantage is that the method is "non destructive" , so it can be used to test various production steps on the same piece of material.

CONCLUSIONS

Since the start of CUORICINO the comprehension of the different sources of our background consistently improved. Presently the only open problem for CUORE is the surface contamination of Copper. Besides the R&D work in order to direct measure this surface contamination great effort is spent in order to find new cleaning techniques. Preliminary tests based on the laser ablation measurements indicate that the plasma cleaning of Copper reduces the surface contaminants with respect to the standard etching/passivation procedure adopted for CUORICINO.

ACKNOWLEDGMENTS

Thanks are due to M. Perego, G. Galotta, S. Parmeggiano, R. Gaigher, A. Rotilio, B. Romualdi, and especially to P. Nuvolone for continuous and constructive help in various stages of this measurements.

REFERENCES

1. C. Arnaboldi *et al. hep-ex/0501034 , submitted to Phys. Rev. Lett.*
2. A. Alessandrello *et al. Physics Letter B* **486** 13–21 (2000).

Surface Contamination From Radon Progeny[1]

Michael Leung

Physics Department, Princeton University
Princeton NJ 08544 USA

Abstract. This paper examines methods the Borexino experiment uses to reduce surface contamination from radon progeny. The first section describes the deposition of airborne radon progeny. It includes measurements of the deposition rate in clean room conditions and describes steps that were taken to minimize the exposure of critical components in the experiment. The second part covers the adsorption of contaminants on surfaces through liquid contact. It summarizes measurements of the partitioning constant, cleaning effectiveness, and the time scales for the desorption of contaminants from a surface.

Keywords: Borexino, solar neutrino, low background, surface contamination, radon progeny, adsorption, deposition
PACS: 01.30.Cc

INTRODUCTION

The sensitivity of physics experiments looking for neutrino oscillations, dark matter, and neutrinoless double beta decay is limited by radioactive backgrounds. These backgrounds are often difficult to distinguish from real events and must be suppressed by many orders of magnitude for a successful measurement to be made. The source of radioactivity arises from either bulk impurities in the materials, external sources, or surface contamination. This paper focuses on surface contamination that results from gas and liquid adsorption of radon progeny. It also discusses steps that can be taken to minimize these surface impurities. The studies are directed towards meeting high purity requirements in the Borexino experiment.

The Borexino solar neutrino experiment is designed to measure the ^{7}Be solar neutrino flux[1]. A diagram of the Borexino detector is shown in figure 1. The detection method is neutrino-electron scattering in 100 tons of organic liquid scintillator. Taking into account the measured neutrino oscillation parameters, Borexino is expected to observe about 35 neutrino events per day in an energy window from 250-800 keV. Since it is desirable to have a large signal to noise ratio, the total rate of background events in this volume must be less than several per day from all sources. Achieving material and scintillator purity that meet this requirement is the main technical challenge facing the experiment.

RADON

One of the most prevalent radioactive backgrounds is from radon gas which is naturally present in air. Typical indoor radon activities are 10-100 Bq/m^3 though they can increase to several thousand Bq/m^3 in mines and caves where low background experiments are usually located to shield cosmic rays. The decay chain of radon is shown in figure 2. The half life of radon is 3.8 days but it is continually replenished by emanation from the ground where deposits of uranium reside. To reduce radon activity, buildings are typically ventilated with clean outside air.

Radon is particularly difficult to control due to its high mobility. In practical terms this means low background detectors must be designed to be leak tight to air. As an additional constraint, surface contact of detector materials with radon progeny must also be minimized.

Properties of radon and its progeny have been extensively studied to understand human radiation exposure from their inhalation, which is a cause of lung cancer. Many of the associated measurements and models are also of interest to low background experiments that must deal with surface contaminants. An excellent review of the literature can be found in *Radon And Its Decay Products In Indoor Air* by W. Nazaroff and A. Nero [2].

---

[1] Proceedings from Low Background Techniques 2004 at SNOLAB and Laurentian University, Sudbury, Canada, December 12-14, 2004.

CP 785, *Topical Workshop on Low Radioactivity Techniques,* edited by B. Cleveland, R. Ford, and M. Chen

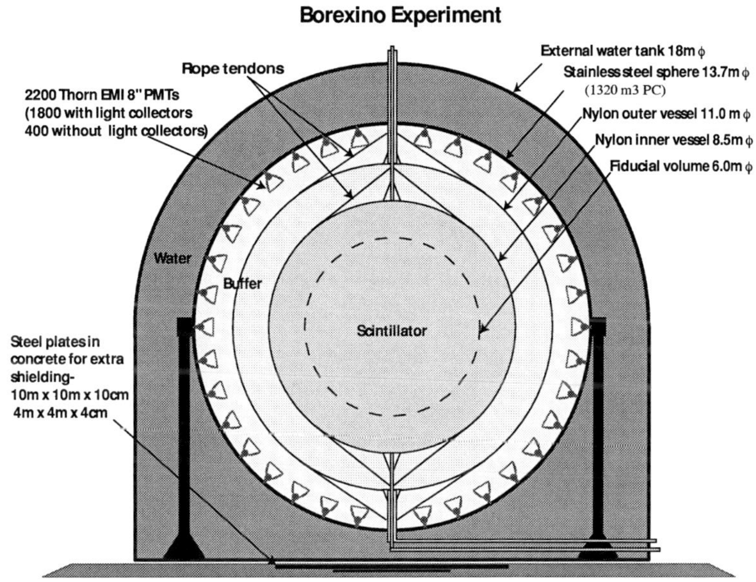

FIGURE 1. A diagram of the Borexino detector. The sensitive volume of the detector is a liquid scintillator solution composed of 1,2,4-trimethylbenzene (pseudocumene) with 1.5 g/l of 2,5-diphenyloxazole (PPO), a wavelength shifter. It is shielded by an inactive buffer of pseudocumene with dimethyl 1,2-benzenedicarboxylate (dimethylphthalate) which quenches scintillation. Events are observed with 2200 8" PMTs mounted on a stainless steel sphere. A high purity water shield outside the stainless steel sphere acts as a muon veto.

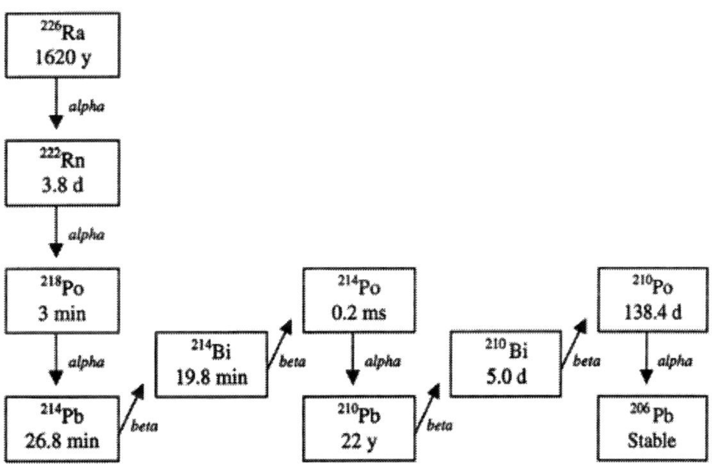

FIGURE 2. The radon decay chain. Radon is produced directly from the α decay of radium. When radon decays is it produces a series of progeny with short half lives of less than an hour until ^{210}Pb is formed. ^{210}Pb with a half life of 22.3 years produces ^{210}Bi (β decay, $Q = 1.17\,\mathrm{MeV}$) and ^{210}Po (α decay, $Q = 5.3\,\mathrm{MeV}$) before reaching the stable isotope ^{206}Pb. It is the long half life of ^{210}Pb and the high energies of the two subsequent decays which makes radon exposure such a problem for Borexino.

DEPOSITION OF RADON PROGENY

Though radon is a noble gas and remains airborne, its progeny readily stick to surfaces. Over time these progeny buildup ^{210}Pb which has 22 year half life. Since this half life is long compared to the lifetime of a typical experiment it is infeasible to wait for ^{210}Pb to decay away. The ^{210}Pb produces ^{210}Bi and ^{210}Po which β and α decay respectively, adding to the radioactive background. Thus the deposition of radon progeny on surfaces results in the presence of three radioactive isotopes long after the surface is isolated.

Radon progeny are formed in energetic decays allowing them to easily ionize and react with molecules the environment. The exact chemistry is difficult to predict as it is strongly dependent on the concentrations of various pollutants in the atmosphere.

The concentration of radon progeny in a room is relatively constant when measured sufficiently far from the walls. Near the walls, the concentration of the progeny decreases because the walls act like a sink. The surface concentration σ is the product of flux J towards the walls and the time of exposure t, for a stable element. The flux is found from the diffusion equation

$$J(x,t) = -D\nabla C(x,t) \tag{1}$$

where D is the diffusion constant and $C(x,t)$ is the concentration at a given position and time. However, if the concentration is constant in time and space (away from surfaces) it is convenient to instead represent the flux J as the product of the room concentration C and a deposition velocity v_d. The deposition velocity is the diffusion constant D divided by the boundary layer l near the surface

$$v_d = \frac{D}{l} \tag{2}$$

The boundary layer l is the region near a surface where the air is essentially static and convection is negligible. Thus the deposition velocity can be thought of as the rate at which particles diffuse towards a surface. The deposition of radon progeny can then be described by

$$\sigma(^{210}\text{Pb}) = \sum_{i=1}^{5} C_i v_d t \tag{3}$$

where:
$\sigma(^{210}\text{Pb})$ = the surface density of ^{210}Pb atoms [atoms/m^2];
C_i = the concentration of ith radon daughter [atoms/m^3];
i = index denoting specific progeny $\{1,2,3,4,5\} \rightarrow \{^{218}\text{Po}, {}^{214}\text{Pb}, {}^{214}\text{Bi}, {}^{214}\text{Po}, {}^{210}\text{Pb}\}$;
v_d = the deposition velocity [m/s];
t = the exposure time [s].

The ^{210}Pb does not appear immediately but is produced when the short lived precursors that stick to the surface decay. Since the time scale of low background experiments is on the order of years, the time for these quickly decaying isotopes to become ^{210}Pb can be safely neglected.

Studies have found that for progeny unattached to aerosols the deposition velocity $v_d = 4 \times 10^{-3}$ m/s in an average room[3]. This number can fluctuate by a factor of two in either direction depending on the local conditions. Usually the deposition velocity is measured for a particular situation since it depends strongly on variables which are not easily determined.

The deposition velocity is strongly influenced by the attachment of progeny to aerosols, which reduces it by roughly a factor of 10. Generally the fraction of progeny attached to aerosols increases with aerosol density though it is also dependent on the size distribution of the aerosols. A partial list of the factors influencing the deposition velocity is given in table 1.

The concentration of each radon progeny is determined by its half life, the concentration and half life of its parent, the ventilation rate of the room, the surface area of the room which acts as a sink, filtration, and surface recoils which can free attached progeny. The Jacobi Model[4] (which is not discussed here) uses these parameters to describe the steady state progeny concentrations with a set of coupled differential equations.

Deposition Measurements

The nylon vessels in Borexino remain in direct contact with the liquid scintillator and are the nearest component to the fiducial volume, making a high level of vessel cleanliness essential. Measurements were performed to test the build up of ^{210}Pb on a nylon surface in a clean room environment[5]. A small test clean room (2 m×2 m×3 m) was built to simulate the full scale class 100 clean room where the Borexino nylon vessels were fabricated. A clean room environment differs significantly from a typical room because the aerosol concentration is small enough that the attachment of radon progeny is negligible. The air flow due to the filters is also significantly higher than normal. Though the clean room filters do little to remove radon gas, they remove the radon progeny quite efficiently. It is worth

TABLE 1. Factors influencing the deposition velocity. A higher air velocity reduces the boundary layer near a surface, increasing v_d. Surface orientation is also a factor. Electric forces on charged progeny (a fraction of which are produced as positive ions when they decay) can substantially increase the deposition velocity.

Variable	Deposition Velocity
Air Velocity	Increased
Aerosol Attachment	Decreased
Charged Fraction	Increased*
Negative Ion Density	Decreased†
Surface Roughness	Increased

* if static charges are present on surfaces
† reduces charged fraction

noting that the test clean room was different from a standard clean room in that it was well sealed to retain radon during the exposure. This resulted in a small ventilation rate while maintaining a high filtration rate.

To measure the effective deposition velocity, nylon film samples (5 cm×5 cm) were placed in the test clean room. A 2 MBq radon source was used to expose the room and nylon samples to a high radon concentration. The radon concentration in the test clean room was monitored over time using a RAD7 radon detector. The samples of nylon were then removed and the ^{210}Pb surface contamination was measured. The actual measurement was of the ^{210}Po α decays with a silicon detector. The ^{210}Pb rate was calculated by accounting for the build up time of ^{210}Po. Since only the radon concentration was measured and not that of the progeny, the ^{210}Pb surface contamination is expressed as

$$\sigma(^{210}\mathrm{Pb}) = [\sum_{i=1}^{5} C_i] v_d t = [kC_0] v_d t \tag{4}$$

where we take the sum over progeny to be proportional to the initial radon concentration C_0. The constant k, which is a small number dominated by the high filtration rate, is incorporated into an effective deposition velocity $v_0 = kv_d$. The ^{210}Pb surface contamination is then

$$\sigma(^{210}\mathrm{Pb}) = C_0 v_0 t \tag{5}$$

where C_0 is the radon concentration and v_0 is an effective deposition velocity. The test clean room experiment provided the integrated radon exposure and ^{210}Pb surface contamination. It showed that in a clean room environment the effective deposition velocity is

$$v_0 = 3 \times 10^{-8}\,\mathrm{m/s} \tag{6}$$

This effective deposition velocity is roughly three orders of magnitude smaller that the average deposition velocity for an unattached radon progeny in a typical room. The most likely explanation is that the high filtration rate removes the radon progeny from the air before they can reach the nylon surface. If this is indeed the case then the effective deposition velocity should scale with the filtration constant λ_f, the volume normalized filtration rate (number of volumes filtered per time).

Borexino Nylon Vessel Fabrication Precautions

The Borexino nylon vessels were fabricated in a customized class 100 clean room in Princeton, NJ. This clean room used a radon filter to provide low radon air for ventilation. The filter was a vacuum swing adsorption system which could supply up to 170m^3/h of air with a radon activity of $<$0.3 Bq/m^3[6]. This clean room was sealed to minimize the makeup air required to keep an over pressure and to prevent the back diffusion of radon. Inside the clean room the radon activity was reduced from $\sim$30 Bq/m^3 to 1 Bq/m^3 during fabrication. The clean room specifications are shown in table 2.

During the vessel fabrication the nylon film was discharged with anti-static rods to prevent charged progeny from being attracted to the vessel. It was also kept well covered when not in use. Only the edges where the joints were made

TABLE 2. Specification for the class 100 clean room where the Borexino nylon vessels were fabricated. Modifications were made to reduce the deposition of radon progeny. The filtration constant $\lambda_f = 0.036/\text{s}$.

Parameter	Value
Volume	$770\,\text{m}^3$
Filters	84 HEPA
Flow Type	Laminar
Relative Humidity	50%
Temperature	290 K
Ventilation Rate	$0.024\,\text{m}^3/\text{s}$
Filtration Rate	$28\text{m}^3/\text{s}$
Particulates $d > 0.5\,\mu\text{m}$	$< 3530/\text{m}^3$
Radon Activity	$1\,\text{Bq/m}^3$

saw significant air exposure. Since the nylon vessels will be filled with water for many weeks during commissioning, it is expected that much of the deposited contamination will be removed. Details can be found in reference [7].

ADSORPTION

Data from the Counting Test Facility (CTF), a small scale prototype of the Borexino detector, initiated the group's interest in adsorption from liquids. The CTF uses a nylon vessel with a one meter radius that is filled with the same liquid scintillator as Borexino. It is surrounded by a water shield and PMTs that observe the scintillator. The total CTF surface activity on the nylon vessel is several times the total internal activity of the scintillator. The surface activity has shown a slow decay over time roughly consistent with the presence of ^{210}Pb, uncorrelated with changes in the bulk activity.

Adsorption describes processes where molecules from a gas or liquid bind to a surface. Adsorption is an exothermic process that is broadly classified as either physical, involving Van der Waals binding, or chemical, where chemical bonds are formed. Physical adsorption is a reversible process that approaches an equilibrium whereas chemical adsorption is an irreversible process often involving the dissociation of molecules.

A flat surface typically has a surface binding site density $n^s = 10^{-5}\text{moles/m}^2$ or $6 \times 10^{18}\text{sites/m}^2$. This number reflects how many molecules can be attached in a monolayer in a given surface area. When a surface is in contact with a liquid all of the binding sites are occupied. The mole fractions of species on a surface N_i^s are generally different from the species' mole fractions in the liquid or gas.

The following describes physical adsorption at the interface of a binary solution and a solid surface. The analysis is simplified when the solution is dilute[8]. Let the solvent be component 1 and the solute be component 2. The moles of solute n_2^s adsorbed per area is related to the mole fraction of solute on the surface N_2^s and the moles of adsorption sites n^s per area by

$$n_2^s = N_2^s n^s \tag{7}$$

The equilibrium constant for adsorption to a surface is

$$K = \frac{N_2^s a_1}{N_1^s a_2} = e^{-\Delta G^o/RT} \tag{8}$$

where a_i is the concentration of component i in solution and the N_i^s are the mole fractions on the surface. ΔG^o is the adsorption free energy. Since the solution is dilute, a_1 is constant and

$$N_2^s = \frac{n_2^s}{n^s} = \theta = \frac{ba_2}{1+ba_2} \tag{9}$$

where we define $b = K/a_1$ and use the fact that $N_1^s + N_2^s = 1$. θ is the fractional surface coverage of the solute. If ba_2 is much less than one then the mole fraction of solute on the surface is simply proportional to the concentration in the liquid.

TABLE 3. ^{210}Pb removal from various cleaning methods on stainless steel. These results are representative of the relative efficiencies of each method. They are also indicative of the desorption rate for each process.

Soak Treatment	Temperature	Time	Percent Remaining
Pseudocumene	20° C	10 Hours	96.1±1.1%
Pseudocumene	20° C	1 Day	95.9±0.8%
Water	50° C	1 Day	63.7±0.9%
Brulin Detergent	50° C	1 Day	16.0±0.6%
Brulin Ultrasonic	20° C	1 Hour	11.7±0.5%

Partitioning Measurements

When pseudocumene was delivered for the experiment it contained relatively high concentrations of contaminants including ^{210}Pb. Before it can be purified the pseudocumene comes in contact with pipes, storage tanks, and column packing material. These stainless steel surfaces have an area of hundreds of square meters. This steel had the potential to adsorb a significant fraction of the contaminants. These surface contaminants could later be released into the pseudocumene after it had been purified.

A laboratory test was done to determine the equilibrium constant for the partitioning of ^{210}Pb between an electropolished stainless steel surface and pseudocumene. Pseudocumene was exposed to a high concentration of radon gas which was allowed to decay producing ^{210}Pb. The electropolished stainless steel sample (3 cm×3 cm) was then soaked in the spiked pseudocumene (~500 ml). The activity of the ^{210}Pb in the pseudocumene and on the stainless steel sample was then measured to find K. Two 2" photomultiplier tubes were used to measure the ^{210}Po content of the scintillator using a pulse shape discrimination technique from which the ^{210}Pb concentration was derived. The steel sample was counted with a planar germanium detector which observed the 46 keV ^{210}Pb gamma.

The results of the measurements are as follows. The measured ^{210}Pb rate of 36 kBq/m^3 in the scintillator implied a lead concentration $a_2 = 6.1 \times 10^{-11}\,\text{mol/m}^3$. The surface density of ^{210}Pb with a rate of 1.2 Bq/cm^2 is $n_2^s = 2.0 \times 10^{-11}\,\text{mol/m}^2$. Thus the fraction surface coverage $N_2^s = 1.3 \times 10^{-6}$ and $N_1^s \approx 1$. The concentration of PC is $a_1 = 7.3 \times 10^3\,\text{mol/m}^3$, and stays essentially constant since the solution is dilute. Thus we find

$$K = 1.6 \times 10^8 \tag{10}$$

and $b = K/a_1 = 2.1 \times 10^4\,\text{m}^3/\text{mol}$.

Wash-Off Measurements

The reaction rates at which a liquid and adsorbed phase approach equilibrium are not necessarily the same for adsorption and desorption. While contaminants are adsorbed quite rapidly, desorption is a much slower process. This has implication for various cleaning methods. A set of measurements was done to determine the time scale for desorption of ^{210}Pb on steel.

The results of the desorption measurements are shown in figure 3. The graph shows that the desorption appears to be a function of log time. This is unfavorable for cleaning processes that involve flushing or soaking pipes and tanks with water because removal of surface contaminants is slow. Measurements of the desorption rate of ^{210}Pb from nylon into water also show a logarithmic dependence on time, though the rate is faster. It is suspected that the desporption rate is generally logarithmic for radon progeny bound to flat surfaces. The rate is a function of the substrate, adsorbed species, contact fluid, agitation, and temperature.

Several measurements have been done to test the effects of aggressively cleaning and not cleaning steel samples with a detergent. Data are shown in table 3. With no detergent cleaning of lines and tanks, desorption is a problem for recontamination in Borexino. These measurements show that the desorption rate of ^{210}Pb from steel into pseudocumene is slower than desorption into water.

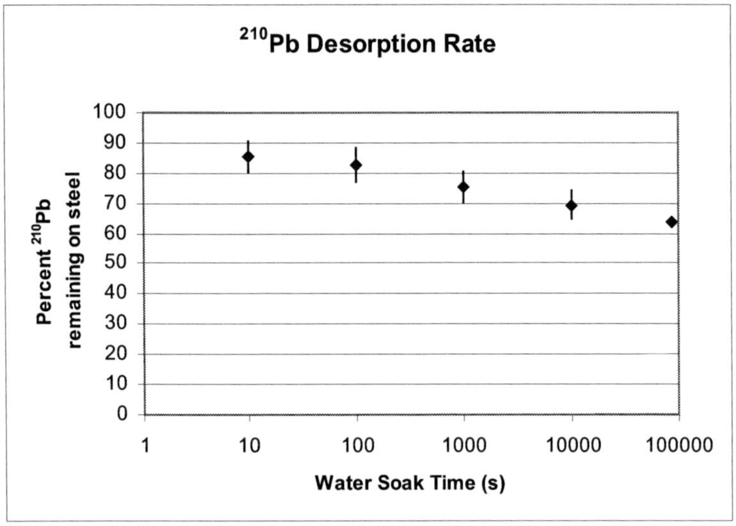

FIGURE 3. The desorption of ^{210}Pb from stainless steel into water as a function of time. The steel (3 cm×3 cm) was soaked in one liter of 10 MΩ de-ionized water at 50° C. The first four data points were made with the same sample that was repeatedly counted and soaked. A separate sample was measured with higher statistics after a one day soak. The 10 second data point exhibited higher ^{210}Pb removal due to agitation during the water soak.

SUMMARY

Radon progeny are a source of contamination that low background experiments must contend with. ^{210}Pb in particular which is produced from ^{222}Rn has a long half life and is followed by two high energy decays before a stable isotope is reached. While bulk contamination of materials from radon and its progeny must be dealt with, adsorption of these impurities on surfaces can also be significant. When considering surface contamination in designing a low background experiment it would be useful to know the following:

1) The partitioning constants for radioactive elements in a liquid between the different surfaces they contact.
2) The desorption rates to determine potential recontamination and to understand the effectiveness of cleaning.
3) The radon concentration, effective deposition velocity and time of exposure during fabrication. Better would be the concentration of each progeny concentration, deposition velocity, and exposure time.

This information can be attained though small-scale laboratory measurements if not previously avaliable.

ACKNOWLEDGMENTS

I would like to thank Frank Calaprice, Jay Benziger, Cristian Galbiati, Tom Shutt, and Kevin McCarty for their ideas and help with the measurements. The CTF measurements were made possible by the Borexino collaboration and staff working in Italy. This work was made possible by National Science Foundation Award Number 002961-001.

REFERENCES

1. The Borexino Collaboration, G. Alimonti et al., *Astropart. Phys*, **16**, 205-234, 2002.
2. W. Nazaroff and A. Nero, *Radon And Its Decay Products In Indoor Air*, John Wiley & Sons Inc., New York, 1998.
3. G. Schiller, *Ph.D. Thesis*, University of California Berkeley, Princeton NJ, 1984.
4. W. Jacobi, *Activity and potential alpha energy of* ^{222}*Rn and* ^{220}*Rn daughters in different air atmospheres*, Health Physics, **22**, 441, 1972.
5. T. Shutt, *Radon Plateout Program*, Internal Borexino Memo, Princeton NJ, 2000.
6. A. Pocar, *Ph.D. Thesis*, Princeton University, Princeton NJ, 2003.
7. A. Pocar, *Low background techniques for the Borexino nylon vessels*, Low Radioactivity Techniques 2004 proceedings, 2005.
8. A. Adamson and A. Gast, *Physical Chemistry of Surfaces*, John Wiley & Sons Inc., New York, 1997.

VII – LIQUID SCINTILLATORS AND WATER PROPERTIES AND PURIFICATION

Liquid Scintillator Purification

Y. Kishimoto

Research Center for Neutrino Science, Tohoku University

Abstract. The KamLAND collaboration has studied background requirements and purification methods needed to observe the ^{7}Be neutrino from the sun. First we will discuss the present background situation in KamLAND where it is found that the main background components are ^{210}Pb and ^{85}Kr. It is then described how to purify the liquid scintillator. The present status and results on how to remove ^{210}Pb from the liquid scintillator are discussed. Specifically, the detailed analysis of the effects of distillation and adsorption techniques are presented.

Keywords: Liquid scintillator, purification
PACS: 29.40.Mc

INTRODUCTION

Liquid scintillator is widely used in neutrino physics. One example is KamLAND, which recently reported results on the observation of neutrino oscillations from nuclear power reactors [1]. Also, many future neutrino experiments, such as the proposed $\theta_{1,3}$ [2] experiments, will use liquid scintillator in the detectors. Currently one of the most stringent applications of liquid scintillator purification is the program for ^{7}Be neutrino observation with KamLAND. The liquid scintillator must be reduced to extremely low levels of naturally accurring radio-isotopes to succeed in a real time observation of ^{7}Be neutrinos from the sun[3]. Neutrino experiments as a whole, such as dark matter and double β decay experiments, demand ultra-low backgrounds to reach the required sensitivities to observe the physics. This in turn requires new novel techniques to purify the liquid scintillator. It is thus worthwhile to discuss how to purify liquid scintillator. The status and results of such studies for use in KamLAND ^{7}Be neutrino observation is presented in this report.

KAMLAND BACKGROUND STATUS

In this section, we will discuss what the most important backgrounds to reduce are in order to observe ^{7}Be neutrino. Figure 1 shows the low energy singles spectrum of KamLAND. In this figure the expected ^{7}Be and pp neutrino spectra, with no oscillations, are also plotted.

It is clear that the β spectrum from ^{210}Bi and ^{85}Kr provided the largest contribution of singles events in the region above 0.4 MeV. It is also found that the peak around $E \sim 0.3$MeV fits nicely to the α spectrum of ^{210}Po.

If the rather short lifetimes of ^{210}Bi($T_{1/2} = 5.0$d) and ^{210}Po($\mathrm{T}_{1/2} = 138$d) are taken into consideration, this provides evidence their mother nuclear ^{210}Pb is feeding the stable rate of these isotopes seen in KamLAND. The concentrations of ^{238}U and ^{232}Th in KamLAND are measured by observing the β and α decays of ^{214}Bi $-^{214}$Po with a delayed coincidence technique. Assuming radiochemical equilibrium with the parent nuclei, the ^{238}U and ^{232}Th concentrations are found to be 3.4×10^{-18}g/g and 5.6×10^{-17}g/g, respectively. The concentration of these isotopes are already low enough for a ^{7}Be measurement.

The concentration of ^{40}K was estimated from the 1.4MeV γ-ray peak in the singles spectrum of KamLAND. The concentration is found to be on the order of 10^{-16}g/g. For other radio-isotopes, we have found no clear evidence of their presence.

The status of backgrounds in KamLAND is summarized in TABLE 1. The methods used to determine these values are described in [3].

As shown in TABLE 1, we have to reduce ^{210}Pb and ^{85}Kr five or six order of magnitude which is a very challenging task.

CP 785, *Topical Workshop on Low Radioactivity Techniques,* edited by B. Cleveland, R. Ford, and M. Chen

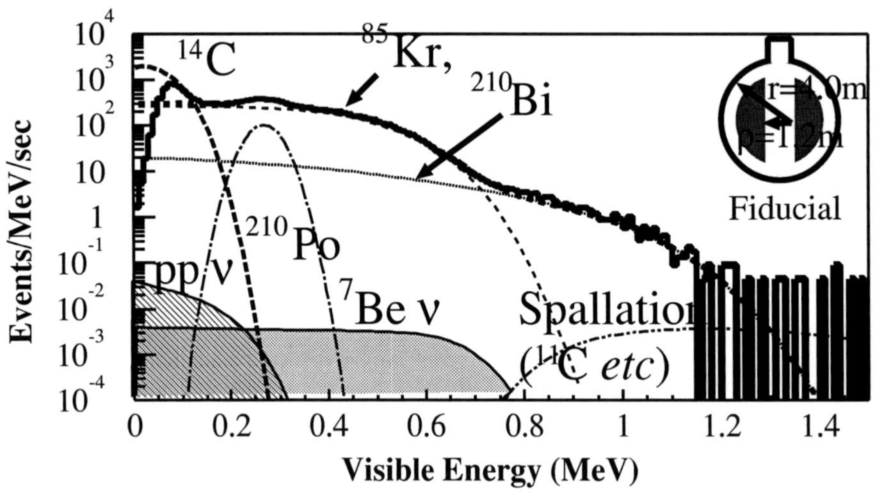

FIGURE 1. Low energy spectrum of KamLAND. The thick solid and thick long-dashed lines indicate the actual data and the estimated background contributions, respectively. The main background components used in this estimate are, *ie,* ^{85}Kr, ^{210}Bi, and ^{210}Po, which are drawn in the figure in the thin dashed, the thin doted, and the thin dash-doted lines, respectively. The expected ^{7}Be and *pp* neutrino events, without oscillation, are also indicated with the shaded and the hatched regions, respectively.

TABLE 1. Background concentration in KamLAND

species	concentration	target level
^{238}U	3.4×10^{-18}g/g	10^{-18}g/g
^{232}Th	5.6×10^{-17}g/g	10^{-17}g/g
^{40}K	2.7×10^{-16}g/g(Max.)	10^{-17}g/g
^{210}Pb	1.1×10^{-20}g/g	5×10^{-25}g/g
^{85}Kr	0.46Bq/m^3	10^{-6}Bq/m^3

LIQUID SCINTILLATOR PURIFICATION

As clearly seen in TABLE 1, the important issue to address in order to observe ^{7}Be neutrinos are the removal of ^{85}Kr and ^{210}Pb. This report focuses on how to remove ^{210}Pb. The possible purification methods of ^{85}Kr, such as ultra-pure-nitrogen gas-purge, are discussed in [4].

Chemical form of ^{210}Pb

^{210}Pb is believed to have originated from ^{222}Rn in the air. The ^{222}Rn came in to contact with the liquid scintillator during the filling of KamLAND. ^{222}Rn is highly soluble in organics and it would not take much radon to see the level of activity we observe in KamLAND. The daughter nuclei,^{210}Pb, has accumulated in KamLAND due to it having the longer lifetime, $T_{1/2} = 22.3$y, of the radon daughters.

The chemical form of ^{210}Pb is believed to be mainly ionic or metallic. There also seems to be an organic-lead component in the liquid scintillator. Organic-lead can be made in the decay process of ^{222}Rn: ^{210}Pb, the mother nucleus of ^{210}Pb, decays with a $Q = 7.7$MeV and emits one α-particle. The recoiled particle, ^{210}Pb, is energetic enough to ionize the ambient liquid scintillator. In this ionization process, there are a lot of organic radicals and highly ionized lead produced, Pb^{+n}. Pb^{+n} is unstable, so almost all of these isotopes are neutralized to Pb or settle into the Pb^{+2} state. However, a very tiny amount interact with organic-radicals and make organic-lead.

Many forms of organic-lead disintegrate because they are unstable at room temperature, and a small portion are stable enough to remain in the KamLAND scintillator. There are experimental results which imply the presence of an organic-lead component, but this will be discussed elsewhere.

Candidates and issues

As discussed above, the chemical forms of ^{210}Pb in the liquid scintillator are metallic, ionic, and organic. The former two are the dominant components, however, due to the very small event rate of the ^{7}Be neutrino the last form cannot be ignored.

We have to be able to remove *all* of the forms of ^{210}Pb. One other important point is that the liquid scintillator is not a pure chemical, but chemical mixture: KamLAND scintillator is made of 80% of dodecane and 20% of pseudo-cumen(PC) with $1.5\text{g}/\ell$ of diphenyloxazole(DPO). This chemical form has unique properties for our detector capabilities that must be retainted after purification.

A list of candidates that could reduce ^{210}Pb in liquid scintillator are: distillation, adsorption, water-extraction, and so on. Each technique has its merits and demerits.

Which technique is adopted will depend on how the optical properties,i.e. attenuation length and light yield, are preserved after purification and what the reduction efficiency of ^{210}Pb is for this method.

Distillation technique

The merits for the distillation technique in removing ^{210}Pb are that we expect high reduction efficiencies for *all* chemical components, if their boiling points are different from that of dodecane, PC, or PPO. One of the demerits is that the distillation process separates dodecane, PC and PPO. The most serious demerit is the safety: The distillation temperature is above the flash-point of the liquid scintillator.

The experiments performed on finding the reduction efficiencies with the distillation technique are carried out using dodecane soaked with ^{212}Pb from the ^{232}Th series, not ^{210}Pb. These techniques operate under the assumption that process in which the ^{212}Pb isotope is born into the liquied scintillator is assumed to be the same in which the ^{210}Pb was introduced into the KamLAND scintillator.

The experimental procedure is as follows:

1. Dodecane with ^{212}Pb is prepared by bubbling with ^{220}Rn from ^{228}Th.
2. The soaked dodecane is separated.
3. One portion is distilled and then the distilled dodecane is separated for another distillation step.
4. Pure PC and PPO are then added to each sample to make the liquid scintillator.
5. The activities of ^{212}Pb from the original and distilled samples are measured as a function of time.
6. The individual activities are then compared with each other.

The measurements are performed with PMTs, and the delayed coincidence of ^{212}Bi-^{212}Po is applied to reduce the ambient backgrounds. The time variation in the delayed α-particles are plotted in FIGURE 2.

We tried performing distillation many times on the same sample to see if we could achieve higher purification efficiencies. The results are shown in TABLE 2. We can expect a reduction of 10^{-4} in the first distillation. We can remove still more ^{212}Pb after a multi-step distillation, even if the overall reduction efficiencies get worse between each step. We have obtained the reduction efficiency of 3×10^{-5} after a total of seven distillations.

The reduction efficiency achieved with the distillation technique is rather good, so the next studies are performed on the liquid scintillator properties after the distillation. To investigate them, we prepared a sample of ordinary liquid scintillator and one made of distilled dodecane, distilled PC and distilled PPO. We measured the light attenuation as a function of the length of the path and calculated the attenuation length. The results are listed in TABLE 3. No significant changes are seen between the samples. We also obtained the light-yield by comparing the PMT charge output from the Compton edge of γ-rays from ^{137}Cs. We found 1392 ± 4[ch] for the ordinary scintillator and 1401 ± 7[ch] for the distilled one. (The unit is ADC channel.)

These results are consistent with each other.

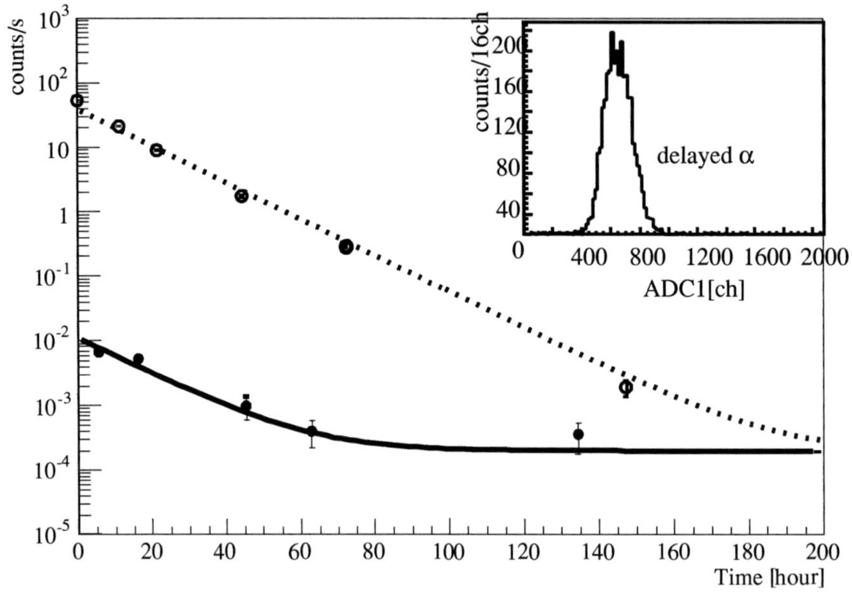

FIGURE 2. Example of distillation result. The delayed α peaks are measured as a function of time to estimate the activities before and after the distillation process. The open and closed circles show the counting-rates of liquid scintillator before and after a distillation, respectively. The delayed alpha spectrum is also shown in the upper right corner of the figure.

TABLE 2. Reduction efficiency after distillation. Dodecane samples are distilled many times, and the activities are measure after each distillation.

Run	**First**	**Second**	**Fifth**	**Seventh**
Run 1	2.7×10^{-4}	7.0×10^{-5}	4.0×10^{-5}	NA
Run 2	1.6×10^{-4}	4.1×10^{-5}	4.1×10^{-5}	NA
Run 3	1.3×10^{-4}	2.9×10^{-5}	4.3×10^{-5}	NA
Run 4	1.3×10^{-4}	NA	4.3×10^{-5}	3.1×10^{-5}
Run 5	5.3×10^{-4}	NA	6.4×10^{-5}	4.3×10^{-5}
Run 6	2.7×10^{-4}	NA	3.4×10^{-5}	3.2×10^{-5}

We found no difference between the properties of the ordinary scintillator and those of the distilled sample. Combining these results with the high reduction efficiencies after multi-step distillation, the distillation technique is quite promising for the purification of ^{210}Pb.

TABLE 3. Attenuation lengths of the liquid scintillator

Sample	**L[m] at 365nm**	**L[m] at 436nm**
Ordinary LS	1.4 ± 0.5	12.7 ± 0.4
Distilled LS	1.0 ± 0.3	11.2 ± 0.4
Adsorbed LS *	1.3 ± 0.4	9.3 ± 0.2

* The tested dsobent is SiO_2

TABLE 4. Best reduction efficiencies in various tested adsorbents.

Sample	**Reduction**
Selecto Si-Gel (32-63μm)	$(5.24 \pm 0.13_{stat.} \pm 0.40_{sys.}) \times 10^{-2}$
Selecto Alusil 70	$(3.7 \pm 0.1_{stat.} \pm 0.4_{sys.}) \times 10^{-2}$
Crushed Cu/Mn Ctalyst	$(3.8 \pm 0.1_{stat.}) \times 10^{-2}$

Adsorption technique

The merits of the adsorption technique are the safety and the ease of operation. This technique has been studied and the studies show that it is effective for ionic and metallic lead, which are the dominant components in the scintillator.

To study the reduction efficiency of ^{212}Pb, we prepared ^{212}Pb-soaked liquid scintillator by bubbling ^{220}Rn. Then the soaked scintillator is put through a column containing an adsorbent, or is mixed with it in a flask. After separating the liquid scintillator form the adsorbent, the scintillator is counted using the γ-rays of ^{212}Pb and the daughters' with a high purity Ge detector. We tested various adsorbents and the three best reduction efficiencies obtained so far are presented in TABLE 4. The experiments show that more than 95% of ^{212}Pb is removed and that the adsorption technique is effective.

We also investigated the attenuation length and light yield of the scintillator after adsorption. The results of the attenuation length are shown in TABLE 3. We see no serious change in the attenuation length measurement, but we found that the light yield after the adsorption is quite small. This is because the PPO is also removed by the adsobent: The 1.5 ± 0.2g$/\ell$ of PPO concentration is reduced to 0.19 ± 0.03g$/\ell$ after adsorption. Faint light output results from this PPO reduction.

One of the other drawbacks in using adsorption is the radioactive concentration in adsorbents. Especially potassium, the concentration of natural potassium is on the order of 1000ppm in commonly available SiO_2 and this can be a source of ^{40}K.

In summary, the adsorption technique removes more than 95% of Pb, and it is effective to some extent, but it also remove PPO which changes the chemical ratios in the scintillator. There are also the potential concerns about emanating radioactivity back into the scintillator.

CONCLUSIONS

We discussed the relative amounts and chemical forms of the background present in KamLAND. The studies performed on the distillation and the adsorption techniques used to purify the liquid scintillator have also been described. The distillation reduces one of the main background components, ^{210}Pb, very effectively. The efficiency is about 10^{-4} after the first distillation and 3×10^{-5} after the seventh distillation. The properties of the liquid scintillator after the distillation were also studied and no problems have been found so far. The adsorption technique is also effective for reduction of lead in the liquid scintillator: 3-5 $\times 10^{-2}$ reduction are achieved in several different adsorbents. The issues on the adsorption technique are that it removes PPO and that there is the concern of emanating radioactive impurities.

From the results so far, it can be stated that the distillation is promising. We are now going to install a larger size distillation apparatus for in depth studies of purifying the liquid scintillator. We have plans to start the actual KamLAND liquid scintillator purification process in 2006 and look to observe the ^{7}Be neutrino in 2007.

ACKNOWLEDGMENTS

All studies are supported by the KamLAND collaboration.

REFERENCES

1. T. Araki, *et al.*, *Phys. Rev. Lett.*, **94**, 081801 (2005)

2. For example, Double CHOOZ (http://doublechooz.in2p3.fr/), KASCA(http://kaska.hep.sc.niigata-u.ac.jp/) *etc.*.
There are other groups to measure θ_{13}. See http://neutrinooscillation.org/ for other information.
3. Y. Kishimoto, talk in *LowNu2003*, May 19-21, 2003, Paris http://cdfinfo.in2p3.fr/LowNu2003/Transp/kishimoto.pdf
4. K. Furuno, takl in *LowNu2002*, May 22-24, 2002, Heidelberg http://www.mpi-hd.mpg.de/nubis/www_lownu2002/transparency/furuno_lownu_2002.pdf

Assay of Radon and Radium in Water: Techniques Developed at SNO

Jacques Farine
for the SNO Collaboration

Department of Physics and Astronomy, Laurentian University, Ramsey Lake Road, P3E 2C6 Sudbury, Canada

Abstract. Water assay techniques developed for measuring ^{222}Rn, ^{226}Ra and ^{224}Ra in the SNO detector are presented. Recent upgrades to improve the performance of the techniques and to increase the sensitivity to lower levels are discussed.

Keywords: trace radon, radium and thoron monitoring, electrostatic counters, hydrous titanium oxide, manganese oxide
PACS: 29.30.Aj,29.40.-n,29.40.Cs,29.40.Wk

INTRODUCTION

The SNO detector [1] records solar neutrino interactions using Cherenkov light produced in the following reactions: (1) elastic scattering (ES): $\nu_x + e^- \rightarrow \nu_x + e^-$, (2) charged current (CC): $\nu_e + d \rightarrow e^- + p + p$, and (3) neutral current (NC): $\nu_x + d \rightarrow \nu_x + n + p$, where ν_x refers to any active flavor of neutrino. In order to make a sensitive measurement of the NC reaction, the production of free neutrons from background radiation by photo-disintegration of the deuteron, $d + \gamma \rightarrow p + n$, must be kept below one neutron per day in the one kilotonne D_2O target. The isotopes ^{208}Tl and ^{214}Bi in the Thorium and Uranium chains emit a gamma ray with energy above the binding energy of the deuteron, 2.2 MeV (see Fig. 1). This background target sets the maximum concentration of progenies of ^{232}Th and ^{238}U in the heavy water to 3.8×10^{-15} g Th_{equiv}/gD_2O and 3×10^{-14} g U_{equiv}/gD_2O. Another background at low energy comes from β and γ decays in the natural radioactive chains. These events add to the Cherenkov light and possibly contaminate the neutrino data set. Water purification and monitoring methods are essential to obtain a good understanding of the backgrounds. The strategy chosen by SNO to determine the activity of ^{208}Tl and ^{214}Bi parents is two-fold: (1) *ex-situ* techniques extract the radioactivity from the water (^{224}Ra, ^{226}Ra, and ^{222}Rn), and count it outside of the detector, and (2) *in-situ* techniques use PMT data and rely on the different average isotropy of ^{208}Tl and ^{214}Bi decays. The first techniques benefit from relatively larger statistics (thousands of atoms extracted), but are not strongly correlated with neutrino runs. The second technique suffers from lower statistics, but this is compensated with the background being measured with the same data set as used for neutrino analysis. The two measurements are therefore complementary, and are merged to infer the number of background events in the neutrino candidates data set – an input for the final stages of neutrino signal extraction of CC, NC, and ES interaction rates.

Details on water purification techniques can be found in [1]. Recent progresses on background monitoring techniques are discussed here.

RADIOCHEMICAL ASSAYS (*EX-SITU*)

At the level of 3.8×10^{-15} g Th_{eq}/g there are about 4 atoms of ^{224}Ra per tonne of D_2O. Two competitive radium techniques have been developed that are as dissimilar as possible. They are referred to by the chemical used to extract the radium from the water: HTiO for hydrous titanium oxide and MnOx for manganese oxide. In preparation for counting, the former technique further concentrates the radium, from an initial 200 tonnes or more into 10–30 mL, mixes the concentrate with a scintillator and uses Bi-Po $\beta - \alpha$ coincidence to determine the activities from the Th and U chains respectively. The latter technique counts the radon and thoron emanating from the MnOx compound using electrostatic chambers (ESCs). Results from these two techniques are merged and produce the *ex-situ* radioactivity measurement for ^{224}Ra and ^{226}Ra.

CP 785, *Topical Workshop on Low Radioactivity Techniques,* edited by B. Cleveland, R. Ford, and M. Chen

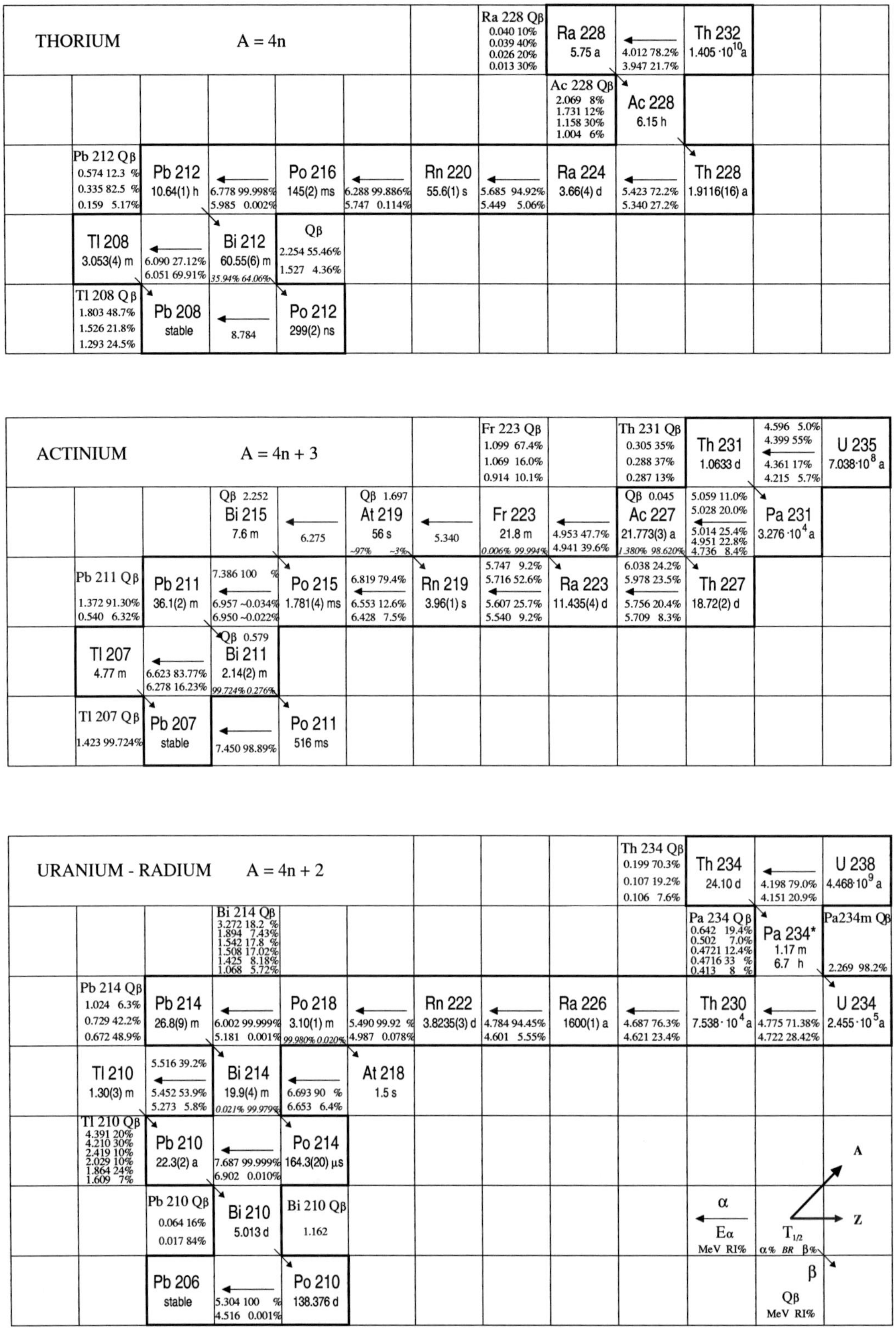

FIGURE 1. Selection of isotopes in the natural radioactive chains of relevance to low background experiments. The nuclei ^{214}Bi and ^{208}Tl induce photo-disintegration of the deuteron and form the principal source of background in SNO. Updated from [2].

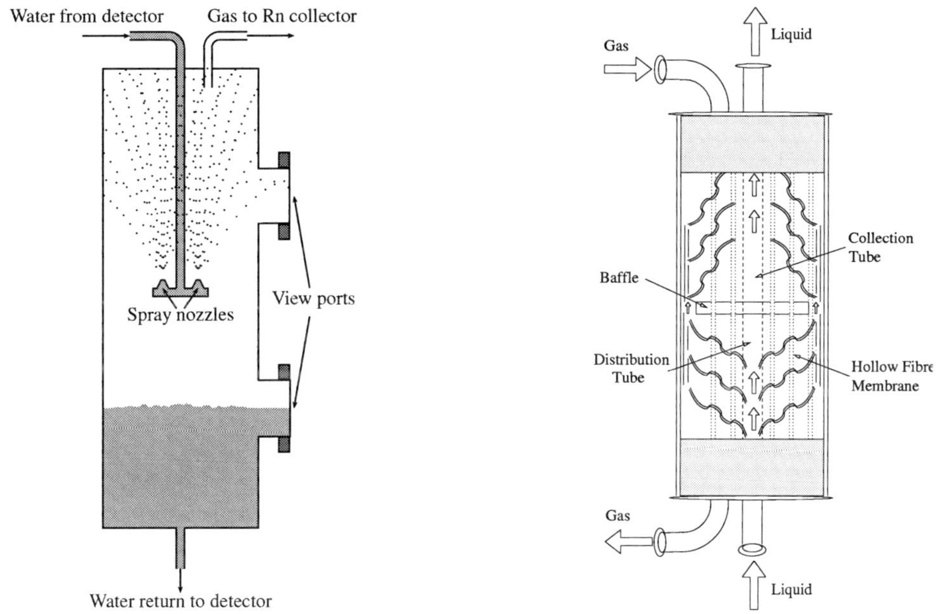

FIGURE 2. Schematics of the radon degassers used on the H_2O (left) and D_2O systems (right).

Because of its 3.82-d half-life and solubility in plastics, ^{222}Rn is not necessarily in equilibrium with the ^{226}Ra dissolved in water, and must be assessed separately. This technique is discussed next.

The Radon Assay Technique

Water from the SNO detector is passed through a vacuum degasser (in the light water system) or a membrane contact degasser (in the heavy water system) where dissolved gases, including radon, are liberated. The degasser is connected to a vacuum system which collects the radon on a cold trap and removes most other gases, such as water vapor and N_2. After roughly 0.5 tonnes of H_2O or 6 tonnes of D_2O have been sampled, the accumulated radon is transferred to a Lucas cell. The cell is mounted on a photomultiplier tube which detects the α-particles from the decay of ^{222}Rn and its progeny. The overall degassing and concentration efficiency is about 38% and the single-α counting efficiency is approximately 75%. The sensitivity of the radon assay system for D_2O is equivalent to $\sim 3\times 10^{-15}$ g U/g water.

At the target concentration given above, there are 0.2 ^{222}Rn atoms/L of D_2O (0.4 mBq/m^3).

Before concentration, ^{222}Rn is extracted from the water systems using different degassers: in the H_2O, water is projected into a vacuum vessel by three nozzles creating a very fine mist. The radon diffuses to the vacuum phase, where water vapour acts as a carrier gas. In the D_2O systems contactors were chosen to facilitate operations during the salt phase. The latest measurements of the degassers' efficiency are: 58$\pm$10% at 19 LPM in H_2O and 62^{+11}_{-9}% at 21 LPM in D_2O. The radon is then transferred by vacuum pumping to a concentrator equipped with a LN_2-cooled trap, with an efficiency of 100.5$\pm$2.3%. The transfer efficiency from the concentrator to the Lucas cell is 62$\pm$3%, essentially from volume sharing. The best Lucas cell background achieved to date is 5 counts per day. The counting efficiency is 74% per alpha. Three useful alpha particles are emitted for every Rn atom. Table 1 lists the contributions to the systematic uncertainty for this assay technique.

See [3] for more details about the technique.

The MnOx Assay Technique

A water volume of 50 m^3 (H_2O) to 400 m^3 (D_2O) from the detector is passed through columns which contain beads coated with a compound of manganese oxide onto which the Ra dissolved in the water is adsorbed. The columns are removed, dried, and mounted below an electrostatic chamber into which the Rn from the decay of trapped Ra is

TABLE 1. Contributions to systematic uncertainty in radon assays in percent. All uncertainties are symmetric unless otherwise indicated. The values for background are evaluated for a water sample whose radon content is 25% of the maximum allowable level.

Source	Inner H_2O	D_2O
Flow rate F	14	17
Flow rate F (stroke counted)	10	11
Degassing efficiency ε_{degas}	17	$^{+18}_{-14}$
Assay system background R_{back}	3	$^{+10}_{-21}$
Trapping efficiency ε_{trap}	2	2
Transport efficiency to Lucas cell $\varepsilon_{transfer}$	3	2
Cell counting efficiency ε_{count}	10	10
Cell background rate B_{Lucas}	7	2
Electronic noise	<0.1	<0.1
Photomultiplier stability	4	4
Combined quadratically	26	$^{+29}_{-32}$
Combined (stroke counted)	24	$^{+26}_{-30}$

continuously flowed by a stream of N_2 gas. The subsequent decay of Rn gives charged Po ions which are swept by the electric field onto a solid-state α counter. The content of Ra in the water is inferred from the measured decay rates of ^{212}Po, ^{214}Po, ^{216}Po, and ^{218}Po. The Ra extraction efficiency is >95%, the overall counting efficiency is 24% for ^{214}Po and 6% for ^{216}Po, and the method can detect a few atoms of ^{224}Ra per m^3 and a few tens of thousands of atoms of ^{226}Ra per m^3. Converted to equivalent equilibrium values of the topmost elements of the natural radioactive chains, the detection limit in a single assay is a few times 10^{-16} g Th or U/cm^3, as indicated in Table 2. For the actinium chain the emanation and transport efficiencies of 4 s ^{219}Rn have not been determined precisely. The combined value is estimated to 10%.

TABLE 2. Sensitivity of the MnOx radium assay technique.

Chain	Isotope	Sensitivity
Thorium	^{224}Ra	5×10^{-16} g Th/g
Uranium	^{226}Ra	2×10^{-16} g U/g
Actinium	^{223}Ra	$\simeq 10^{-14}$ g ^{235}U/g

Background Reduction

The initial design of the ESC assumed that the radium concentrated from large amounts of water into a small volume would result in activities high enough that extreme care was not necessary in the choice of O-rings, even at the SNO target levels. As the measured concentrations came below that limit, we could increase the counters' sensitivity by a factor of 3–4. This was achieved by grouping four columns from the same experiment on a single counter (the activity measured in a blank being mainly in the counter), and by sampling larger quantities of water (the integrity of the columns was shown to be preserved at 130 tonnes per column). With lower levels being pursued, of order of 1 ^{224}Ra per tonne of D_2O and below, it became of interest to try and further increase the counters' sensitivity by lowering their background. Our approach included the following actions:

Replace seals with custom-made teflon gaskets Most joints on the ESCs are made from Viton O-rings (NW25, NW16, and NW10 – only the diode feedthrough is sealed with a butyl O-ring). We replaced these O-rings with gaskets cut from a sheet of teflon and conditioned their surface with clean sand paper. Seals are very good after the conversion, but leaks tend to develop after some time. We are now designing an upgrade of the teflon seal using compression washers.

Surface contamination removal Some counters used during the development of the technique were exposed to ^{228}Th spikes, and exhibit background levels about a factor of two higher than the other ESCs. The spectrum of these higher background counters shows the 5685 keV α from ^{224}Ra, indicating that the internal walls are plated with ^{228}Th. Attempts were made to remove this activity by stripping 3 μm of the aluminum walls by chemical

TABLE 3. Evolution of backgrounds (counting rates and inferred activities) of the electrostatic chambers 7, 9, 3, and 8 after various improvements. For all runs on ESC#9 the counter inlet was sealed with a blank cap. Others had their input valve closed to isolate the chamber. Note how the replacement of this valve on ESC#7 was a determining factor.

	Date (yymmdd)	Pressure range (mb)	Counting time (d)	Counting rate (cpd) ^{214}Po	Counting rate (cpd) ^{216}Po	Inferred activity (dpd) ^{224}Ra	Inferred activity (dpd) ^{226}Ra
ESC#7							
Before actions	040519	25–30	19.25	61.1±2.2	9.1±0.8	40 (36–44)	292 (283–301)
Teflon conversion	041008	25–26	9.67	26.5±1.7	8.2±0.9	43 (40–46)	117 (109–125)
EDTA cleaning	041110	25–26	10.88	27.2±1.6	10.2±1.0	46 (43–49)	97 (90–105)
Inlet valve exchange	050513	25–27	18.10	7.1±0.6	2.7±0.4	14 (12–16)	33 (31–36)
ESC#9							
Before actions	040408	26–26	19.25	13.9±0.8	2.5±0.4	9 (8–10)	60 (57–62)
EDTA cleaning	040828	24–39	17.75	12.5±0.9	1.2±0.3	6 (5–7)	29 (26–33)
Teflon conversion	040917	26–28	13.03	6.1±0.7	1.5±0.3	5 (4–6)	18 (15–21)
ESC#3							
Before actions	050309	26–30	9.68	22.5±1.5	5.4±0.8	25 (23–28)	118 (110–126)
EDTA and Teflon conversion	050322	24–34	13.06	18.4±1.2	2.7±0.5	12 (10–13)	75 (70–80)
ESC#8							
Before actions	050517	26–27/8	14.78	16.1±1.0	5.3±0.6	23 (21–25)	73 (69–77)
EDTA and Teflon conversion	050601	26–26	18.50	6.5±0.9	5.6±0.5	26 (24–27)	25 (23–27)

attack. EDTA was dissolved in UPW to prepare a 0.1 M solution, corresponding to a pH=10. The chamber was disassembled, wiped with methanol, and threads were protected using polypropylene bolts. The chamber was rinsed twice with UPW then immersed for 2 hours in the EDTA solution under constant agitation. After fast draining the chamber was rinsed three times with EDTA, then washed and dried using methanol. It was then put back together and a background run started to measure the background of the chamber alone.

Replacement of contaminated elements Thorium spikes used in development work have probably deposited ^{228}Th all over in those counters. In the course of normal operations, most parts have been replaced over the years, except for a few valves. The example of ESC#7 in Table 3 shows that one particular valve was a major contributor to the background. We also investigated if the Tygon tubing used for the recirculation loop was contributing significantly. Brand new tubing has been finely cut to expose a total area of 1800 cm^2, about ten times more than exposed to a typical counter. The pieces were ultrasonically cleaned, dried, and counted. The activity was rather high, of order of hundreds of decays per day. But, interestingly, this turned out to be outgassing of dissolved ^{222}Rn. After the run the Tygon was kept under vacuum for two weeks with periodic flushes using clean dry nitrogen. After this conditioning, barely detectable levels of Th and U were observed, making Tygon a very small contributor to the background, at the level of a few %. The inferred emanation rate for Tygon is 11^{+22}_{-17} ^{220}Rn m^{-2}day^{-1} and 139 ± 61 ^{222}Rn m^{-2}day^{-1}.

These actions were combined in various ways on different counters. Their influence on background levels is given in Table 3. One observes that in general the chemical attack did not reduce the Th background substantially. The U rates went down but they are more difficult to interpret, due to interference with Rn ingress from leaks. A leak of order of 1 mbar/day contributes at most an equivalent activity in ^{226}Ra of 60 dpd, depending on the Rn level in the laboratory air. Replacing the gaskets reduced the backgrounds in U in all counters, but only in half the cases for Th.

Clearly there is uncertainly associated with these attempts, but the background could be reduced by a factor of two in three counters out of four for Th and by a factor of two to almost ten in all counters for U.

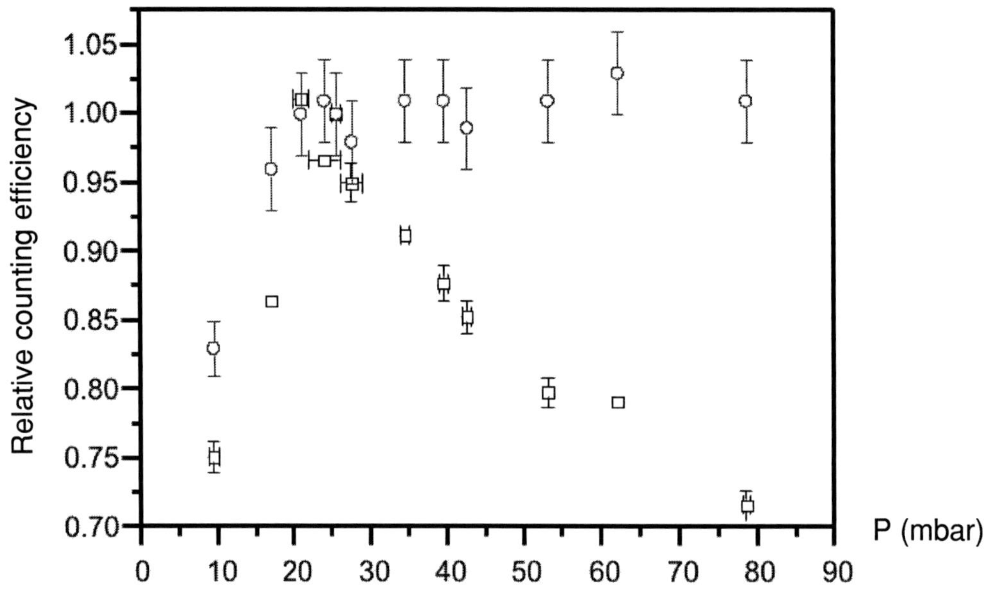

FIGURE 3. Measurement of the relative ESC counting efficiency versus pressure for ^{216}Po (squares) and ^{212}Po (circles), using an MnOx column spiked with ^{228}Th. The counting efficiency is normalized at 25 mbar, where it amounts to 22.5% for ^{216}Po.

Calibrations

A number of calibration sources with known activities of ^{224}Ra, ^{222}Rn, ^{228}Th are used periodically to check the relative counting efficiency between counters, between columns configurations on the counters, reproducibility, and the dependence of the counting efficiency with pressure or drift high voltage for the various α-emitters used in the analysis.

Figure 3 shows a recent measurement of the relative ESC counting efficiency versus pressure for ^{216}Po and ^{212}Po, using an MnOx column spiked with ^{228}Th. The counting efficiency is normalized at 25 mbar, where it amounts to 22.5% for ^{216}Po. Even if the pressure increases enough that the counting efficiency of ^{216}Po drops substantially, it does not affect the ability to count ^{212}Po, and the assay can be evaluated,

See [4] for more details about this technique.

The HTiO Assay Technique

This radium assay technique uses hydrous titanium oxide (HTiO) adsorbent on a filtration membrane together with a $\beta-\alpha$ delayed coincidence counting system. For a 200 tonne assay the detection limit for ^{232}Th is a concentration of $\sim 3\times10^{-16}$g Th/g water and for ^{238}U of $\sim 3\times10^{-16}$g U/g water.

Hydrous titanium oxide is an inorganic ion-exchanger that has been used in the nuclear industry for the extraction of heavy ions. It is formed as a white colloidal suspension by the hydrolysis of titanium salts, which forms particles of sizes ranging from 1 to 100 microns, with a mode at around 10 microns.

In the HTiO assay procedure there are five steps: deposition, extraction, elution, secondary concentration, and counting. The HTiO ion-exchanger is first deposited onto microfiltration membranes: 0.25 m long filters for a light water assay and 1.0 m long filters for a heavy water assay. Then the columns containing the loaded filters are taken underground to extract ^{224}Ra from the Th chain and ^{226}Ra from the U chain from a large volume of D_2O ($\sim$200 tonnes) or H_2O ($\sim$30 tonnes). After extraction, the columns are brought back up to the surface laboratory, and the radium is eluted with 15 L of dilute nitric acid. In the secondary concentration stage, the radium in the 15-L nitric eluate is further concentrated down to $\sim$10 mL of final eluate. This is then mixed with $\sim$40 g of liquid scintillator and counted using a $\beta-\alpha$ delayed-coincidence liquid-scintillation counter. Details about this technique can be found in [5].

The procedure has been modified since to offer the following advantages (see Figure 4): (1) less interference from

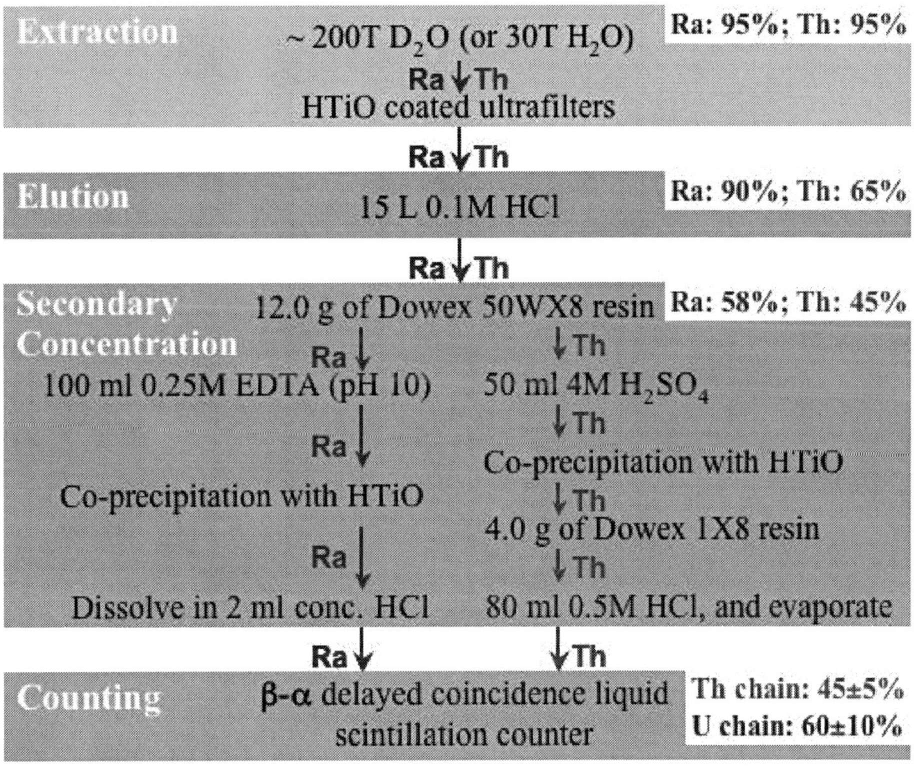

FIGURE 4. Modified HTiO procedure for ^{228}Th, ^{224}Ra and ^{226}Ra in SNO water. See text for details.

Mn and Ni in the D_2O (the previous secondary concentration step uses the HTiO co-precipitation method to concentrate Ra, and its efficiency could be significantly reduced in the presence of Mn and Ni); therefore, the new procedure is much more reliable for the heavy water assay in NCD phase; (2) less occupancy of the $\beta - \alpha$ counters (only 2 samples are produced for the modified procedure: one assay sample plus one blank; while 6 samples were taken for the previous method: three assay samples plus three blanks).

This modified HTiO procedure shows total chemical efficiencies of 50$\pm$8% for Ra and 28% for Th. The total efficiencies are 30$\pm$7% for ^{226}Ra, 22$\pm$4% for ^{224}Ra, and 12% for ^{228}Th. This procedure could have a variety of applications in the measurement of low-level natural radioactivity in different environmental and biological samples.

The following describes a variation optimized for the assay of radium and thorium from leaching tests, in which the sample to be measured is immersed into 15 L of UPW for between 4 days and a week:

Extraction 1-2 mL of 15% $Ti(SO_4)_2$ solution are added and NaOH is used to titrate to pH 9. Ra and Th co-precipitate with HTiO. The efficiencies are 98$\pm$2% for Ra and 95$\pm$5% for Th

Elution the HTiO precipitate is trapped onto small ultrafilters and Ra and Th are eluted into 10 mL of 0.5 M HCl. The efficiencies are 90$\pm$10% for both Ra and Th

The total, chemical efficiencies are 86$\pm$10% for Ra and 88$\pm$10% for Th. The same counting efficiencies apply, leading to overall efficiencies of 51$\pm$11% for ^{226}Ra, 38$\pm$6% for ^{224}Ra, and 40$\pm$6% for ^{228}Th. The procedural blanks return 0.3$\pm$0.1 cph for ^{226}Ra and $<$0.05 cph for ^{224}Ra and ^{228}Th.

HTiO can also be used to measure ^{238}U directly in the water. The measured extraction efficiency is 95$\pm$5%. The U is eluted into 0.03 M HNO_3 with an efficiency of 90$\pm$10%. The solution is then sent for ICP-MS analysis. The detection limit for a 200-tonne assay is lower than $\sim 10^{-16}$ g U/g.

Another possible use of HTiO is the purification of Ra, Pb, U, and Th isotopes, from a number of media that include water, salt and liquid scintillator. The purification methods rely on HTiO co-precipitation, HTiO-loaded ultrafiltration, and HTiO-loaded resin.

TABLE 4. ^{224}Ra assay results for elements of the D_2O systems, compared to the total activity measured in the complete system.

Element of heavy water system	Th bkgd. (^{224}Ra/day)
Ultrafiltration unit (assay system)	28^{+27}_{-23}
Heat exchanger	233^{+50}_{-48}
Ultrafiltration unit (main stream)	78^{+37}_{-31}
Main recirculation pump	72^{+33}_{-30}
Process degasser	75^{+31}_{-29}
Filtration unit (main stream)	< 36
Piping	330 ± 184
Complete system	816 ± 165

Link of Assay Results to Neutrino Data

The *ex-situ* assays return the concentration of sources of background in the water feeding the extractors. These results must be translated into effective concentrations a) in the fiducial volume used for PMT data analysis and b) during the times neutrino data was being acquired.

Both effects are taken into account with the development of a multiple sources model, that identifies sources of activity in the water systems, and uses the system's history (flow rates, paths and times). During a normal assay, the activity measured on the MnOx assay columns is the sum of contributions from the D_2O and from the piping leading to the columns. These sources were found by a combination of assays of the water system with the SNO detector bypassed and so-called 'Peristaltic assays' – extractions of small, isolated subsets of the systems using a peristaltic pump. For the latter extractions, advantage was taken of long, idle periods of the D_2O systems, during which all valves were kept closed (the normal sleep state). The radium leach rate of components like filters, degassers, heat exchangers, and pumps were studied. Table 4 shows the observed activities during the salt phase. These measurements do not include the small piping section in the detector. Prior to salt addition the total activity measured in the systems was <16 dpd. The salt brine was assayed, showing that no Th had been added. Measurements taken recently in pure D_2O showed that most of the activity is gone with the salt. Our interpretation is that the presence of Cl and Na in the water has displaced the equilibrium between ^{224}Ra fixed to its sources in the system and ^{224}Ra in solution.

IN-SITU TECHNIQUE: USING LOW ENERGY PMT DATA

The *in-situ* technique uses pattern recognition on the Cherenkov light distribution to determine the equivalent concentration of ^{232}Th and ^{238}U in the water. The decays of ^{208}Tl produce a more isotropic light distribution than ^{214}Bi decays because of a more complex decay scheme. In the energy window $4 < T_{\rm eff} < 4.5$ MeV, decays of ^{208}Tl and ^{214}Bi in the Th and the U chains are the dominant components of the Cherenkov signal. By studying events that reconstruct with $R < 450$ cm, ^{208}Tl and ^{214}Bi are separated statistically by their differences in the light isotropy β_{14}, as defined below. In this energy interval, the Cherenkov light from ^{214}Bi decays is primarily from the direct ground state β decay with an endpoint of 3.27 MeV, while almost every ^{208}Tl decay emits a 2.614-MeV γ, accompanied by one or more low-energy γ-rays and a β with an endpoint energy up to 1.8 MeV. Therefore ^{208}Tl decays produce a more isotropic light distribution than ^{214}Bi decays. The *in-situ* technique also has the advantage of providing direct determination of the background levels during data taking without any assumptions regarding temporal variation between assays. Figure 5 shows the *in-situ* determination of the low-energy background in the D_2O. The data points represent low-energy PMT events. The data isotropy (β_{14}) distribution is fit to a combination of ^{208}Tl and ^{214}Bi distributions. Also shown are the ^{24}Na background and solar neutrino contributions that were constrained in the fit. The fit result is shown as the sum histogram.

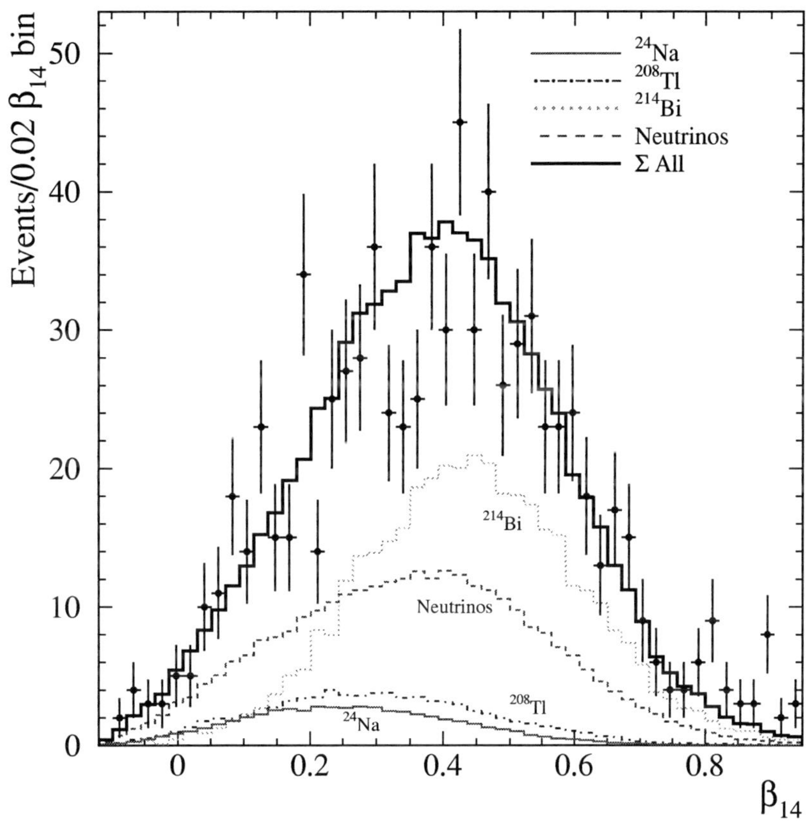

FIGURE 5. *In-situ* determination of the low-energy background in the D_2O. See text for details.

Event Isotropy

In the context of SNO event analysis and signal extraction, isotropy refers to the uniformity of the distribution of triggered PMTs on the PMT array.

Several variables constructed to measure isotropy are found to have comparable separation power between the single electron (CC and ES) and the neutron (NC) signals. The variable used, which could be simply parameterized and facilitates systematic uncertainty evaluations, is $\beta_1 + 4\beta_4$ where $\beta_l = \frac{2}{N(N-1)} \sum_{i=1}^{N-1} \sum_{j=i+1}^{N} P_l(\cos\theta_{ij})$. In this expression P_l is the Legendre polynomial of order l, θ_{ij} is the angle between triggered PMTs i and j relative to the reconstructed event vertex, and N is the total number of triggered PMTs in the event.

MERGING *IN-* AND *EX-SITU* RESULTS

Figure 6 shows Th (a) and U (b) backgrounds (equivalent equilibrium concentrations) in the D_2O deduced by *in-situ* and *ex-situ* techniques. The MnO_x and HTiO radiochemical assay results, the Rn assay results, and the *in-situ* Cherenkov signal determination of the backgrounds are presented for the salt phase on the left-hand side of frames (a) and (b). The right-hand side shows time-integrated averages including an additional sampling systematic uncertainty for the *ex-situ* measurement. The weighted average of the ex-situ Th measurements is decreased by 0.65×10^{-15} to allow for the effect of source terms in the piping external to the acrylic vessel, as discussed in the text. The large ^{222}Rn excess near day 580 is the decay of Rn that was added for calibration purposes, and was excluded in calculating the

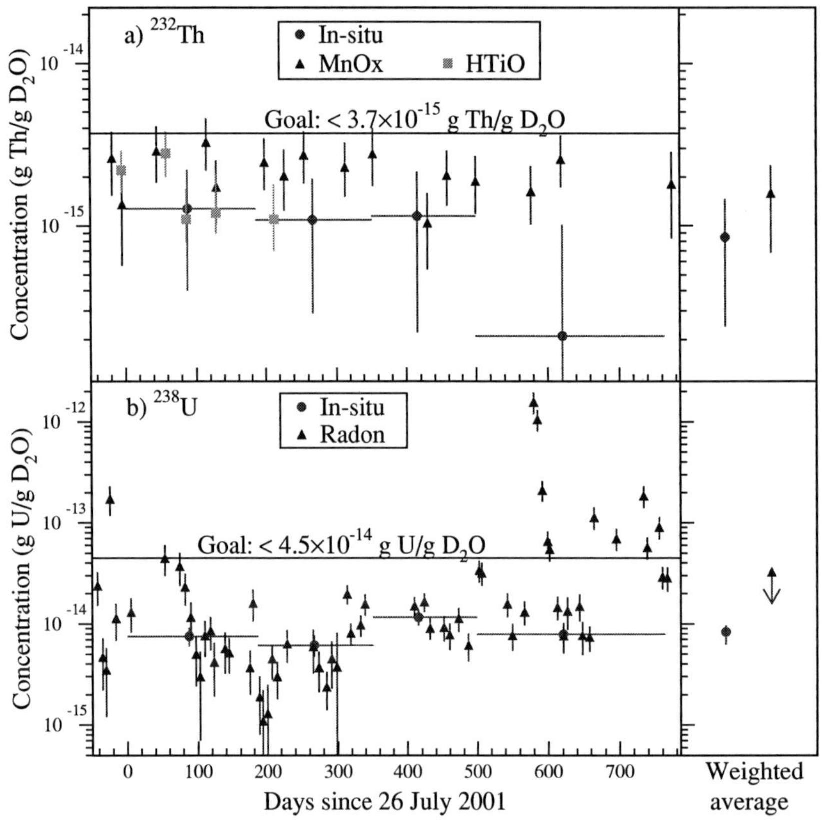

FIGURE 6. (a) Th and (b) U backgrounds (equivalent equilibrium concentrations) in the D_2O deduced by *in-situ* and *ex-situ* techniques. See text for details.

mean *ex-situ* results. The weighted average of the *ex-situ* ^{222}Rn measurements appears as an upper limit only. This is due to intermittent background appearing in the radon extraction process.

CONCLUSIONS

Specific aspects of the measurement of Ra, Rn, and Th in the SNO water systems were presented and discussed. During the course of the experiment the sensitivity of all techniques was improved and reached well below target levels.

REFERENCES

1. J. Boger et al., "The SNO Experiment," in *Nucl. Instrum. Meth. A* **449** 172–207 (2000).
2. J. Farine, "New low background techniques for neutrino experiments at low energy. First Results," *PhD Thesis*, Neuchâtel University, 1996.
3. I. Blevis et al., "Measurement of ^{222}Rn dissolved in water at the Sudbury Neutrino Observatory," in *Nucl. Instrum. Meth. A.* **517** 139–153 (2003).
4. T. Andersen et al., "Measurement of radium concentration in water with Mn-coated beads at the Sudbury Neutrino Observatory," in *Nucl. Instrum. Meth. A.* **501** 399–417 (2003).
5. T. Andersen et al., "A radium assay technique using hydrous titanium oxide adsorbent for the Sudbury Neutrino Observatory," in *Nucl. Instrum. Meth. A.* **501** 386–398 (2003).

Studies of Doped Scintillator at BNL: A Generic Method for Neutrino Measurement

M. Yeh, R. L. Hahn, A. Garnov, and Z. Chang

Brookhaven National Laboratory, Chemistry Department, Upton, NY 11973

Abstract. Using multiple liquid-scintillatior detectors at distances between 0.1 and 3 km from a nuclear reactor has the potential to determine the yet unknown neutrino-mixing angle θ_{13} via inverse ß-decay reaction on protons. The coincidence tag between the emitted prompt positron and delayed neutron provides a clear signature of anti-neutrino event. The neutron capture signal is greatly enhanced by loading gadolinium into the liquid scintillator, due to the large (n, γ) cross sections of Gd isotopes. BNL has developed a procedure to synthesize Gd-loaded scintillator with long attenuation length (> 10 m), high light output (> 95% of pseudocumene), and long-tern stability (> 5 months to date). The purification and chemical quality of BNL-synthesized Gd-LS are discussed in the paper.

Keywords: theta-13, doped scintillator, gadolinium scintillator
PACS: 98.85.Ry

INTRODUCTION

Recent experiments provide unambiguous evidence that neutrinos oscillate between their weak eigenstates. This implies that neutrinos have small but finite masses and mix among generations. The oscillation of neutrinos can be illustrated by angles describing the couplings between the neutrino flavor ($\nu_{e,\mu,\tau}$) and mass eigenstates ($\nu_{1,2,3}$). Besides the individual masses, which are not known, all but two of the mixing parameters have been measured: the coupling of the electron neutrino flavor to the third mass eigenstate, Ue3, which is commonly parameterized through the mixing angle θ_{13}, and the value of the CP-violating phase, δ_{CP}.

Nuclear-reactor experiments have played a key role in the history of neutrino physics, from the first direct detection of the antineutrino by Reines and Cowan in 1956 [1] to Palo Verde, CHOOZ, and KamLAND. Experiments with reactors led to the discovery of the neutrino, to the first observation of reactor $\overline{\nu_e}$ disappearance, and recently to the measurement of spectral distortion, a unique signature of neutrino oscillation. Reactor neutrino experiments have also allowed us to place the best constraints on the neutrino magnetic moment. Furthermore a relative measurement of antineutrinos with identical multi-detectors at a nuclear reactor has the potential to measure the yet unknown θ_{13}. Since all CP violating effects in the neutrino sector rely heavily on θ_{13}, the value of $\sin^2 2\theta_{13}$ (small or bigger than 0.01) could be used to determine the feasibility of searches for δ_{CP} and would have a profound impact on the future of neutrino physics.

Reactors produce an antineutrino ($\overline{\nu_e}$) spectrum with an average energy of 3.7 MeV from the nuclear fission in the reactor core. Such low energy $\overline{\nu_e}$ can usually be detected through the inverse ß-decay reaction on protons:

$$\overline{\nu_e} + p \rightarrow e^+ + n$$

The coincidence signal from the prompt positron and the delayed neutron capture on another proton allows a clear identification of $\overline{\nu_e}$ events. The most recent experiments measured the absolute reactor neutrino flux to search for neutrino oscillations. Among them, the best upper limit of θ_{13} comes from the CHOOZ reactor antineutrino disappearance experiment, as $\sin^2 2\theta_{13}<0.17$ for $\Delta m_{13}=2.5\times10^{-3}$ eV^2 [2].

Research sponsored by the Office of Nuclear Physics, Office of Science, US DOE, under contract DE-AC02-98CH10886 with BNL.

CP 785, *Topical Workshop on Low Radioactivity Techniques,* edited by B. Cleveland, R. Ford, and M. Chen

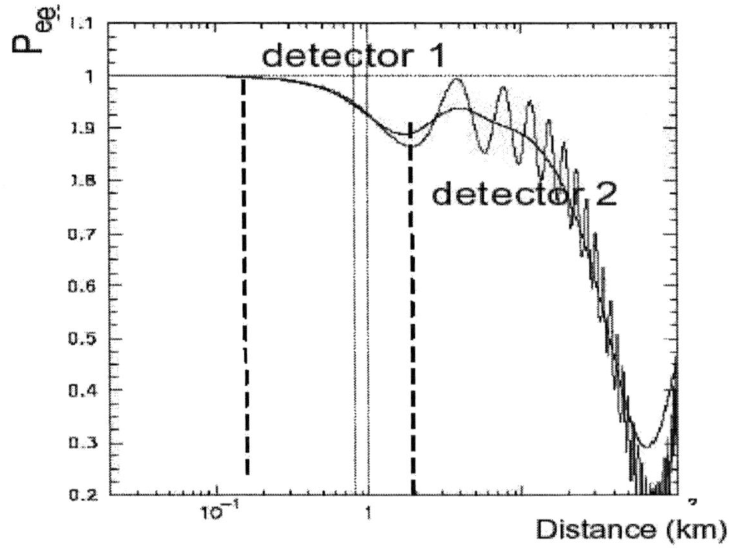

FIGURE 1: Multiple detectors reactor experiment for future neutrino oscillation measurement [4].

The idea of a 2-detector reactor neutrino experiment to measure θ_{13} was first discussed by Mikealyan et al [3]. Since then, a number of groups have investigated the concept of multi-detector experiments for a relative measurement of the $\overline{\nu_e}$ interaction rate at different distances from a reactor to determine the neutrino-mixing angle θ_{13}. General consensus is to use two or more liquid-scintillator detectors placed at distances between 0.1 and 3 km from a nuclear reactor to measure the rate and energy spectrum of $\overline{\nu_e}$ as a function of distance from the reactor core. The relative measurements between two or more detectors largely eliminate the biggest systematic uncertainties that limit absolute measurements, such as the detection efficiency uncertainty, the fiducial volume uncertainty, and the reactor flux uncertainty, and thus have the potential to do a high-precision measurement of mixing angle θ_{13}.

The unique sensitivity of reactor anti-neutrinos to $\sin^2 2\theta_{13}$ and the recent technical successes of both CHOOZ and KamLAND provide the basis for a new reactor neutrino experiment dedicated to the measurement of the important parameter θ_{13}. In 2004, the APS-sponsored Study on The Future of Neutrino Physics [5] recognized this complementary nature of neutrino experiments at reactors and accelerators, and gave one of its two "high priority" recommendations to an "expeditiously done" reactor-based, high-precision ($\sin^2 2\theta_{13}$~1%) measurement of θ_{13} that would then be followed by accelerator-based experiments (the other "high priority" recommendation was for new double beta-decay experiments).

LIQUID SCINTILLATOR AS THE NEUTRINO DETECTOR

Organic liquid scintillators (LS) have been the detection medium of choice for antineutrinos since the early discovery experiment of Reines and Cowan. For the delayed neutron capture signal, one of the main signatures in the reactor antineutrino experiment, the advantages of adding an element such as Gd to the LS (to form "Gd-LS") are significant. Only a small concentration of Gd in the liquid is required, ~0.1-0.2% by weight, because some naturally occurring Gd isotopes have huge (n,γ) cross sections, ~10^4 barns. Also the energy released by the neutron capture of Gd is large, ~8 MeV, and is distributed among a few emitted γ rays, giving a distinctive angular distribution. Last, the delayed coincidence time between the e^+ and n signals is shortened from ~200 μs for n-capture in H to ~30 μs (depending on the concentration of Gd in the LS), which could reduce the accidental background by a factor of ~7.

However there are challenges of adding inorganic salts of Gd, such as $GdCl_3$ or $Gd(NO_3)_3$, directly to the LS. It is well known that the CHOOZ experiment was forced to shut down when the optical properties of its LS + $Gd(NO_3)_3$ mixture deteriorated, degrading the quality of its output signal. The observed yellow color of the CHOOZ Gd-LS was likely produced by oxidation of the organic liquid by the nitrate. It is not trivial to synthesize a chemical complex of Gd that will be soluble in the organic LS, which is usually a non-polar aromatic compound (i.e., contains phenyl groups and primarily C and H). This type of LS is generally immiscible with aqueous solutions.

Once the Gd-LS is synthesized, several requirements must be satisfied in a multi-ton LS neutrino detector. The doped scintillator must (1) be optically transparent, i.e., have a large attenuation length, (2) have high light output, (3) contain ultra-low concentrations of contaminants, chemical as well as radioactive, (4) be chemically compatible with the vessel in which it is contained, and (5) be chemically stable for very long time periods, ~ years, as opposed to the shorter times needed for the more traditional uses of Gd-LS, for detecting high fluxes of neutrons from nuclear reactors and spallation neutron sources. "Chemical stability" means the absence of (slow) chemical reactions, such as hydrolysis and/or polymerization, which can lead over time to cloudy suspensions in the LS or formation of gels or precipitates.

In-LS for LENS-Sol

The Solar-Neutrino/Nuclear-Chemistry Group in the BNL Chemistry Department has a long history of neutrino research beginning with the pioneering HOMESTAKE experiment initiated by Ray Davis in 1962. Since 2000, the BNL Group has been involved in R&D of chemical techniques for loading metallic elements, such as In^{3+} and Yb^{3+}, up to concentrations of several % by weight, into organic LS (in collaboration with R. S. Raghavan of Bell Labs). This research, in a project called LENS-Sol, has as its main goal the development of a detector capable of observing low-energy neutrinos from the solar pp and ^{7}Be branches, via neutrino capture on the In (with a Q-value of 0.118 MeV). The principle of the chemistry is straightforward, namely to prepare an organometallic complex of In that is soluble in the LS. But the execution has been difficult, with many problems to solve and details to master.

Several organic complexing agents come to mind, for example (i) carboxylic acids (RCOOH) that can be neutralized with inorganic bases such as NH_3 to form metal carboxylates, (ii) organic phosphorus-oxygen compounds, such as organic phosphates or phosphine oxides, that can complex neutral inorganic species such as $InCl_3$, and (iii) organic diketones, such as acetyl acetonate. Compounds from groups (i) and (ii) were selected as candidates for testing proposed chemical procedures at BNL. Much of their organometallic chemistry has already been developed in the fields of separations chemistry and chemical treatment in the nuclear fuel cycle.

After extensive laboratory R&D, procedures have been worked out for the synthesis of the metal-LS by solvent-solvent extraction. Much of the focus has been on the carboxylic acids, because they are produced in bulk by the chemical industry and have lower cost and ease of chemical disposal, as compared with the phosphorus-containing compounds. Several carboxylic acids, with side chains containing from 1-8 carbons, have been studied. The best candidate found to date is methylvaleric acid, MVA. Reproducible results with MVA have been found for samples of In-LS, where In concentrations are in the 5-10% range and the organic solvent, pseudocumene (PC), is the LS. Liters of doped scintillator can be synthesized directly in one step within a few hours, compared to other methods, which need to form organometallic solids first, and then dissolve them into LS.

Many techniques, instrumental and chemical, have been developed at BNL to analyze the resulting In-LS samples. Among these are (1) measurement of the attenuation length both by a UV-Visible absorption spectrophotometer with 10-cm optical cells and by a dual-beam, long-path (>1 meter) blue laser system; (2) measurement of the light yield, S%, relative to the pure PC; and several chemical determinations of the In-LS, such as of the concentrations (3) of the In^{3+} by a colorimetric method, (4) of the total RCOOH by acid-base titrations, (5) of the uncomplexed RCOOH by IR spectroscopy, (6) of the H_2O by Karl-Fischer titrations, (7) of the NH_4^+ and Cl^- by electrochemistry, and (8) of different In species in the organic liquid by IR spectroscopy. A summary of transparency, long-term stability, and light yield for the present In-LS at BNL is shown in Figure 2, with the several features of: high In concentration (~7%), good light yield (~40% of PC), long attenuation length (L(1/e)~9m without wavelength shifter), and chemical stability for over 13 months since synthesis [6].

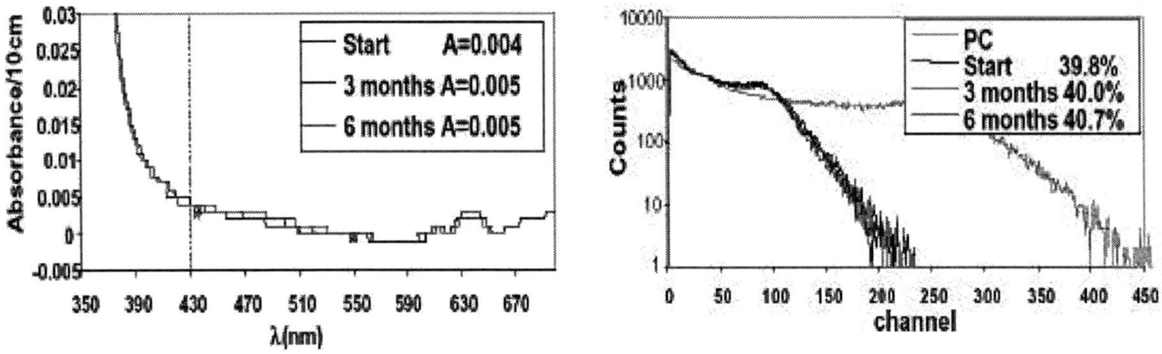

Stability: The UV-VIS absorbance (430 nm) with time over six months (BNL#115, In%=6.77)

Light yield with time over six months (BNL#115, 3g PBD/L , 15mg bis-MSB/L)

FIGURE 2. In-doped Scintillator for LENS-Sol

Gd-LS for High-Precision θ_{13} Experiments

Early in 2004, because of the developing interest of the BNL Group in θ_{13} experiments, the procedures that had been developed for In-LS were successfully applied to the preparation of some initial samples of Gd-LS. There are two approaches to making batches of the Gd-LS: (a) preparing each batch at the desired final Gd concentration, 0.1-0.2%, or (b) preparing more concentrated batches, at least 1-2% Gd, and then diluting with the organic LS to the desired concentration. The two approaches are not identical, with regard to possible long-term effects such as hydrolysis, polymerization, and effects on the optical properties of the final Gd-LS. Approach (b), however, may simplify preparation of larger volumes of Gd-LS. The scenarios of chemical preparation at BNL are to (1) develop a reproducible procedure of Gd-LS synthesis, (2) dope Gd into different scintillating solvents, such as PC, phenyl cyclohexane (PCH), and mixtures of PC with mineral oil or with other organic-chemical, inert solvents, such as dodecane, and (3) to establish chemical assays to remove and measure chemical and radioactive impurities.

Reproducibility

The synthesis of Gd-LS is very consistent, and its reproducibility is presented in Figure 3. The samples of 0.2% Gd-LS are diluted from different batches of ~2% concentrated Gd-LS prepared under different chemical conditions. The synthesis procedure is very reproducible in terms of Gd extraction (> 90%) and physical property (for instances, similar attenuation length that is dominated by the optical property of scintillator, PC in this case).

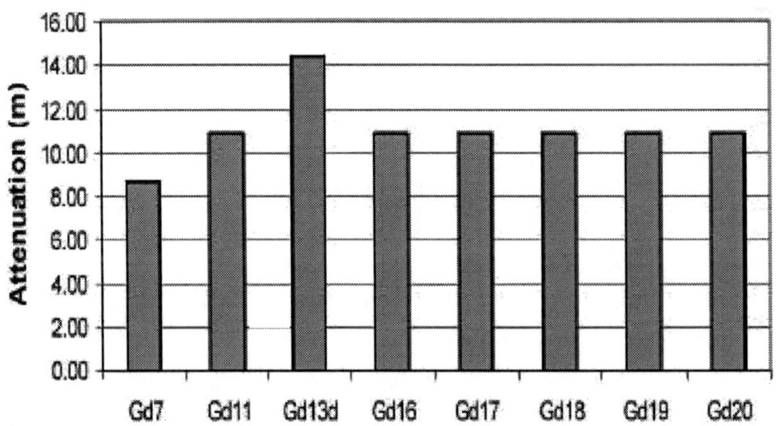

FIGURE 3. The reproducibility of Gd-LS Synthesis estimated from the 10-cm UV Spectrometer.

It should be noted that the absorption of Gd-doped scintillator at UV region is very close to the spectrometric baseline (see Figure 4), thus it's very difficult to precisely estimate the attenuation length by short pathlength experiments and indicates that a longer pathlength measurement is necessary. The attenuation lengths of Samples Gd16 - Gd20 presented in the above figure were later measured as ~16 m by the 1-m blue laser system.

Long-Term Stability of Gd-LS

To test the long-term stability of Gd-doped scintillator, the attenuation length and light yield of Gd-LS are measured periodically since synthesis. In Figure 4, the UV absorption of 1.2% Gd-LS is presented as a function of time. The transparency of Gd-LS is slightly better at few weeks after synthesis. This is probably due to the settlement of fine particles or bubbles formed at the loading of just synthesized Gd-LS. After that period of time, the Gd-LS is very stable for over 4 months (the length of the monitoring period to date).

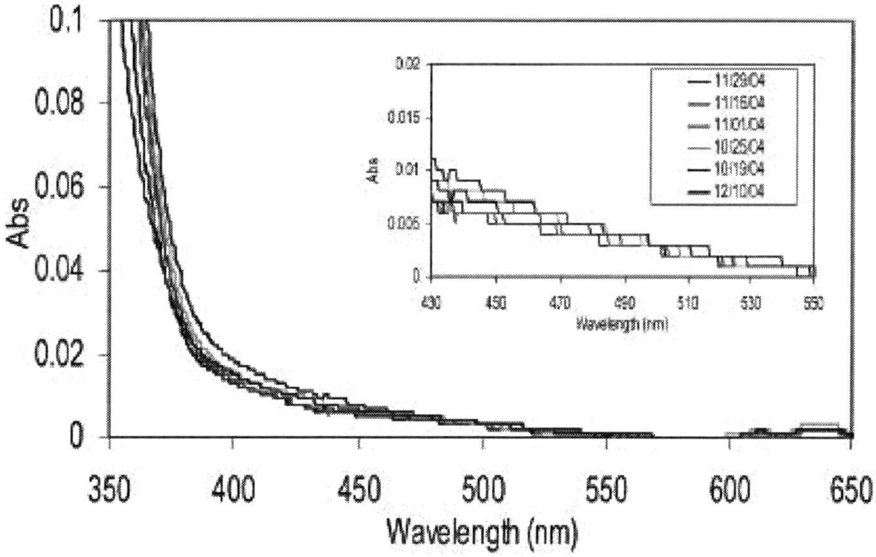

FIGURE 4. Stability test of 1.2% Gd-LS as a function of time.

Light Yield of Gd-LS

The light yield of doped scintillator is determined by the Compton Scattering curve of an external γ-source irradiating the Gd-LS samples and is a function of the liquid scintillator of interest (see Figure 5, where Cs-137 is the γ-emitter). For 1.2% Gd-LS, the light yield is ~81% of PC, while 0.2% Gd-LS has light yield better than 95% of PC.

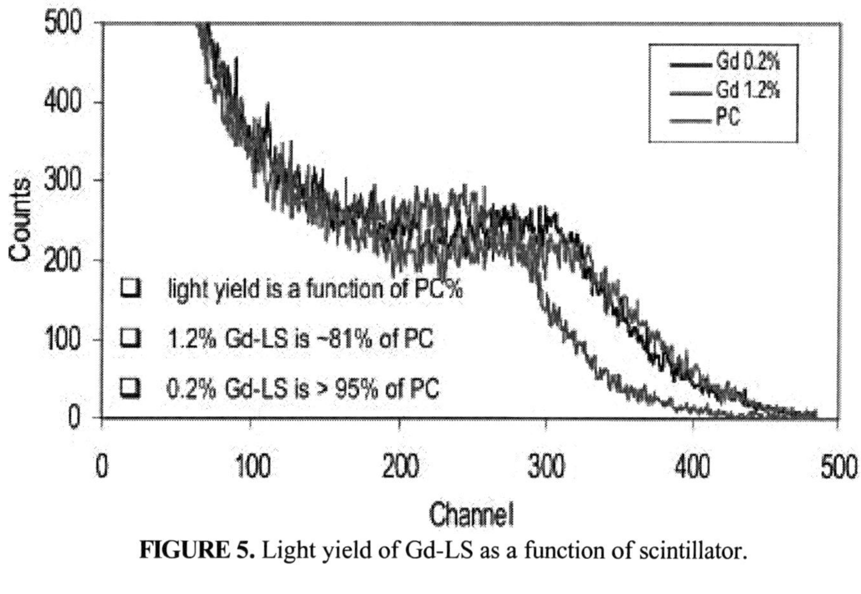

FIGURE 5. Light yield of Gd-LS as a function of scintillator.

Purification of Gd-LS

Most of the envisaged purification steps will have to be applied before and during the synthesis of the Gd-LS. Most chemical separation schemes that would be used after the Gd-LS has been put into the detector vessel are unsuitable because they would likely remove some of the Gd as well as the other inorganic impurities. Even vacuum techniques for removing Rn from the Gd-LS could cause problems by changing the concentrations of the volatile organic liquid components in the LS. The requirement of preparing Gd-doped scintillator for measurement of $\sin^2 2\theta_{13}$ to a 1% or better precision magnifies the challenges of purification inherent in preparing the Gd-LS. For example, the concentrations not only of Gd but also of LS (i.e., the concentrations of target H atoms for the antineutrino capture) must be identical in the near and far detectors. Special care will thus have to be given to the batch-wise preparation of the Gd-LS and to the chemical quantitative analyses, particularly the C/H ratio, of these batches.

The purification techniques developed at BNL are divided into removals of chemical and radioactive steps, respectively. The starting material of $GdCl_3$ is purified separately to remove the radioactive impurities; this topic will be discussed in a later section. Concentrations of other non-radioactive chemical species that can adversely affect the optical properties of the Gd-LS will also have to be strictly controlled. Indeed, the removal of chemical impurities not only can reduce the attenuation of the light, but also enhance the long-term stability of the Gd-doped scintillator. Primary investigations found that some impurities could induce slow chemical reactions, which would later cause the transparency of Gd-LS to deteriorate gradually.

Removal of Chemical Impurities

Several Chemical purification steps are developed on basis of synthesis phase. Purification of the aqueous phase (NH4-carboxylate) is done with mixtures of organic chelating solvents; while the organic phase (carboxylic acid and liquid scintillator) is purified by dry column absorption and by temperature-dependent, vacuum distillation, depending on the types of organic acid and scintillator used. The purity of organic solvents to be used in the synthesis was measured by UV spectrometer, as illustrated in Figures 6-8.

Removal of Radioactive Impurities

The water used for synthesis is purified by several techniques that have been extensively studied at SNO, where our group participated actively in the early development stages. For the organic solvents of use, it is expected that the radioisotopes, if any, will be separated out from purified solvents after distillation. Lanthanides, such as Gd^{3+} of the starting materials, often contain naturally occurring radioactive impurities, such as Th^{4+} and U^{4+} and U^{6+} (as UO_2^{2+}). An example of radioisotopes existing in gadolinium oxides, prepared and measured by our colleagues at the Institute of High Energy Physics (IHEP) in Beijing, is presented in the following table.

TABLE 1. The Concentrations of Radioisotopes in Gd_2O_3 Prepared at IHEP

	U-238	*Th-232*	*K-40*
ppb in Gd_2O_3	< 13	440±32	< 2.3

Purification steps will have to be developed to reduce these radioactive species to concentrations of ~10^{-12} g/g, ~0.3 Hz for 10 tons, in 0.1% Gd-LS with an external 45-cm γ-catcher. At BNL, 99.999% $GdCl_3$ is used as the starting materials and the contents of radioactive impurities are expected to be less. Two purification methods, cation exchangers and solvent extraction, are being considered, but the latter technique is preferred. Since there is a step at the purification of chemical impurities by organic solvents, as mentioned in the above section, it will be useful to modify the combinations of organic solvents used for purification of chemical impurities to remove the radioactive impurities as well. Thus the cleanings of chemical and radioactive impurities could be done simultaneously. Several organic solvents used in chemical reprocessing in the nuclear industry, such as TBP, TOPO, TBPO, and TTA, are under investigation. The results will be reported in the future.

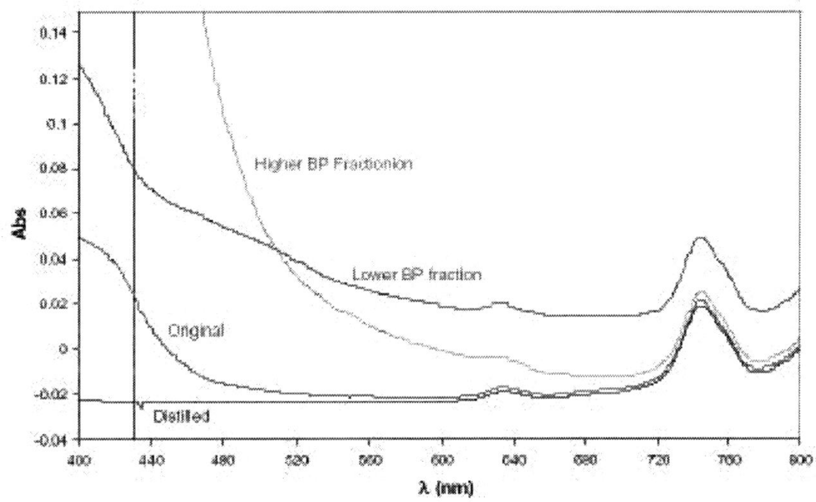

FIGURE 6. Purification of carboxylic acid by temperature-dependent, vacuum distillation.

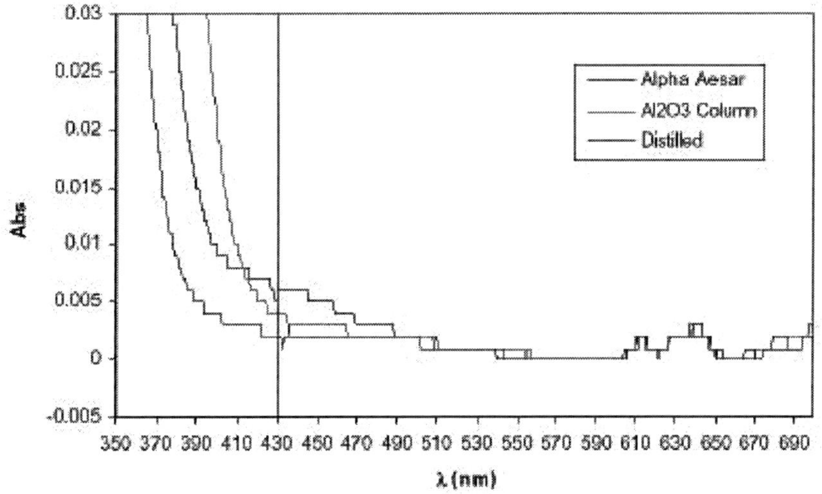

FIGURE 7. Purification of PC by dry column and vacuum distillation.

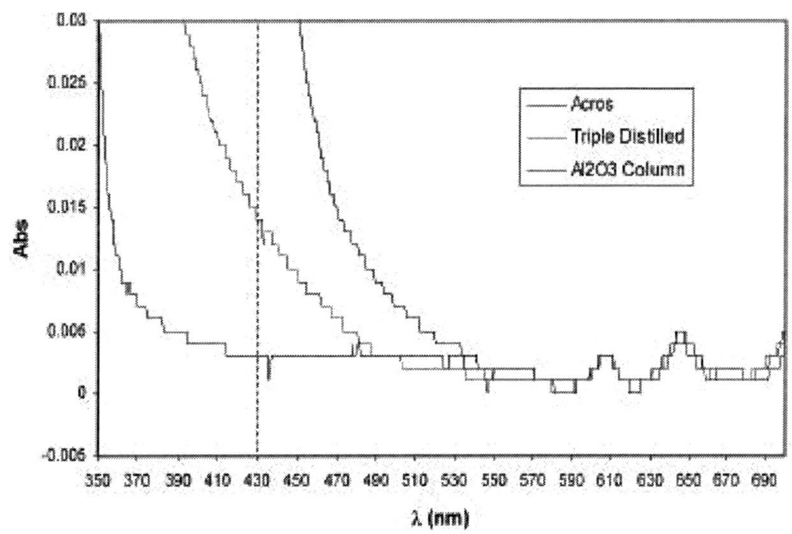

FIGURE 8. Purification of PCH by dry column and vacuum distillation.

Comparisons of Different Scintillation Systems

Another important issue is to study the chemical compatibility of the organic Gd-LS and of the unloaded LS that will be used in the buffer region with the material of the detector vessel. Acrylic is known from SNO R&D to be attacked by many chemical liquids. Nylon, as used in Borexino and in KamLAND, is known to be resistant to some LS, such as PC and mineral oil. Another aspect of this problem that should be realized is that chemical attack of the vessel by the LS could introduce unwanted impurities, organic as well as inorganic, from the vessel into the liquid and adversely affect the LS properties.

In addition to PC, other organic liquid scintillators, such as PCH and various types of mineral oil, are being studied. It has been reported that a mixture of 20% PC – 80% mineral oil, and possibly even 40% PC – 60% mineral oil, will not attack acrylic. These claims will have to be verified. At this stage, it is our best interest to investigate different mixtures of scintillators that might be used for future neutrino detector. The comparisons of different scintillation systems are shown in Figures 9 & 10, in terms of attenuation length and light yield. It should be noted that the light yield of 23.8%, relative to PC, for mineral oil might include some of the lights produced from the background, such as the scintillating impurities existing in the glass sample holder that was used during the measurement.

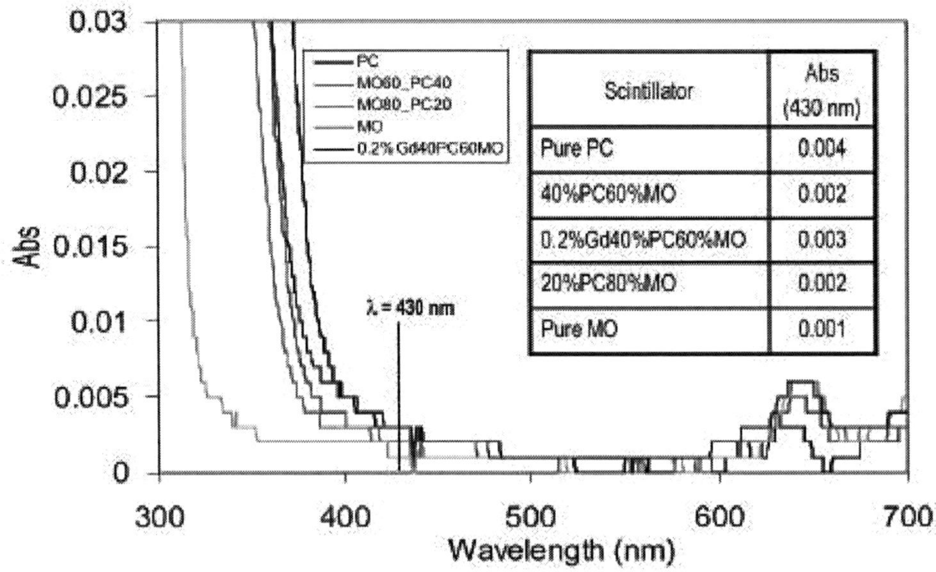

FIGURE 9. Absorption of PC and MO mixtures.

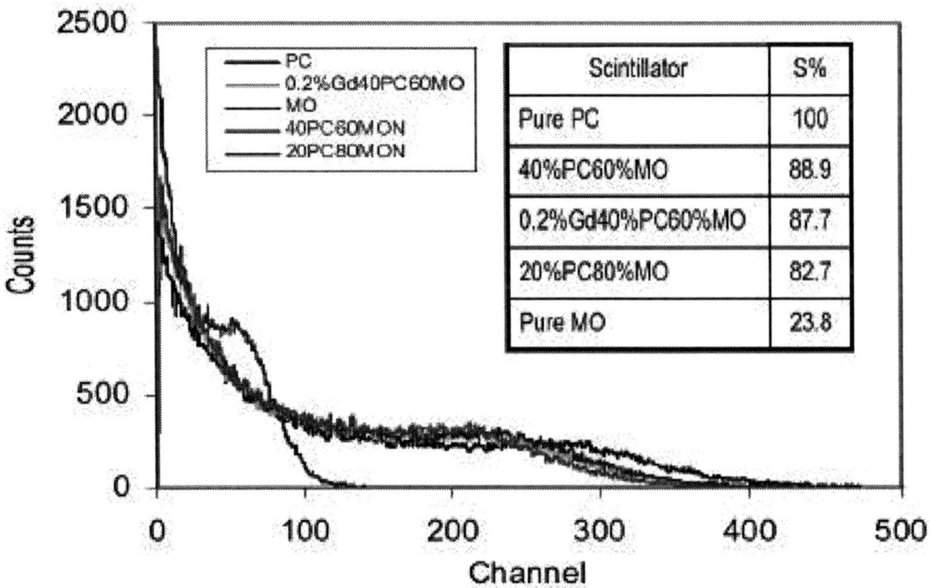

FIGURE 10. Light yield of PC and MO mixtures.

Mineral oil has very good light transmission compard to that of PC at the UV region of 350 to 550 nm. The addition of mineral oil to PC significantly enhances the transparency of diluted scintillator. On the other hand, mineral oil produces no or very little scintillating light, which reduces the photon efficiency of Gd-LS. The general consensus is to add the mineral oil as little as possible to maintain the good photon production of Gd-LS. The minimum level of mineral oil required in the PC depends on the degree of chemical aggressiveness of final Gd-LS mixtures to acrylic. As mentioned, the compatibility of Gd-LS with the detector vessel needs to be studied. At the same time, testing of other diluting agents, such as dodecane, is also underway.

Comparisons with Commercial Gd-LS

Some samples of commercially available Gd-LS, BC-521 with 1% Gd in PC, and BC-525 with 0.5% Gd in mineral oil, purchased from Bicron, are compared with the BNL samples. It should be noted that BC-521 is the concentrated Gd-doped scintillator that was used at the Palo Verde experiment. The light yields of the respective BNL and Bicron samples were found to be comparable, but the attenuation lengths of the BNL Gd-LS were a factor of 2 longer than the BC samples, probably reflecting the care put into the BNL pre-synthesis purification steps. The details of comparisons are listed in Table 2.

TABLE 2. The Comparisons of BNL and Bicron Gd-LS Samples

	BNL Gd-LS	BC-521	
Measured by	BNL	BNL	Bicron
Gd %	~1.2%	~1%	1%
Attenuation Length (m)	13.5 (1-m laser) 6.23 (10-cm UV)	2.55 (10-cm UV)	> 4.00
Light Yield (s%)	81.9% PC	85% PC	57% anthracene
Stability	2 months and continuing	N/A	long term

SUMMARY AND FUTURE AT BNL ON GD-LS

The BNL Group has established a very consistent method for Gd-LS synthesis, in terms of over 90% Gd extraction, ~16 m of long attenuation length and better than 95% relevant to PC of light yield. These initial results on Gd-LS are very encouraging for the planned θ_{13} experiments and merit further R&D. We plan to increase our efforts in the short term. One major focus will be to optimize, simplify and finalize the chemical procedures to synthesize the Gd-LS. Issues of chemical purifications and of the control and assay of radioactive impurities are also under investigation.

In the long term, consideration will have to given to designing and building chemical systems that (i) are closed to prevent ingress of air, which can degrade the LS, and (ii) are automated to replace many of the procedures that are currently done in the laboratory by hand. After that, one will have to increase the scale of the syntheses, from the present 0.5-1 liter per batch to at least 5-10 liters per batch. A goal has been set of accumulating ~200 liters for prototype tests.

As the BNL R&D continues, all of the analytical methods that have been developed for In-LS will be used to determine the chemical and physical properties of the Gd-LS. The timing characteristics of the light signals produced by the Gd-LS may also be investigated, such as the pulse widths and decay times.

Once the project gets into planning for the full-scale experiment, or possibly for an intermediate-size prototype, the volumes of liquids being produced will be on the level of tons. Industrial cooperation will likely be required. At that juncture, it will be very important to have a quality control program already developed and in place in the collaboration so as to verify the quality of the materials that will be supplied by the vendors, e.g., pure solvents, starting materials for the Gd-LS syntheses, or even commercially manufactured Gd-LS.

The BNL Group is funded by DOE's Office of Nuclear Physics, for its ongoing research in SNO and for exploring new research directions, such as LENS-Sol and the θ_{13}-reactor experiment. At the moment, the Group is relatively small and may have more ambitious plans than it can accomplish in a reasonable time span. It intends to cooperate with other groups to leverage its efforts. It will also provide samples of Gd-LS for measurements that will be done at other institutions and/or participate in exchanging samples between different institutions ("round robins") to get independent measurements of key quantities, such as attenuation lengths and light yields.

REFERENCES

1. F. Reines and C. Cowan, Phys. Rev. 113 (1959) 273-279.
2. M. Apollonio et al., Eur. Phys. J. C27 (2003) 331-374.
3. L. Mikaelyan, e-print arXive: hep-ex/0008063 (2000).
4. Daya Bay NSF Proposal: Collaborative Research: Towards a Precision Measurement of θ_{13} with Reactor Anti-Neutrinos (2004).
5. The Neutrino Matrix: Study on the Future of Neutrino Physics, e-print arXiv: physics/0411216 (2004).
6. R. S. Raghavan et al., LENS-Sol: A Detector for measuring the Neutrino Luminosity of the Sun, NOON Proceeding, Japan (2004).

VIII – BACKGROUND STUDIES AND SIMULATIONS

Background Models for Muons and Neutrons Underground

Joseph A. Formaggio

Center for Experimental Nuclear Physics and Astrophysics, and Department of Physics University of Washington, Seattle, WA 98195

Abstract.
Cosmogenic-induced activity is an issue of great concern for many sensitive experiments sited underground. A variety of different arch-type experiments – such as those geared toward the detection of dark matter, neutrinoless double beta decay and solar neutrinos – have reached levels of cleanliness and sensitivity that warrant careful consideration of secondary activity induced by cosmic rays. This paper reviews some of the main issues associated with the modeling of cosmogenic activity underground. Comparison with data, when such data is available, is also presented.

Keywords: underground experiments, spallation, muon-nuclear interactions

INTRODUCTION

For the past 40 years, sensitive experiments performed deep underground have revealed a wealth of knowledge about some of the most elusive phenomena of nature. The environment beneath the Earth's surface often provides the necessary conditions for experiments dealing with these rare or exotic phenomena. The dominant background sources remaining in underground environments differ in character from those normally found on the surface. Special techniques have been developed to shield out, discriminate against, or accurately measure the background present in these experiments. However, the level of sensitivity projected for the next-generation of underground experiments will push the limits of those techniques. As a result, careful accounting of all backgrounds must take place to ensure the given experiment reaches the level of sensitivity desired. With a number of proposals presently coming forward for one or more underground laboratories in the North American continent, emphasis on backgrounds –and how to model them– has subsequently increased.

In this paper, we consider the models and simulation tools employed to estimate the level of activity due to cosmogenic activity. For underground environments, the cosmogenic activity involves mainly high energy muons created in the upper atmosphere and any subsequent activity due to muon-nuclear interactions. The paper discusses the models used to estimate the muon flux, energy spectrum, and, more importantly, the secondary interactions that result from muon interactions. A more extensive review of this subject can be found in Reference [1].

COSMIC RAYS

Muons in and of themselves do not usually constitute a serious background to underground experiments, since the muons are easily identified and/or vetoed as they pass through the detector. Muon-induced spallation, on the other hand, produces short charged tracks at the event site, possible secondary neutrons, and delayed decays of radioactive products. Unwanted events from the first and last categories can be suppressed greatly by rejection of multiple interactions and of interactions with too great an energy deposition occurring nearby in time and space, and by recoil nucleus discrimination.

However, muon-induced backgrounds are still troublesome in several ways. First, the muon veto rate contributes to the experiment's dead time. This, however, is only important at relatively shallow sites, where muon rates are high. Second, low-energy negative muons can stop and undergo nuclear capture in any

CP 785, *Topical Workshop on Low Radioactivity Techniques,* edited by B. Cleveland, R. Ford, and M. Chen

material, often giving rise to neutrons and unwanted radioactive isotopes. Such backgrounds are also more important at shallow sites, where the rate of stopping muons is relatively high.

High-energy muons often create spallation products, such as neutrons, photons, and radioisotopes, which can cause serious backgrounds for low-energy experiments. The secondary neutrons produced by high-energy muons interacting in the surrounding rock, in the local shielding, and in the detector itself are not easily tagged. Particularly difficult are the neutrons produced in the adjacent rock, which are not necessarily in time coincidence with a muon veto that covers the entire experiment and its shielding. These neutrons may represent a fundamental limiting background that only deeper sites [2, 3] can reduce sufficiently to allow the coming generation of ton-scale experiments to succeed.

The flux of muons underground has been studied for many years. Most recently, measurements of the muon flux as a function of depth have been reported by the Frèjus [4], MACRO [5], and LVD [6] experiments. For muon energies between 10 GeV and 100 TeV, one can use a simple relation to describe the energy and angular dependence of the muon flux at the Earth's surface [7]:

$$\frac{dN_\mu}{dE_\mu d\Omega} \simeq 0.14 E_\mu^{-(\gamma-1)} \left(\frac{1}{1+\frac{1.1 E_\mu \cos\theta}{115\mathrm{GeV}}} + \frac{0.054}{1+\frac{1.1 E_\mu \cos\theta}{850\mathrm{GeV}}} \right) \mathrm{cm}^{-2}\mathrm{s}^{-1}\mathrm{sr}^{-1}\mathrm{GeV}^{-1}, \tag{1}$$

where $\gamma - 1 = 2.78 \pm 0.05$. This formula describes the spectral shape with relative errors around 5%. The error on the total flux, however, is much larger, near 20%. As the muon traverses the rock, it loses energy through ionization, bremsstrahlung, e^+e^- pair production, and photoproduction. By neglecting fluctuations in the energy loss, one can use a simple relation to determine the average energy loss of muons passing through matter:

$$-\left\langle \frac{dE(E_\mu)}{dX} \right\rangle = \alpha + \beta E_\mu \tag{2}$$

At large depths, however, fluctuations become large and the above approximations no longer adequately characterize the muon spectrum. Simulation codes such as PROM_MU [8] and MUSIC [9] are well tested against existing data and appear to adequately describe the rare, high-energy physics processes that dominate at large depths. Typical differences in the energy spectrum between the simple Gaisser parameterization using Equation 2 and full simulations are around 5% [10].

SECONDARY PRODUCTS OF COSMIC RAY INTERACTIONS

Cosmic ray muons produce neutrons through several different mechanisms, each of which dominates at different energies and depths. The processes that contribute to this neutron production include the following:

1. Negative muon capture on nuclei.
2. Electromagnetic showers generated by muons.
3. Muon interactions with nuclei via the exchange of virtual photons.
4. Muon-nucleon quasielastic scattering.
5. Secondary neutron production from any of the above processes.

Each of these processes is described in greater detail below.

Muon Capture

Once a negative muon is of sufficiently low energy, it can be attracted into the Coulomb field of a nucleus, forming a "muonic atom" bound state. The bound muon quickly cascades down to the 1s state, where it will either decay:

$$\mu^- \rightarrow e^- + \bar{\nu}_e + \nu_\mu, \tag{3}$$

or else undergo nuclear capture via the weak charged-current process:

$$\mu^- + \mathrm{A(Z,N)} \rightarrow \nu_\mu + \mathrm{A(Z-1,N+1)}. \quad (4)$$

The number of neutrons from muon capture can be computed from the following expression:

$$N_n(h) = I_\mu(h) \otimes P_{\mu^-} \otimes P_c \otimes P_n. \quad (5)$$

Here, $N_n(h)$ is the total number of neutrons produced by stopped muons at a depth h, $I_\mu(h)$ is the flux of μ^- stopping at a given depth, P_{μ^-} is the relative charge fraction of cosmic ray muons, P_c is the capture probability, and P_n is the neutron multiplicity following capture.

The muon stopping rate as a function of depth (I_μ), as well as the muon charge fraction (P_{μ^-}), has been studied in detail by Charalambus [11] and Bilokon [12]. The principles used to compute the muon stopping function are the same as are needed in general for muon transport through matter, and thus are described by well-known cross sections and fluxes. Uncertainties that dominate the direct interactions in deep sites, such as photonuclear cross sections at high energy, do not significantly affect calculation of the muon energy loss.

Likewise, the μ^-/μ^+ charge ratio has been measured in numerous experiments. The ratio grows logarithmically with muon momentum from 1.2 (at 1 GeV/c) to 1.4 (at 1 TeV/c). Below 1 GeV/c, geomagnetic effects become important and P_{μ^-} therefore varies with position on the Earth [7].

The muon capture probability has also been well studied, both theoretically and experimentally [13, 14, 11], and is described briefly here. It is possible to relate the capture probability $\mathrm{P_c}$ to the capture and decay rates:

$$P_c = \frac{\Gamma_c}{\Gamma_c + Q\Gamma_d} \quad (6)$$

where

$$\Gamma_d = \frac{1}{\tau_{\mu^+}}. \quad (7)$$

The symbol Q refers to the Huff factor, which corrects the free muon decay rate for the phase space change due to the muonic atom binding energy.

The muon capture rate Γ_c rises as Z^4_{target}. Comprehensive measurements have been made at TRIUMF for a variety of targets [13], and results for a selected sample of those targets are shown in Table 1. Phenomenological models also exist to estimate the muon capture rate. For low-Z targets, one can use the Primakoff model [15]:

$$\Gamma_c(A,Z) = Z^4_{\mathrm{eff}} X_1 (1 - X_2 \frac{A-Z}{2A}), \quad (8)$$

where Z_{eff} is the effective nuclear charge for a given atom, X_1 represents the capture rate in hydrogen, and X_2 takes into account Pauli suppression for the given nuclear environment. Phenomenological fits to available data from TRIUMF yield $X_1 = 170\ \mathrm{s}^{-1}$ and $X_2 = 3.125$ [13]. For heavier elements, Pauli suppression becomes more important and higher-order Pauli corrections are necessary [16].

Finally, the average muon-capture neutron multiplicity (P_n) has been measured for a variety of elements. Table 1 shows some of those measurements.

The energy spectrum of neutrons emitted by muon capture contains direct and evaporation components, as shown in Equations 9 and 10 [14]:

$$\frac{dN_{\mathrm{evap}}}{dE_n} \propto E_n^{5/11} e^{-\frac{E_n}{\theta}}, \quad (9)$$

$$\frac{dN_{\mathrm{direct}}}{dE_n} \propto e^{-\frac{E_n}{T}}, \quad (10)$$

where $\frac{dN}{dE_n}$ is the neutron energy spectrum, E_n is the neutron kinetic energy, T is obtained from an exponential fit to the high-energy portion of the neutron spectrum, and θ is the neutron evaporative temperature. The parameters T and θ vary with target atomic number, but typical values are $T \sim 10$ MeV and $\theta \sim 1$ MeV.

TABLE 1. Huff factors, average neutron multiplicities P_n, and muon mean lifetimes, for negative muons stopping in various targets (multiplicity data from Reference [17] and mean lifetime from Reference [13])

Material	Z (Z_{eff})	Huff factor	Multiplicity	Mean lifetime (ns)
Al	13 (11.48)	0.993	1.262 ± 0.059	864 ± 2
Si	14 (12.22)	0.992	0.864 ± 0.072	758 ± 2
Ca	20 (16.15)	0.985	0.746 ± 0.032	334 ± 2
Fe	26 (19.59)	0.975	1.125 ± 0.041	206 ± 1
Ag	47 (27.95)	0.925	1.615 ± 0.060	87.0 ± 1.5
I	53 (29.27)	0.910	1.436 ± 0.056	83.4 ± 1.5
Au	79 (33.64)	0.850	1.662 ± 0.044	74.3 ± 1.5
Pb	82 (34.18)	0.844	1.709 ± 0.066	74.8 ± 0.4

Electromagnetic Interactions

As a muon passes through matter, real photons and electrons from bremsstrahlung and pair production are released and produce particle showers. Typically the contribution from such processes is small. However, the cross sections grow as $\langle Z^2 \rangle$, and thus become more important for high-Z targets, such as lead. In general, the cross sections associated with electromagnetic showers are very well known. In certain cases, the electromagnetic component of showers produced by cosmic rays were explicitly measured and shown to rise linearly with muon energy [18].

Muon Spallation

Neutrons produced in muon spallation involve the exchange of a virtual photon as the muon passes through matter. As it stands, muon spallation via virtual photon exchange is one of the least understood processes in muon interactions at high energies. Theoretically, it is possible to calculate the muon photonuclear cross section by using the equivalent photon approximation. This technique, originally proposed by Fermi [19] and later developed by Weizsäcker [20] and Williams [21], relates the real photonuclear cross section to the virtual cross section by assuming the longitudinal component of the γN cross section (which would apply only to virtual photons) to be negligible:

$$\sigma_{\mu N} = \int \frac{n_\gamma(\nu)\sigma_{\gamma N}(\nu)}{\nu} d\nu. \tag{11}$$

Here, $n_\gamma(\nu)$ is the virtual photon spectrum associated with the passage of the muon, ν is the photon energy, and $\sigma_{\gamma N}(\nu)$ is the real photon-nucleon cross section. For low energy processes, where the photon interacts with the nucleus collectively, the virtual photon spectrum can be written[22] as a simple energy-dependent expression :

$$n_\gamma(\nu) = \frac{\alpha}{\pi}\left[\frac{E^2+E'^2}{p^2}\ln\frac{EE'+PP'-m^2}{m_\mu \nu} - \frac{(E+E')^2}{2P^2}\ln\frac{(P+P')^2}{(E+E')\nu} - \frac{P'}{P}\right], \tag{12}$$

where E (E') and P (P') are the initial (final) muon energy and momentum, respectively.

Bezrulov and Bugaev [23, 24] have further developed the model of virtual photon exchange within the framework of a generalized vector-meson-dominance model. This model gives a good description of virtual photon exchange in the diffractive region (low 4-momentum transfers and large photon energies). These authors also took into account nuclear shadowing effects that influence muons at high energies. The modified photon spectrum equation is:

$$\begin{aligned} n_\gamma(\nu) &= \frac{\alpha}{2\pi}\omega^2\left\{\frac{1}{4}H(\omega)\ln(1+\frac{m_2^2}{t}) - \frac{m_\mu^2}{2t}[1-\frac{m_2^2}{4t}\ln(1+\frac{t}{m_2^2})]\right. \\ &\left. +3G(z)[\frac{1}{4}H(\omega)\left(\ln(1+\frac{m_1^2}{t}) - \frac{m_1^2}{m_1^2+t}\right) - \frac{m_\mu^2}{2t}(1-\frac{1}{4}\frac{m_1^2}{m_1^2+t})]\right\} \end{aligned} \tag{13}$$

Here $\omega \equiv \frac{\nu}{E_\mu}$ and:

$$H(\omega) = \frac{1+(1-\omega)^2}{\omega^2}$$
$$G(z) = \frac{3}{z}[\frac{1}{2} + \frac{(1+z)e^{-z}-1}{z^2}]$$
$$z = 0.00282A^{1/3}\sigma_{\gamma p}(\nu)$$
$$t = \frac{m_\mu^2\omega^2}{1-\omega}$$

A is the atomic number of the target, and m_1^2 and m_2^2 are the vector meson masses (with values of 0.51 GeV^2 and 1.80 GeV^2, respectively). Note that the expression is no longer linear in A, but rather has a subtle dependence via the expression $G(z)$, which is due to nuclear shadowing. The trend in the literature has been to use expression 12 for reactions that take place below the pion threshold, while using that of Bezrukov & Bugaev for reactions above it.

Real and virtual photons from the cosmic ray muons produce neutrons through various mechanisms at different energies. In order of increasing energy these include contributions from giant dipole resonance (GDR) absorption quasideuteron photoabsorption [25], $\pi-N$ resonance [26], and deep-inelastic scattering [23, 24]. These various components of the $\sigma_{\gamma N}$ cross section are described in more detail in the next section.

Photonuclear Cross-Sections

Because the virtual photon spectrum falls only as $1/E_\mu$ rather than $1/E_\mu^2$, both high-energy and low-energy portions of the γN cross section are important. At low energies (below $\sim$ 30 MeV), photon absorption is dominated by a nuclear collective excitation, the giant dipole resonance (GDR), which decays predominantly by neutron emission. The GDR photo-absorption spectrum is often described by a Lorentian shape:

$$\sigma_{\mathrm{GDR}}(E_\gamma) = \Sigma_i A_i \frac{(E_\gamma\Gamma_i)^2}{(E_\gamma - E_i)^2 + (E_\gamma\Gamma_i)^2}, \tag{14}$$

where A_i, E_i, and Γ_i are respectively the empirical peak cross section, energy, and width of the excitation for the nucleus in question. The summation denotes whether one is dealing with spherical (i=1) or deformed (i=2) nuclei. The International Atomic Energy Agency has recently compiled a comprehensive review of photonuclear cross sections from threshold to 140 MeV for a variety of nuclear targets and isotopes (http://iaeand.iaea.or.at/photonuclear/).

Above the GDR peak but below the photopion threshold—that is, for photon energies in the range of $40 \leq E_\gamma \leq 140$ MeV—the dominant mechanism of nuclear photoabsorption is described by the quasideuteron model [25]. This model relates the nuclear photoabsorption cross section (σ_{qd}) to that of the free deuteron σ_d:

$$\sigma_{qd} = \frac{L}{A} NZ\sigma_d(E_\gamma)f(E_\gamma), \tag{15}$$

where L is the Levinger parameter, and $f(E_\gamma)$ is the Pauli suppression function. It is possible to parameterize the free deuteron photodisintegration cross section as follows:

$$\sigma_{\mathrm{d}}(E_\gamma) = 61.2\frac{(E_\gamma - 2.224)^{3/2}}{E_\gamma^3}. \tag{16}$$

The cross section and energy are given in units of mb and MeV, respectively. Chadwick et al. have evaluated the effect of the Pauli suppression term $f(E_\gamma)$ explicitly, and show good agreement with experimental data [27].

Above photon energies of 140 MeV, pion production—particularly through the Δ isobar resonance—becomes the dominant feature in the photonuclear cross section. The rapid decay of the Δ produces πN or

$\pi\pi N$, contributing to the neutron production from muons underground. Also, stopping π^- particles form pionic atoms and undergo nuclear absorption, producing energetic neutron pairs by the ipseudo-deuteron capture mechanism $\pi^- + d \rightarrow n + n$. Even π^+ have a several-percent probability of nuclear absorption as they range out in matter. It has been suggested that correlated double neutron events could serve as a tag for pion production by cosmic muons. Such pairs have been experimentally measured. The cross section and contribution of photo-pion production from muons has been explicitly calculated [26], both with and without the equivalent photon approximation, using the cross-section functional forms of Chanfrey et al. [28].

Finally, at photon energies above the Δ resonance, one must consider more general interactions between the photon and the actual participating partons. As mentioned above, Bezrukov & Bugaev [23, 24] have calculated the muon-nucleon cross section within the framework of a vector meson dominance model, taking into account the Q^2 dependence of the meson exchange, as well as nuclear shadowing effects. The original work parametrized the total muon cross section in the form:

$$\sigma_{\gamma N} = 114.3 + 1.647 \ln (0.0213 E_\gamma)^2, \tag{17}$$

with the cross section $\sigma_{\gamma N}$ in μb and the photon energy E_γ in GeV. A more recent treatment of this process [29] used a Regge-type parameterization of the cross section [30]:

$$\sigma_{\gamma N} = 67.7 s^{0.0808} + 129 s^{-0.4525}. \tag{18}$$

Here, $s \equiv 2 m_n E_\gamma$ and the energy and cross section again are expressed in GeV and μb, respectively.

Although the bulk of neutron production comes from the high-energy portion of the cross section, the low-energy GDR and quasideuteron components of the muon-nucleon interaction are not negligible ($\sim$ 10–30% of the total neutron rate). Thus, all components must be considered to accurately model neutron/hadron production from cosmic ray muons.

MONTE CARLO MODELING OF NEUTRON PRODUCTION UNDERGROUND

Even though different components of the muon cross section are both experimentally and theoretically well understood, the prediction of neutron and hadron production underground remains largely uncertain. There are several issues associated with such theoretical estimations. First of all, the Weizsäcker-Williams approach (and its subsequent adaptations) explicitly relies on the premise that the longitudinal polarization state—which exists only for virtual photons—can be ignored. However, at low energies the virtuality of the photon becomes comparable to its energy and thus longitudinal polarization contributions become non-negligible. Second, the low-energy interaction occurs with the nucleus as a whole (giant dipole resonance) rather than individual partons, further complicating the interaction. Finally, the simple factorization of high-energy photoabsorption cross sections into a product of the virtual photon spectrum with the photon-nucleon cross section stems from the relative Q^2-independence of the photon cross section, which is spoiled at higher energies.

Even if the above issues affecting the primary neutron production processes were negligible or surmountable, the primaries themselves interact further in dense media to produce secondary neutrons, which quickly become dominant at larger depths. Secondary neutron contributions become as large as 75% at high energies or deep sites. Thus, although the theoretical models provide useful guidance for estimating underground backgrounds, their predictive power is still limited without detailed simulation tools.

Recently, several new Monte Carlo models have been developed in order to deal with both the full muon-neutron production cross section and the secondary neutron production. Codes such as FLUKA [31] and GEANT ([32]; see also the GEANT4 Web page at CERN, http://geant4.web.cern.ch/geant4) have recently been adopted to simulate high-energy muon interactions taking place in underground experiments. These programs include many different physical processes, including electromagnetic, weak, and nuclear interactions, neutron transport, and so forth. Calculations using these tools are thus better suited than analytical calculations to deal with the problem of particle production from cascade processes. In the next few sections, we examine how such codes compare with existing experimental data.

A number of measurements have been carried out to explicitly measure the neutron and hadron production by muons at various depths. Most of the data available are on neutron production, typically involving

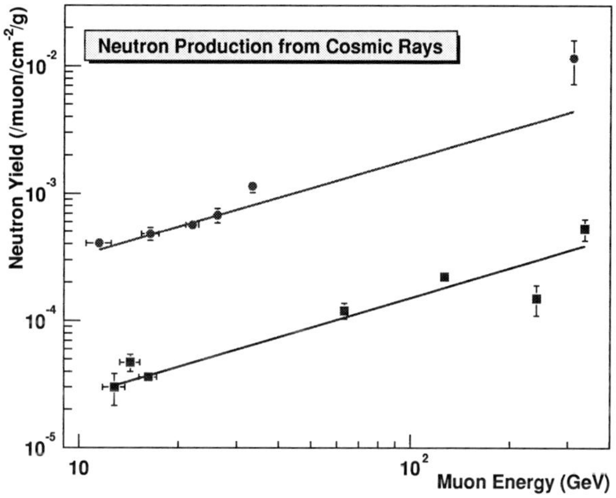

FIGURE 1. Neutron production rate as a function of average muon energy. Points taken for scintillator (*filled squares*) and lead (*filled circles*) targets at various depths. Solid lines indicate phenomenological fits to the functional form αE_μ^β. Reprinted with permission from the Annual Review of Nuclear and Particle Science, Volume 54 (c) 2004 by Annual Reviews

scintillator targets. A few experiments, however, have also been able to furnish information on neutron spectrum, neutron multiplicity, and pion production.

Several measurements have been made at shallow sites, such as Palo Verde [33], the Stanford Underground Facility [34], and facilities in Russia [23]. All have used gadolinium-loaded liquid scintillator as the main target. Gadolinium-loaded scintillator quickly captures neutrons and emits a gamma cascade with a total energy of 8 MeV. As such, the coincidence signature between the entering muon and the subsequent gamma cascade allows the neutron yield from cosmic rays to be measured.

Measurements at deeper sites include those from the Artemovsk Scientific Station [18] (depth 750 wme); the Large Volume Detector (LVD) in Gran Sasso [35] (depth 3650 mwe); and the Large Scintillation Detector (LSD) in Mont Blanc [36] (depth 5200 mwe). Each of these detectors also uses liquid scintillator as the main target. Results from all these experiments are illustrated in Figure 1, along with data on lead targets, which will be discussed below.

Theoretical models [37] predict that, based on the shower evolution, the neutron production rate should grow as $\sim E_\mu^{0.75}$, where E_μ is the average muon energy at a given depth. A simple fit to the existing scintillator target data to the function αE_μ^β yields $\alpha = (5.12 \pm 0.62) \times 10^{-6}$ and $\beta = 0.73 \pm 0.03$ for scintillator targets. These results can be compared with those predicted from detailed simulations [38]:

$$N_n = 4.14 \times 10^{-6} E_\mu^{0.74} \text{ n / muon/g/cm}^{-2}. \tag{19}$$

Although the energy dependence of muon-induced neutron production is in good agreement with the data, the Monte Carlo predictions tend to underestimate the overall neutron production rate by about 20%.

Target Dependence

Although most muon-induced neutron measurements have been made using liquid scintillator, a few earlier measurements also explored neutron production with other targets. Gorshkov & Zyabkin [39, 40, 41] measured the neutron production for aluminum, iron, cadmium, and lead at a variety of depths. Similarly, Bergamasco et al. [42] have reported on neutron production from lead targets as deep as 4300 mwe. Figures 1 and 2 summarize these results.

Gorshkov & Bergamasco's data for lead targets at a variety of depths [39, 40, 41, 42] gave a combined global fit to the neutron production parameter β for both scintillator and lead data of $\beta = 0.77 \pm 0.03$, again

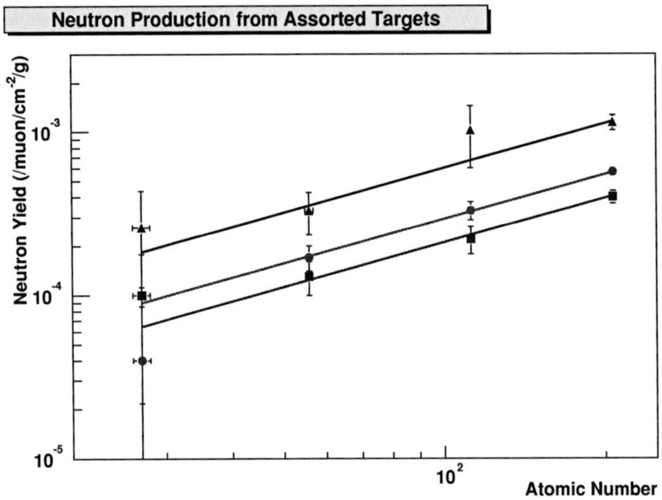

FIGURE 2. Behavior of neutron production at various depths as a function of atomic weight $\langle A\rangle$ for Al ($\langle A\rangle = 27.0$), Fe ($\langle A\rangle = 55.9$), Cd ($\langle A\rangle = 112.4$), and Pb ($\langle A\rangle = 207.2$). Data taken at 40 mwe (circles), 80 mwe (squares), and 150 mwe (triangles). Lines indicate fit to the function $\alpha_i\langle A\rangle^{\gamma}$, with a best fit of $\gamma = 0.90 \pm 0.23$. Reprinted with permission from the Annual Review of Nuclear and Particle Science, Volume 54 (c) 2004 by Annual Reviews

consistent with Monte Carlo and analytical predictions.

The data collected by Gorshkov and colleagues allows further study of the target dependence of neutron production by muons. One can fit the data with the simple relation $\alpha_i\langle A\rangle^{\gamma}$, where α_i is allowed to vary for different depths but γ is fixed. The best fits are obtained for $\gamma = 0.90 \pm 0.23$. One can compare this extracted A-dependence to that predicted by detailed Monte Carlo simulations, where one finds a dependence of $A^{0.76\pm0.01}$ [43, 44]. The results are not dissimilar to experimental measurements, though the thickness of the target may prohibit direct comparison.

The target dependence of neutron production from high-energy muons has also recently been studied in the NA55 experiment at the CERN-SPS facility [45]. Using a mono-energetic beam of 190 GeV/c^2 muons, the group measured neutron production using thin carbon, copper, and lead targets at different scattering angles. The results of this experiment agree with detailed Monte Carlo simulations to within 15%.

CONCLUSION

Underground environments are currently the ideal location for rare event searches across a variety of fields and disciplines. Despite their ideal setting, these underground locations still possess unique issues and challenges for the next generation of experiments. A great deal of effort has recently been taking place in trying to estimate –both analytically and through simulations– the backgrounds that one would encounter underground. These models and techniques are sure to increase in precision and predictive power over the next several years. Data taken specifically at various underground sites will provide further validity to these models.

ACKNOWLEDGMENTS

The author wishes to thank C. J. Martoff for useful discussions with regards to underground backgrounds, in particular with respect to dark matter experiments. This work was supported by the U.S. Department of Energy under Grant No. DE-FG06-90ER40537 and the University of Washington.

REFERENCES

1. Formaggio J A and Martoff, C J *Annu. Rev. of Nucl. and Part. Sci.* 54:351 (2004)
2. Schnee RW, Akerib DS, Gaitskell RJ. *Nuc. Phys. B Proc. Suppl.* 124:233 (2003)
3. Balantekin B (Homestake Collab.) *The national underground science and engineering laboratory at Homestake: project book, reference design stage.* http:int.phys.washington.edu/NUSEL/reference_book.html (2003)
4. Berger C, et al. *Phys. Rev. D* 40:2163 (1989)
5. Ambrosio M, et al. *Phys. Rev. D* 52:3793 (1995)
6. Aglietta M, et al. *Astropart. Phys.* 3:311 (1995)
7. Gaisser TK. *Cosmic Rays and Particle Physics.* Cambridge Univ. Press, UK (1990)
8. Lipari P. *Astropart. Phys.* 1:195 (1993)
9. Antonioli P, et al. *Astropart. Phys.* 7:357 (1997)
10. Hagiwara K, et al. *Phys. Rev. D* 66:010001 (2002)
11. Charalambus S. *Nucl. Phys. A* 166:145 (1971)
12. Biloken H, et al. *J. Geophys. Res.* 94:12,145 (1989)
13. Suzuki T, et al. *Phys. Rev. C* 35:2212 (1987)
14. Measday DF. *Phys. Rep.* 352:243 (2001)
15. Primakoff H *Rev. Mod. Phys.* 31:802 (1959)
16. Goulard B and Primakoff H, *Phys. Rev. C* 10:2034 (1974)
17. MacDonald B, Diaz JA, Kaplan SN, Pyle RV. *Phys. Rev. B* 139:1253 (1965)
18. Enikeev RI, et al. *Sov. J. Nucl. Phys.* 46:1492 (1987)
19. Fermi E. *Physik Z.* 29:315 (1924)
20. Weizsäcker CF. *Z. Phys.* 88:612 (1934)
21. Williams EJ, et al. *Kgl. Dan. Vidensk. Selsk. Mat. Fys. Medd.* XIII:4 (1935)
22. Dalitz RH, Yennie DR. *Phys. Rev.* 105:1598 (1957)
23. Bezrukov LB, et al. *Sov. J. Nucl. Phys.* 17:51 (1973)
24. Bezrukov LB, Bugaev EV. *Sov. J. Nucl. Phys.* 33:635 (1981)
25. Levinger JS. *Phys. Rev.* 84:43 (1951)
26. Delorme J, Ericson M, Ericson T, Vogel P. *Phys. Rev. C* 52:2222 (1995)
27. Chadwick MB, Oblozinsky P, Hodgson PE, Reffo G. *Phys. Rev. C* 44:814 (1991)
28. Chanfray G, Delorome J, Ericson M, Molinari A. *Nucl. Phys. A* 556:439 (1993)
29. Bugaev EV, et al. *Phys. Rev.* 58:05401 (1998)
30. Donnachie A, Landshoff PV. *Phys. Lett.* B296:227 (1992)
31. Fasso A, Ferrari A, Sala PR. *Proc. Monte Carlo 2000 Conf.*, Lisbon, Spain (2000)
32. Agostinelli S, et al. *Nucl. Instrum. Methods A* 498:250 (2003)
33. Boehm F, et al. *Phys. Rev. D* 62:092005 (2000)
34. Hertenberger R, Chen M, Dougherty BL. *Phys. Rev. C* 52:3449 (1995)
35. Aglietta M, et al. *Proc. Int. Cosmic Ray Conf., 26th*, 2:44 (1999)
36. Aglietta M, et al. *Nuovo Cim.* C12:467 (1989)
37. Ryazhskaya OG, Zatsepin GT. *Proc. Int. Conf. Cosmic Rays, IX*, 3:987. London: Institute of Physics and the Physical Society (1966)
38. Wang YF, et al. *Phys. Rev. D* 64:013012 (2001)
39. Gorshkov GV, Zyabkin VA. *Sov. J. Nucl. Phys.* 7:470 (1968); *Yad. Fiz.* 7:770 (1967)
40. Gorshkov GV, Zyabkin VA. *Sov. J. Nucl. Phys.* 12:187 (1971); *Yad. Fiz.* 12:340 (1970)
41. Gorshkov GV, Zyabkin VA, Yakovlev RM. *Sov. J. Nucl. Phys.* 13:450 (1971); *Yad. Fiz.* 13:791 (1970)
42. Bergamasco L, Costa S, Picchi P. *Nuovo Cim.* A13:403 (1973)
43. Kudryavtsev VA, Spooner NJ, McMillan JE. hep-ex/0303007 (2003)
44. Kudryavtsev VA, et al. hep-ex/0301038 (2003)
45. Chazal V, et al. *Nucl. Phys. A* 663:885 (2000)
46. Barton JC. Proc. Int. Conf. Cosmic Rays, 19th, La Jolla, 1985, p. 98. London: Institute of Physics (1985)
47. Khalchukov FF, et al. *Nuovo Cim.* C6:320 (1983)
48. Adams TR, et al. *Phys. Rev. Lett.* 74:5198 (1995). Errata. *Phys. Rev. Lett.* 80:2020 (1998)
49. Wolf J. *The Indentification of Dark Matter: Proceedings*, Edited by Neil J.C. Spooner and Vitaly Kudryavtsev. River Edge, N.J., World Scientific (2003)
50. Wulandari H. Presented at Int. Workshop Identification of Dark Matter, 4th, York UK (2002)

Neutron Backgrounds: $^{13}C(\alpha,n)$ etc

Tadao Mitsui

for the KamLAND collaboration

Research Center for Neutrino Science, Tohoku University, Sendai 980-8578, Japan

Abstract.
$^{13}C(\alpha,n)$ reaction is the main neutron source in an underground large-volume liquid-scintillator detector KamLAND. α sources, targets, cross sections, and neutron transport are studied to estimate the backgrounds of $\bar{\nu}_e$ signal.

Keywords: neutron, alpha-ray, delayed coincidence
PACS: 28.20.-v, 23.60.+e, 13.15.+g

INTRODUCTION

Neutrons are the most serious background in detecting electron anti neutrinos ($\bar{\nu}_e$) via the inverse β decay, $\bar{\nu}_e + p \rightarrow e^+ + n$. In this traditional and the most powerful mode, that have been used in most reactor neutrino experiments, backgrounds are strongly suppressed by requiring a delayed coincidence between prompt e^+ signal and delayed γ from neutron capture on a nucleus. The time and spatial correlations as well as the energy of the delayed γ are characteristic to neutron diffusion and capture in the detector/target (composition of nuclei), and are therefore used for strong tags. As far as concerning the accidental coincidence, those three tags (time and spatial correlations and delayed energy) are essentially independent. On the other hand, background neutrons produce perfect (real) signals of neutron capture (delayed energy), and the correlations from the prompt signals (if existed) are also real ones. This is the reason why removing neutron backgrounds in $\bar{\nu}_e$ experiments has been very difficult or impossible.

In most reactor $\bar{\nu}_e$ experiments before the Kamioka Liquid AntiNeutrino Detector (KamLAND) [1], cosmogenic fast neutrons have been main neutron sources. When those cosmogenic neutrons are strongly screened in deep underground, however, we have to consider other sources, i.e., (α,n), (γ,n) reactions (α and γ from radioimpurities), spontaneous fission (^{238}U etc), and interactions of atmospheric or solar neutrinos with nuclei. For those sources, intensities are essentially independent of the experimental site, but depend on detector volume and purity, so the situation is tough for large-volume experiments. Among those possible sources, we found $^{13}C(\alpha,n)$ reaction is the largest background in KamLAND.

ALPHA-NEUTRON REACTIONS

In this section, possible (α,n) reactions in KamLAND are reviewed [2], by which the effective combination of α source and targets are identified and the precise estimation of it will be shown in the next section.

α sources in KamLAND

Figure 1 shows α spectrum expected in KamLAND scintillator. The spectrum includes α's from ^{238}U and ^{232}Th series. For ^{238}U series, radiative equilibrium could possibly be broken due to relatively long-lived nuclei ^{262}Ra (half life $T_{1/2}$ = 1600 yr), and ^{210}Pb ($T_{1/2}$ = 22.3 yr). Considering that possibility, the decay rate of the series is estimated separately in the three parts, i.e., $^{238}U \rightarrow {}^{226}Ra$, $^{226}Ra \rightarrow {}^{210}Pb$, and $^{210}Pb \rightarrow {}^{206}Pb$. Within each part, radiative equilibrium is assumed. The decay rate of the first part is estimated by single energy spectrum of ^{234}Pa β. No significant contribution of ^{234}Pa β was found in the KamLAND single spectrum, then the upper limit on the decay rate 0.14 Hz / kton has been obtained.

For the second part, tagged ^{214}Bi - ^{214}Po delayed coincidence is used to estimate the decay rate. The mean rate during the livetime is measured to be $\sim 8 \times 10^{-4}$ Hz / kton. We have been monitored this coincidence rate more

CP 785, *Topical Workshop on Low Radioactivity Techniques,* edited by B. Cleveland, R. Ford, and M. Chen

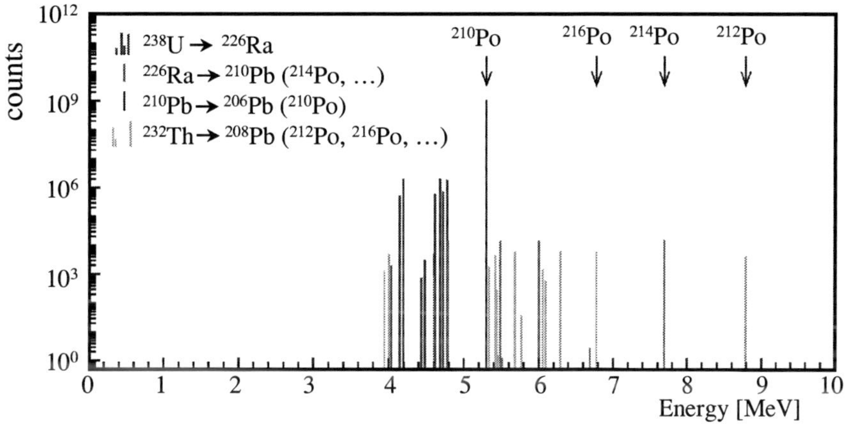

FIGURE 1. α source spectrum in the KamLAND scintillator, estimated[2] from the decay rates of ^{238}U and ^{232}Th series.

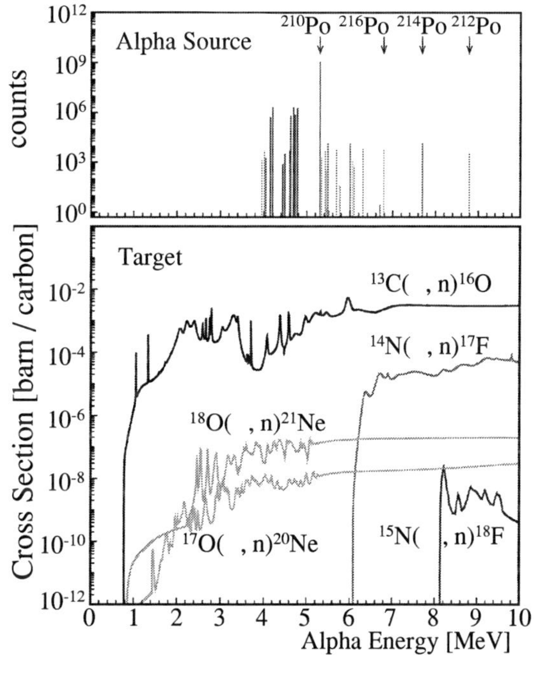

FIGURE 2. Sources and targets of (α,n). Cross sections (lower panel) are shown for each target nucleus with multiplying the abundance in KamLAND.

carefully actually, i.e., monitored as a function of position and time. Excess of the ^{214}Bi - ^{214}Po rate has been observed after some cases of source calibration, followed by clear exponential decay with half life 3.8 day. For these cases, ^{222}Rn should have been let come into the detector unfortunately. This "radon spike" has worked, at the same time, as good monitor for our sensitivity to ^{222}Rn measured with the ^{214}Bi - ^{214}Po coincidence method. We should note that, for most calibration cases performed carefully, no such radon spike was observed and the largest contribution is not from those spikes but from the constant (for position and time) component.

For the 3rd part ^{210}Pb $\rightarrow$ ^{206}Pb, we found, unfortunately, the decay rate is much higher, $\sim$ 60 Hz / kton, measured from the single rate of ^{210}Bi-β and ^{210}Po-α. This implies that, at some stage of the construction, we should have let a lot of ^{222}Rn come into the detector, that decayed down to ^{210}Pb and have been accumulated. As a result this is found to be the largest α source in our scintillator.

The decay rate of ^{232}Th series is estimated from the tagged ^{212}Bi - ^{212}Po delayed coincidence to be $\sim 3 \times 10^{-4}$ Hz / kton. Seeing Figure 1 again, ^{210}Po-α is shown to be the main (more than 99 %) α source.

Targets for (α,n) reactions

In the KamLAND scintillator, possible targets of (α,n) reaction are ^{13}C, ^{14}N, ^{15}N, ^{17}O, and ^{18}O. Carbon is one of the most abundant element in the organic scintillator. Oxygen and Nitrogen are in PPO (fluor) and residual gases dissolved in the scintillator (most are nitrogen, because the scintillator was nitrogen purged). Numbers of oxygen and nitrogen atoms are much smaller than carbon. Isotopic fraction of ^{13}C was measured to be 0.01 for real scintillator sample, in good agreement with natural carbon. Considering elemental and isotopic fractions, ^{13}C is the dominant target of (α,n). Figure 2 shows (α,n) cross sections multiplied with target abundance in the KamLAND scintillator.

ESTIMATION OF RATE AND ENERGY SPECTRUM

Precise estimation of the rate of (α,n) reactions and the energy spectrum of prompt events are shown in this section. As shown in the previous section, the main contribution is ^{13}C(α,n), so this mode is mainly explained here.

Prompt Event of ^{13}C(α,n)

There are 4 types of faked prompt signals; if the final state of ^{13}C(α,n) is the ground state of ^{16}O, i.e., ^{13}C$(\alpha,n)^{16}$O(g.s.), only neutrons can fake prompt signals via (i) elastic scatterings with protons or (ii) inelastic scattering with ^{12}C, which are abundant in the scintillator, with exciting ^{12}C into 4.4-MeV state immediately followed by a γ emission of that energy. (iii) if the final state is the first excited state of ^{16}O, i.e., ^{13}C$(\alpha,n)^{16}$O(6.049 MeV), prompt signal is faked by e^+e^-, emitted from ^{16}O(6.049 MeV), whereas kinetic energy of neutrons are very low, hardly contributing to the prompt signal. (iv) the second excited state of ^{16}O, i.e., ^{13}C$(\alpha,n)^{16}$O(6.130 MeV). γ emission from the 6.13-MeV state fakes a prompt signal.

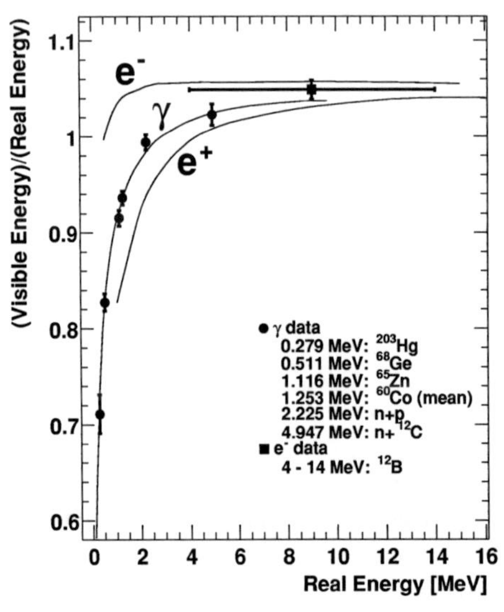

FIGURE 3. Energy calibration. Nonlinearity is due to quenching effect of the scintillator and Cherenkov light emission.

In the cases (ii)-(iv), in which the prompt signals are e^+, e^-, or γ, the quenching effect of the scintillator (energy scale nonlinearity) is relatively small and we understand it very well with calibration sources (^{60}Co, ^{68}Ge etc) and cosmic-ray spallation products (^{12}B-β, and n capture on ^{12}C) as shown in Figure 3. Here scintillator quenching effect is modeled using Birks formula and as the other source of nonlinearity, Cherenkov light emission is taken into account. We use this calibrated energy scale separately for e^+, e^-, and γ to obtain an expected "visible energy" spectra. For the case (i), the signal of low energy recoil protons should show larger quenching effect. We determined this energy scale using Birks constant obtained by α's from ^{210}Po (kinetic energy 5.3 MeV), ^{214}Po (7.7 MeV), and ^{212}Po (8.8 MeV).

Procedure of Precise Estimation

For a precise estimation of the rate and energy spectrum of ^{13}C(α,n) background, real data are used as long as possible, and also used are published cross sections of ^{13}C(α,n) reactions and Geant4 based Monte Carlo. The actual procedure is (i) ^{210}Po decay rate in the fiducial volume is determined from the α peak in the observed energy spectrum. (ii) neutron production rates and energy spectra for ^{13}C$(\alpha,n)^{16}$O(g.s.), ^{13}C$(\alpha,n)^{16}$O(6.049 MeV), and ^{13}C$(\alpha,n)^{16}$O(6.130 MeV) are determined using published cross sections [4] and α source rate determined in (i). (iii) to get energy spectrum that will be observed in our detector, transport of neutrons, electrons, positrons, and gammas are simulated with Geant4 based Monte Carlo with the scintillator quenching effect taken into account.

^{210}Po Decay Rate

Figure 4 shows a typical single spectra of KamLAND with various fiducial volume cut. Clear peaks are seen around 0.3 MeV that are not suppressed by fiducial cut. These are identified as ^{210}Po-α since the energy corresponds to 5.3 MeV (the energy of ^{210}Po-α) taking quenching effect of the scintillator into account. In the fiducial volume $R < 5.5$ m, outside γ's from ^{40}K and ^{208}Tl are suppressed, and the main contributions in the low energies are β's from ^{85}Kr and ^{210}Bi and α from ^{210}Po. Using this spectra, decay rates of ^{210}Po-α are measured as a function of time (Figure 5, quoted from [3]). Also shown is a fitting curve that assumes, at some point T_0 (free parameter), only ^{210}Pb were injected (no ^{210}Bi or ^{210}Po) and the decay chain ^{210}Pb (half life 22.3 yr) $\rightarrow^{210}$Bi (5.0 day) $\rightarrow^{210}$Po (138.4 day) started to converge to radioactive equilibrium after long enough time compared to ^{210}Po life time. For simplicity the decay time of ^{210}Bi is ignored here. As seen, the best-fit T_0 well agrees with the period of initial detector filling (lasted 5 months), that is one of the reasons why we guess those ^{210}Pb are daughters of ^{222}Rn which are unfortunately let to come into the detector from somewhere during filling operation.

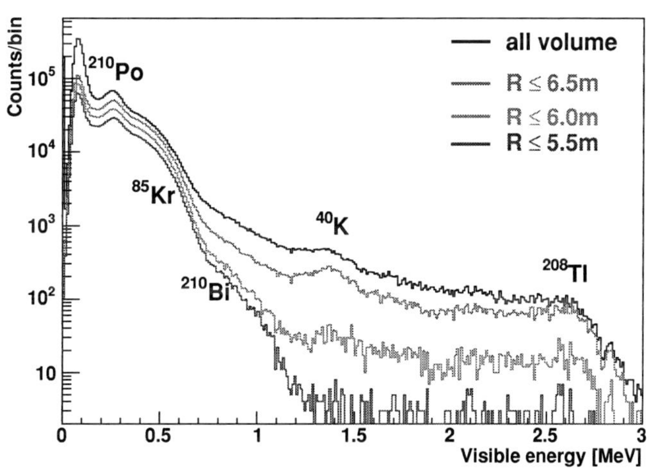

FIGURE 4. KamLAND single spectra with various fiducial volume cut. ^{210}Po-α peaks are clearly seen.

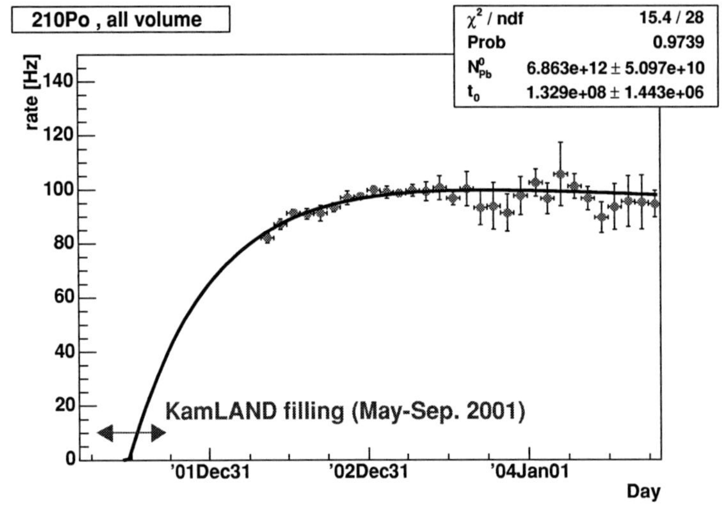

FIGURE 5. Time variation of ^{210}Po decay rate.

^{13}C(α,n) Cross Section and Neutron Spectra

Total and differential cross sections for ^{13}C(α,n) reactions are well measured and parameterized [4]. For the ground-state case ^{13}C$(\alpha,n)^{16}$O(g.s.), both the absolute cross section and its dependence on the α energy are precisely known. Also for the differential cross sections, distributions of neutron emission angle in the center-of-mass system are expressed as series of Legendre polynomials and dependence of those Legendre coefficients on α energy are provided. This enables us to perform precise and reliable calculation for differential cross sections for given α energy.

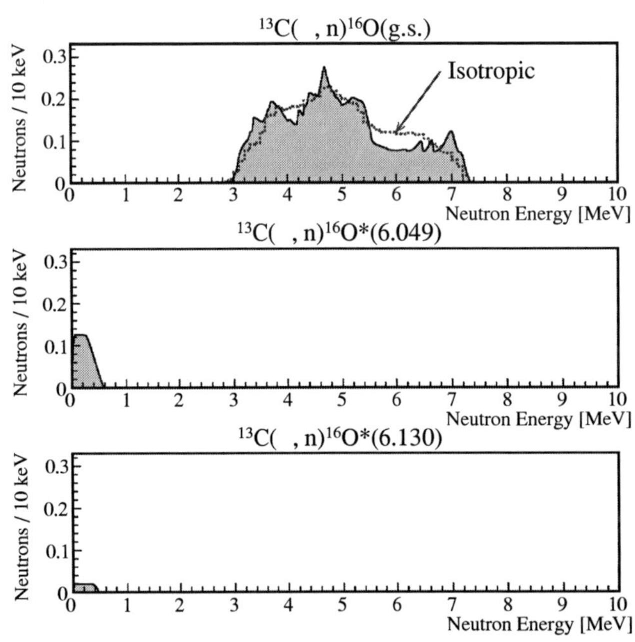

FIGURE 6. Neutron source spectra in the KamLAND scintillator.

We then obtained an expected neutron energy spectrum at emission (Figure 6), by integrating the differential cross sections multiplied by dx/dE from the initial α energy 5.3 MeV to zero. Also shown in the figure is the spectrum calculated with assumption of isotropic neutron emission in the center-of-mass system. The comparison tells us that the differential cross section doesn't change the overall feature of the neutron emission spectrum. This difference is incorporated in the systematic error.

For the excited-state cases $^{13}C(\alpha,n)^{16}O(6.049\,MeV)$, and $^{13}C(\alpha,n)^{16}O(6.130\,MeV)$, we rely only on the α energy dependence, and not on the absolute value of the reaction cross section. Neutrons from those cases are in very low energy as shown in Figure 6, because the most of the available energy (Q value plus kinetic energy of α) are consumed for ^{16}O excitation. In those cases however, e^+e^- and γ emissions from $^{16}O(6.049$ MeV) and $^{16}O(6.130$ MeV) respectively will fake the prompt signals as described already.

Neutron Transport and Expected Prompt Energy Spectrum

To calculate energy spectra of the faked prompt signals, transport of neutrons, electrons, positrons, and gammas are simulated with a Geant4 based Monte Carlo simulation. For electrons, positrons and gammas, the simulation is well tuned with plenty of calibration data, as described already (Figure 3), and a precise energy scales with an estimated systematic error $\sim 2\%$ has been obtained [1]. This error includes uncertainty of Monte Carlo tuning, especially that of Birks constant and amount of Cherenkov light emission allowed within calibration data.

For fast neutrons from $^{13}C(\alpha,n)^{16}O(g.s.)$ case, Monte Carlo tuning includes larger systematic errors, but the precision is enough for the present purpose. Figure 7 shows an expected energy spectrum of the faked prompt events. Low energy part and the 4.4-MeV peak are contributions of $^{13}C(\alpha,n)^{16}O(g.s.)$, and the peak at ~ 6 MeV is from $^{13}C(\alpha,n)^{16}O(6.049$ MeV) and $^{13}C(\alpha,n)^{16}O(6.130$ MeV). Note that spectrum below 2.6 MeV, that is not used in the current analysis [1], is preliminary.

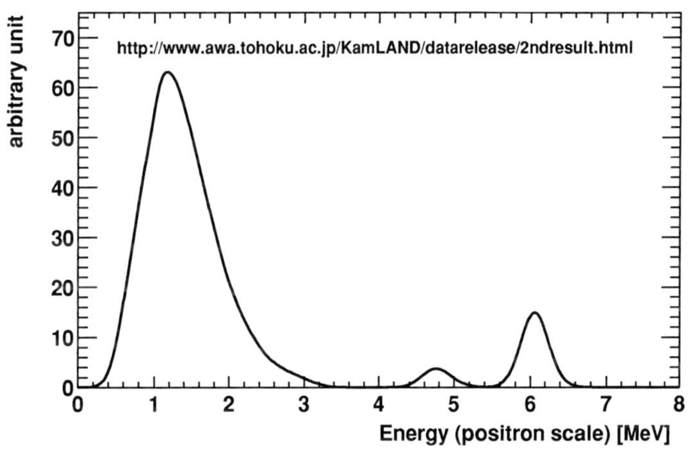

FIGURE 7. Expected prompt energy spectrum of (α,n) backgrounds. All detector response is included for individual particle species, and shown in the positron energy scale for correspondence with $\bar{\nu}_e$ signal.

EFFECT ON OSCILLATION ANALYSIS

Figure 8 shows the prompt energy spectra for selected delayed coincidence events [1]. Data and expectations for the signal (reactor $\bar{\nu}_e$) and backgrounds are shown, including $^{13}C(\alpha,n)$ explained in this report. As seen, $^{13}C(\alpha,n)$ is the largest background above 2.6 MeV, although the effect is small enough if correctly incorporated in the analysis. Number of events in this final sample is 258, while expected $^{13}C(\alpha,n)$ background is about 10 events.

To examine an effect of the existence of $^{13}C(\alpha,n)$, we performed a Monte Carlo study, in which 10^6 of "toy experiments" are generated based on the expected spectra of signal plus background. For $^{13}C(\alpha,n)$ background, expected rate and spectrum are fluctuated based on the systematic errors described above, especially, for 6-MeV events, expected number of events $N^{input}_{event(6-MeV-bg)}$ are varied 0 to 21 with flat probability density. Although the best estimation is $N_{event(6-MeV-bg)} \sim 7$, we regard it as a free parameter in a conservative attitude, because the absolute cross sections for $^{13}C(\alpha,n)^{16}O(6.049\text{ MeV})$ and $^{13}C(\alpha,n)^{16}O(6.130\text{ MeV})$ are rather ambiguous. For each toy experiment, the same analysis as the real data [1] are performed, in which the best-fit number of 6-MeV events $N^{best-fit}_{event(6-MeV-bg)}$ is treated as a free parameter.

Figure 9 shows the result of this Monte Carlo study. The horizontal axis $N^{best-fit}_{event(6-MeV-bg)} - N^{input}_{event(6-MeV-bg)}$ represents a degree of misfit of 6-MeV events. It is shown that Δm^2 changes from the input value (7.9×10^{-5} eV^2) according to the misfit effect (the slanting cluster around the input value means that effect). The thickness of the cluster corresponds to the statistical error, and due to the slanting cluster, distribution will become wider when projected onto the vertical axis. This effect corresponds to the additional error due to 6-MeV background, which does not seem a dominant source of error at present. We will however make the best effort to perform more precise and reliable estimation of this background and more importantly, to purify the scintillator and remove ^{210}Pb [5].

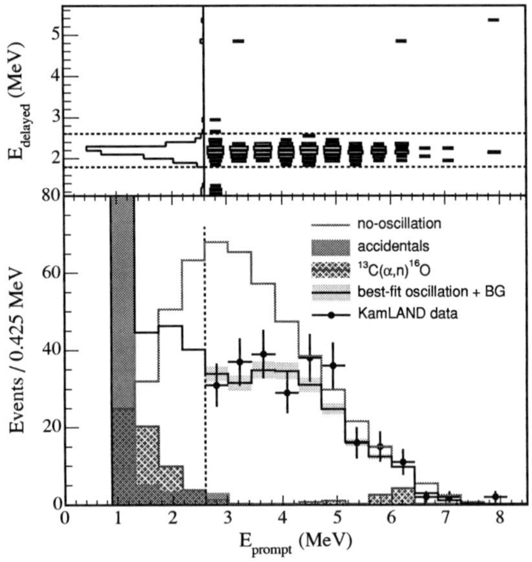

FIGURE 8. Prompt energy spectra: KamLAND data and expectations. Figure quoted from [1].

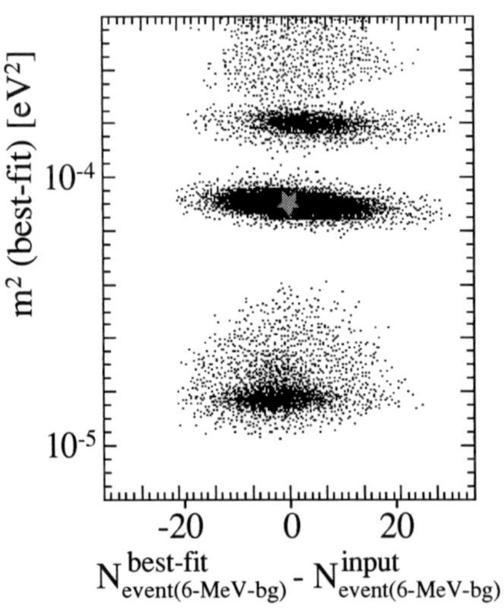

FIGURE 9. Monte Carlo result for oscillation analysis. 10^6 of toy experiments are generated. Star shows the input value.

ACKNOWLEDGMENTS

I am grateful to the organizing staffs of the workshop for warm hospitality at SNOLAB, and also editors for kindness and patience for the delayed manuscript of mine.

REFERENCES

1. T. Araki et al. (KamLAND collaboration), *Phys. Rev. Letters*, **94**, 081801 (2005), `http://www.awa.tohoku.ac.jp/KamLAND/datarelease/2ndresult.html`.
2. S. Enomoto, PhD thesis, Tohoku University (2005).
3. K. Ichimura, Master thesis (in Japanese), Tohoku University (2005).
4. Japan Atomic Energy Research Institute, `http://wwwndc.tokai.jaeri.go.jp/`,
   G.W. Kerr et al., *Nucl. Phys.*, **A110**, 637–656 (1968),
   R.B. Walton et al., *Phys. Rev.*, **170**, 1065–1075 (1957).
5. Y. Kishimoto, *these proceedings*.

Background studies for particle astrophysics experiments

Vitaly A. Kudryavtsev

Department of Physics and Astronomy, University of Sheffield, Sheffield, S3 7RH, UK

On behalf of the UK Dark Matter Collaboration and ILIAS

Abstract. Background radiations typical for the high-sensitivity underground experiments in particle astrophysics are discussed. An emphasis is given to the neutron background coming from spontaneous fission and (α,n) reactions from U and Th traces in rock and detector components, and from cosmic-ray muons. Gammas from radioactivity in various materials are also considered. Special case of a xenon-based large-scale dark matter detector is studied. Several Monte Carlo codes capable of producing, transporting and detecting neutrons are compared with each other and with available experimental data.

Keywords: Underground experiments, Background radiations, Neutrons, Gammas
PACS: 28.20.-v, 24.10.-i, 95.85.Ry

INTRODUCTION

There exist many particle astrophysics experiments located deep underground. Most of them are searching for rare events induced either by particles coming from the space (dark matter WIMPs – Weakly Interacting Massive Particles, astrophysical neutrinos, etc.) or by weak processes predicted by theories beyond the Standard Model (for instance, neutrinoless double-beta decay). More detectors are under construction or in the planning phase. For most (if not all) of them the background radiations are critical in restricting their sensitivity to the expected signal. Knowledge of the background and ability to suppress or reject background events are essential for estimating the detector sensitivity, interpreting experimental results and designing future experiments. Usually the study of background is done through the Monte Carlo simulations based on nuclear and particle physics models. Monte Carlo codes and the corresponding models are checked by the comparison with existing experimental data.

Background studies have been initiated by several collaborations to evaluate the potential of future detectors, in particular, for dark matter searches. A new opportunity for such studies has been offered by the ILIAS – Integrated Large Infrastructures for Astroparticle Science. Several activitites of ILIAS include working groups on background simulations and measurements. Within these activities, the simulations of various types of background in underground laboratories relevant to different experiments are foreseen with goals to test Monte Carlo codes and help to design future large-scale detectors, in particular for dark matter searches, their shielding and active veto systems.

In this paper we discuss various types of background relevant to particle astrophysics experiments located deep underground: dark matter searches, double-beta decay experiments, neutrino detectors. We concentrate mainly on low-energy events in keV-MeV range, although some codes are also relevant to high-energy neutrino astronomy. Useful information on backgrounds can also be found in Ref. [1].

NEUTRON BACKGROUND

Neutrons underground arise from two sources: i) local radioactivity, and ii) cosmic-ray muons. Neutrons associated with local radioactivity are produced mainly via (α,n) reactions initiated by α-particles from U/Th traces in the rock and detector elements. Neutrons from spontaneous fission of ^{238}U also contribute to the flux. The neutron yield from cosmic-ray muons depends strongly on the depth of the underground laboratory. The suppression of the muon flux by a large rock overburden will also reduce the neutron flux but by a smaller factor.

CP 785, *Topical Workshop on Low Radioactivity Techniques,* edited by B. Cleveland, R. Ford, and M. Chen

Neutrons from rock: production

In this Section we will consider production of low-energy neutrons (E<10 MeV) in radioactive decays of U and Th in rock.

Neutrons are produced in rock via spontaneous fission and (α,n) reactions arising from uranium and thorium traces. Spontaneous fission of ^{238}U only gives a significant contribution to the neutron flux, while (α,n) reactions are important for alphas in the U/Th decay chains if the energy threshold of a reaction calculated from the Q-value for any particular nucleus hitted by an alpha particle, is smaller than the alpha energy. Another important effect which affects the (α,n) cross-section is the Coulomb barrier. The Coulomb barrier prevents low-energy alphas approaching the nucleus because of the electric repulsion between positively charged alpha particle and positively charged nucleus. The energy required to overcome the Coulomb barrier increases with the charge of the target nucleus. Below this energy the cross-section is very small even above the energy threshold calculated from the Q-value. The Coulomb barrier energy is about 7 MeV for nuclei in the iron group making the contribution from (α,n) reactions on elements with Z>26 to the neutron flux smaller than that from spontaneous fission.

The only code known to the author, which can calculate neutron yield from (α,n) reactions, is SOURCES [2]. The code also calculates the neutron yield from spontaneous fission (of ^{238}U mainly) using a Watt spectrum [3]. Neutron fluxes and spectra from (α,n) reactions are obtained taking into account cross-sections of reactions as functions of alpha energy, branching ratios for transitions to different excited states, stopping power of alphas in various media, and assuming isotropic emission of neutrons in the centre-of-mass system. The library of cross-sections and transition probabilities can be easily modified by a user.

Original code SOURCES provides a treatment of (α,n) reactions only up to 6.5 MeV α-energies. This restricts significantly the reliability of the results because the cross-section of (α,n) reaction rises with energy and the average neutron energy also increases with the parent alpha energy. Hence, the 6.5 MeV cut reduces the total neutron yield from (α,n) reactions and shifts neutron spectrum to low energies.

The following modifications have been made to SOURCES to overcome the 6.5 MeV limit [4]. Existing cross-sections have been extended to 10 MeV, taking into account available experimental data. For some materials, new cross-sections have been added to the code. If the cross-section for a material has been measured or calculated for low energies only, then it has been extrapolated from low energies up to 10 MeV. Most cross-sections can be found on the web-site of the Nuclear Data Services of the Nuclear Data Centre at the International Atomic Energy Agency [5].

The branching ratios for transitions to the ground and excited states above 6.5 MeV have been chosen to be the same as at 6.5 MeV [4]. This results in a small overestimate of neutron energies for alphas above 6.5 MeV, since the increased probability of transition to the higher states is neglected, but the total neutron flux is not affected. The uncertainties associated with the calculations of such probabilities are not negligible, however. The difference as large as (20-30)% exist between different calculations of the transition probabilities. If the transition probabilities were not present in the code library, as was the case for elements for which the cross-sections were absent too, then the transitions have been assumed to occur to the ground state only.

The modified code SOURCES has been used to simulate neutron production in NaCl in Refs. [4, 6] (see Figure 1). The authors also calculated the neutron flux and energy spectrum of neutrons expected for the Modane rock and compared them to the simulations and measurements reported in Ref. [7] (a revised flux was reported in Ref. [8]). The most dramatic difference concerns the energy spectrum of neutrons, that from SOURCES being significantly softer than measurements and simulations reported in Ref. [7] (see Figure 2 for comparison). The simulated and measured spectra from Ref. [7] both have a peak at around 3 MeV. Note, however, that the spectrum measured in Ref. [7] is not a directly measured neutron spectrum. It was deconvolved from the measured proton recoil spectrum using a complicated procedure involving simulations including neutron propagation through the lead and copper shielding around the detector. Moreover, this spectrum is not a spectrum at production but the spectrum in the laboratory after propagation through the rock. A peak at smaller energies (about 1.5 MeV) is present on the neutron energy spectrum from SOURCES [4, 6] but is smeared out by the propagation of neutrons through the rock (carried out in [4, 6] using GEANT4 toolkit [9]). It is difficult to explain the presence of such a peak in the neutron spectrum at the rock surface (rock/cavern boundary). The simulations carried out in Ref. [7] assumed the final nucleus in the ground state only (this overestimates the neutron energy) and an emission of a neutron at 90° (this means that neutron energy is directly calculated from the alpha energy and the neutron spectrum is a delta-function). These two major effects can be responsible for the difference between these simulations and the results obtained with more accurate simulations with SOURCES.

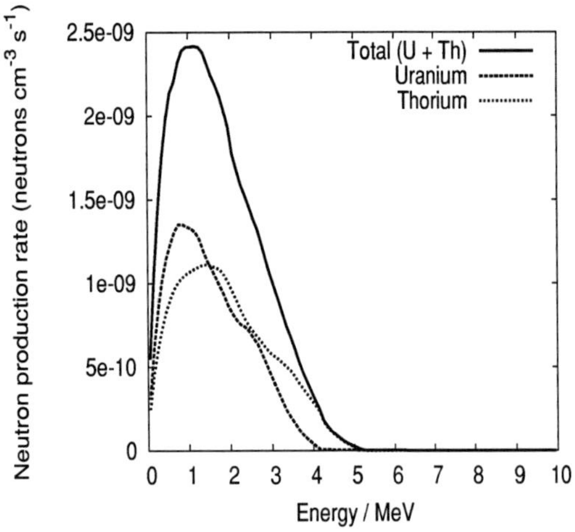

Figure 1: Simulated energy spectrum of neutrons produced in the salt rock, as calculated using modified SOURCES. Solid curve - combined total neutron production rate due to U (60 ppb) and Th (130 ppb) contamination in salt rock; dashed line - neutron production rate due to uranium (60 ppb); dotted line - neutron production rate due to thorium (130 ppb). The graph is taken from Ref. [6].

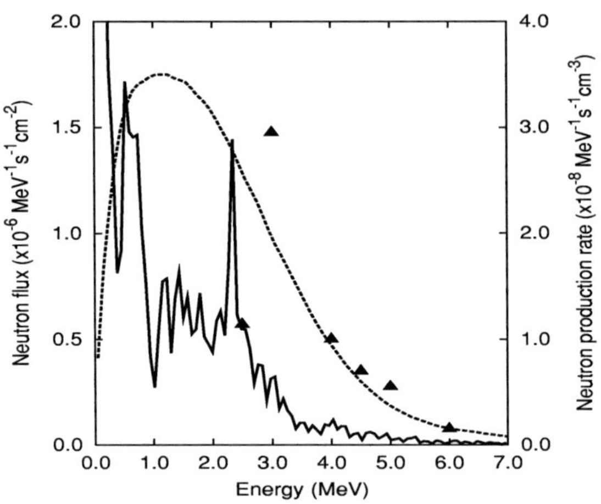

Figure 2: Neutron production spectrum as calculated with modified SOURCES for Modane rock (dashed curve and right y-axis for rate units) is plotted together with the neutron energy spectra at the rock/cavern boundary (left y-axis): solid curve – spectrum from SOURCES propagated with GEANT4; triangles – evaluated spectrum from measurements [7]. The graph is taken from Ref. [4].

The authors of the original code SOURCES carried out extensive tests and comparison of the calculations with existing experimental data [2] which show a remarkably good agreements but the validation of the code was restricted to low alpha energies (below 6.5 MeV) and relatively light targets.

A wide peak (at about 2.2-2.3 MeV) in the measured neutron spectrum in the laboratory has also been reported recently in Ref. [10]. The authors have used the deconvolution of the measured spectrum of proton recoils and no simulation of the neutron propagation through the detector shielding has been performed. No distinctive peaks have been reported for either of the measurements in the Gran Sasso Laboratory above 1 MeV [11, 12], but the evaluated spectra appear to be harder than the typical spectrum from SOURCES, although no precise simulations with SOURCES have been done for the Gran Sasso rock and concrete. Instead, the simulations have been carried out [13] using the assumptions similar to Ref. [7] and neutron spectrum has been corrected to match the experimental data for thick target neutron yield from (α,n) reactions. The resulting calculated neutron spectrum in the Gran Sasso laboratory is softer that that reported in Ref. [12], although it is not very different from Ref. [11].

To conclude, SOURCES provide quite accurate treatment of (α,n) reactions but predict the neutron spectra much softer than all available underground measurements. Keeping in mind that the reported measurements also involve simulations of neutron transport and interactions in the detector, I cannot determine what causes this difference and then cannot favour either experimental results (also affected by simulations) or calculations using SOURCES.

Neutrons from rock: propagation

The neutron transport through the rock has already been discussed in the previous section in connection with the neutron spectra in the underground laboratories. Sensitive experiments are usually shielded from rock gammas by high-Z material such as lead, and from rock neutrons by low-A material such as hydrocarbon: polyethylene or polypropylene. Hydrogen nuclei, having similar weight to the neutrons, are very efficient in reducing the neutron energy below the detection threshold. Optimal configuration of the shielding depends on goals and specific features of the detector. A special case of the xenon-based large-scale dark matter detector has been discussed in Ref. [4], whereas low-pressure gas TPC has been considered in Ref. [6]. It was shown that a thick layer (20-30 cm) of lead reduces significantly the neutron flux above a few MeV due to the high inelastic scattering cross-section [4]. This results in a larger suppression of the neutron flux by the hydrocarbon when the lead is present compared to the case without lead. Note that the hydrocarbon itself is not very efficient in suppressing the neutron flux at high energies (above a few MeV) since the elastic cross-section decreases with energy. These simulations have shown that about 35-40 g/cm^2 behind 20-30 cm of lead is enough to suppress the neutron flux below the level of sensitivity of a tonne-scale dark matter detector. Without lead more hydrocarbon shielding (by about 15-20 g/cm^2) is needed. For a hydrocarbon material with smaller fraction of hydrogen than in polyethylene, the thickness of the hydrocarbon should be recalculated accordingly since only hydrogen contributes to the energy degradation of neutrons.

There exist several Monte Carlo codes able to transport low-energy neutrons. GEANT4 [9] has been used in Refs. [4, 6, 14, 15], MCNP [16] in Refs. [13, 8]. MCNP is a well established package used by many research groups across the world. GEANT4 is a relatively new code but it attracts much attention and becomes more and more popular because of its flexibility, high precision and a large variety of tasks which it can be used for. The comparison between the existing results from the two codes are quite complicated since the simulations quoted above have been set up for different geometries, different initial neutron spectra and different detecting media. I am not aware of any published direct comparison between GEANT4 and MCNP in transporting neutrons through various materials. Only recently such investigation has been started within the ILIAS programme. The first preliminary results for a neutron spectrum and geometry described in Ref. [4], showed a difference in the neutron spectrum above 1 MeV at the rock (NaCl) surface (rock/cavern boundary) as obtained with GEANT4 [4] and MCNP [17], MCNP giving smaller neutron flux at high (> 1 MeV) energies. Similar simulations carried out with GEANT3 [18] support the MCNP results. The observed difference in the neutron spectrum results in the higher suppression factor obtained after propagating neutrons through the hydrocarbon shielding with MCNP. The difference at the rock surface is most likely due to the bug in the neutron inelastic cross-section on chlorine found in the GEANT4 library [19]. Although such a bug should be quite rare and was found in one file only (cross-section on chlorine) [19], its presence indicates that more tests would certainly be useful.

An interesting result has been obtained in Ref. [14]. The authors obtained the neutron flux at the Canfranc Underground Laboratory as the difference between measured event spectra in Ge detectors without and with neutron shielding. Simulating the event spectra with initial soft (spontaneous fission) and hard ((α,n) reactions) neutron spectra, they found no big difference in the shape of the spectrum in the Ge detector. This imposes some restrictions on the accuracy with which the neutron spectrum in a laboratory can be measured to test simulations with SOURCES.

In conclusion, no major differences have been found so far between MCNP, GEANT3 and GEANT4 (with corrected inelastic cross-section for Chlorine) in propagating low-energy neutrons through the salt rock and hydrocarbon shielding.

Neutrons from cosmic-ray muons

At deep underground sites (3 km w. e. or more), the neutron production rate from muons is about 3 orders of magnitude lower than the rate for neutrons arising from rock activity, depending strongly both on the depth and the U/Th contamination. The muon-induced neutron flux can be important, however, for experiments intending to reach high sensitivity to WIMPs or other rare events. There are several reasons for this: 1) the energy spectrum of muon-induced neutrons is hard, extending to GeV energies, and fast neutrons can travel far from the associated muon track, reaching a detector from large distances; 2) fast neutrons deposit larger energies and are visible in the detectors, while many events from α-induced neutrons fall below detector energy thresholds; 3) a detector can be protected against neutrons from the rock activity by hydrocarbon material, possibly with the addition of a thermal neutron absorber; such material, however, will also be a target for cosmic-ray muons.

Muon-induced neutron flux is proportional to the muon flux but it also increases with the increase of the mean muon energy underground (see, for example, Ref. [20] for discussion). The muon flux has been measured for most underground laboratories giving enough of information for testing Monte Carlo codes for muon simulations.

Muon transport through the large thickness of rock can be done with general particle physics codes such as GEANT4 or FLUKA [21], or with specially developed, simple and fast muon propagation codes such as MUSIC [22, 23]. MUSIC was tested against experimental data (see Refs. [24, 20] for discussions) and provide fast and simple way of propagating muons down to large depths. In most cases the muon transport through the rock is not needed and can be substituted with a code which simulates muon energy spectrum and angular distribution at a known depth underground. Such a code – MUSUN [24, 20], has been written and used as a starting point for neutron simulations [20, 4, 6, 25].

Two general-purpose codes GEANT4 and FLUKA have been used so far for production, transport and detection of muon-induced neutrons. The validation of the codes has been done through the comparison of them with each other [26, 27] and with available experimental data [20, 28, 26]. Several models for neutron production in GEANT4 have been tested and compared to FLUKA predictions [27] allowing one to choose the most accurate model. The results of the simulations with the most accurate model agree reasobaly well with most available experimental data [26]. Good agreement has been found for the neutron yields, spectra and lateral distributions in light materials (important measurements have been performed with the LVD detector [29]). There is, however, a number of problems which is worth to mention here.

Both GEANT4 and FLUKA failed to explain the high neutron yields, especially in heavy targets, measured in NA55 experiment at CERN [30]. Earlier experiments carried out underground [31, 32] showed the total neutron yield in lead in reasonable agreement or slightly higher than predicted by FLUKA and GEANT4 [26]. However, the detectors used had quite a small thickness and muon-induced cascades could not develop fully. This effect was not taken into account in the evaluation of the total neutron yield from these experiments. If appropriate correction is done, then the total neutron yield from these experiments may increase significantly exceeding the model predictions from both FLUKA and GEANT4. However, accurate comparison is difficult since it would require detailed simulations of the experimental set-up.

For high-energy muons (>100 GeV) and most targets the neutron yield from GEANT4 has been found to be less (by up to a factor of 2) than that from FLUKA [26, 27]. Direct neutron production in muon spallation reactions is similar in both codes [26]. The neutron production in electromagnetic cascades always dominates in GEANT4 over that in hadronic cascades. In FLUKA hadronic cascades play more important role at high muon energies (>10 GeV) in most materials [20]. GEANT4 shows enhanced neutron photoproduction in electromagnetic cascades in light materials compared to FLUKA. For heavy elements, FLUKA overproduces neutrons in both electromagnetic and hadronic cascades initiated by muons compared to GEANT4. Despite these differences, the simulated total fluxes of fast neutrons (>1 MeV) from salt rock entering the underground cavern were found to agree within 20% [26]. The neutron fluxes generated by FLUKA and GEANT4 behind the lead and hydrocarbon shielding also agree (see Figure 3).

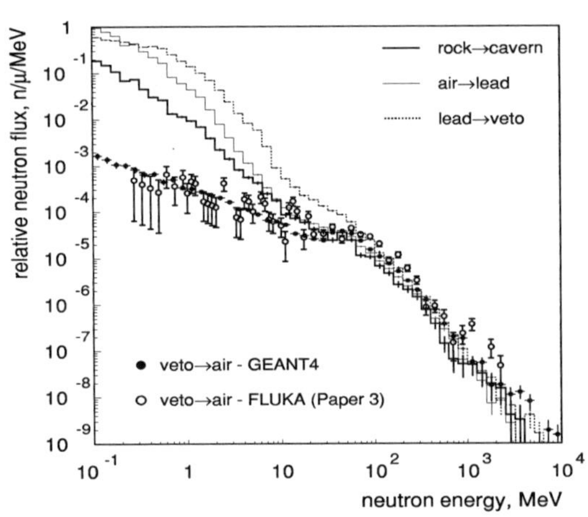

Figure 3: Energy spectra of muon-induced neutron fluxes across several boundaries. The thick line represents the flux at the rock face for an empty cavern. The spectrum of neutrons exiting the hydrocarbon veto obtained with FLUKA (Paper 3 [4]) is also shown. The graph is taken from Ref. [26].

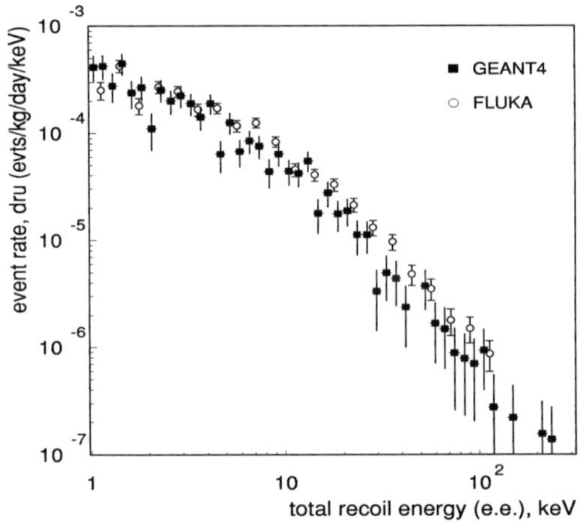

Figure 4: Nuclear-recoil energy spectrum in the liquid xenon detector as a function of the visible energy deposited by all nuclear recoils in each event. The spectra include 'mixed' events involving electromagnetic energy deposits, not just 'pure' nuclear recoils, but only the energy left by nuclear recoils was counted. The graph is taken from Ref. [26].

Although FLUKA uses Kerma factors for generating point energy deposition of neutrons (strictly speaking no nuclear recoils are produced), the spectrum of these depositions in a xenon-based WIMP detector has been found to be very similar to that calculated with more accurate modelling by GEANT4 [26] (see Figure 4). Both codes predict similar rate of neutron-induced events and similar rejection factor for most of these events due to coincidence between nuclear recoils and any energy deposition from other particles in muon-induced cascades. In fact, only a few percent (5-8%) of nuclear recoils in a large-scale (250 kg) xenon detector are 'pure' nuclear recoils without other energy depositions. Note that such coincidences will be missed if neutron production, propagation and detection are separated from those of other particles. So for correct simulations of neutron-induced effects it is very important to produce, trasport and detect neutrons and all other particles associated with muons with a single, accurate Monte Carlo code such as GEANT4 or FLUKA.

Both codes have shown that with an active veto system just around the dark matter detector, the rate of non-vetoed neutron background events from cosmic-ray muons is negligible – less than 0.5 per year in a 250 kg xenon detector.

FLUKA and especially GEANT4 are being improved but a significant change in the modelled neutron rate is not expected. However, the high neutron production in heavy targets as suggested by some experiments has to be checked.

Neutrons from detector components

The source of this background is spontaneous fission and (α,n) reactions as in the case for neutrons from the rock. The problems then are similar to those described in Section 2. The most serious is the existence of only one code SOURCES capable of simulating neutron production via (α,n) reactions. The lack of measured or accurately calculated cross-sections on heavy targets (for example, on most isotopes of iron and other elements in the stainless steel), as well as transition probabilities to various excited states, restrict significantly the accuracy of simulations of neutron production in some targets with Z$\sim$20-30. For example, the neutron production in iron and stainless steel was found to be 50-70% less [6] than the measurements [33]. The contribution from nuclear physicists to this area would be very welcome.

The Monte Carlo codes usually used for the transport and detection of these neutrons are GEANT4 (as in Ref. [6]) and MCNP.

Another restriction on the accuracy of simulations comes from the measurements of contamination levels of radioactive isotopes in different materials. The exact neutron fluxes and associated event rates are detector dependent and cannot be discussed in detail here. It is worth to mention, however, the most common cases.

Hydrocarbon shielding can be made very pure and (α,n) reactions contribute very little to the neutron yield (because of the high threshold of reaction on ^{12}C) [6]. As a good neutron moderator hydrocarbon will protect a detector from neutrons from rock and other shielding (for example, lead, iron or copper) if hydrocarbon is placed just around the detector.

Photomultiplier tubes (PMTs) may dominate in the neutron yield for detectors which use them as a readout [4]. For detectors without PMTs other components (such as detector vessel [6], for example) become important.

GAMMA, BETA AND ALPHA BACKGROUNDS

External gammas are produced in U and Th decay chains and in ^{40}K decay. For salt rock the latter dominates. The suppression of gammas by the lead shielding has been studied in Ref. [34] in connection with a xenon-based large-scale dark matter detector (see Figure 5). The gamma spectra from U/Th/K decay chains have been taken from Refs. [35, 36]. The gamma flux from the rock has also been compared to the expected gamma background from the various detector components. It was found that about 10 cm of lead (plus about 40 g/cm^2 as a neutron shielding) is required to reduce the external gamma background to a level below that of the internal background from the copper vessel. But the dominant contribution comes from the PMTs even if ultra-low-background PMTs are used. Figure 6 shows the spectra of electron-like events from different types of background in 250 kg xenon dark matter detector. The transport and detection of gammas has been carried out with GEANT4. Other Monte Carlo codes to be considered for this job are MCNP, EGS4 and FLUKA.

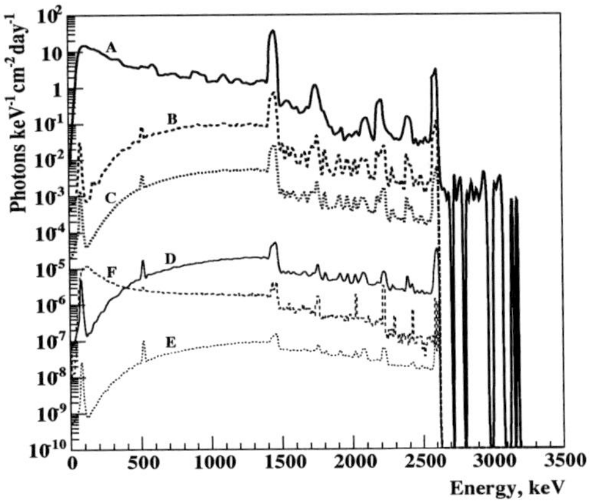

Figure 5: Gamma spectra from Boulby rock/cavern interface (A) and after 5 cm (B), 10 cm (C), 20 cm (D) and 30 cm (E) of Pb shielding. The usual U/Th peaks can be seen as well as the peak at 1460 keV from ^{40}K. The lines at 511 keV and 88 keV are due to e^-e^+ annihilation and lead K-shell emission respectively. Line (F) is the spectrum after 20 cm Pb + 40 g cm^{-2} of CH_2. The graph is taken from Ref. [34].

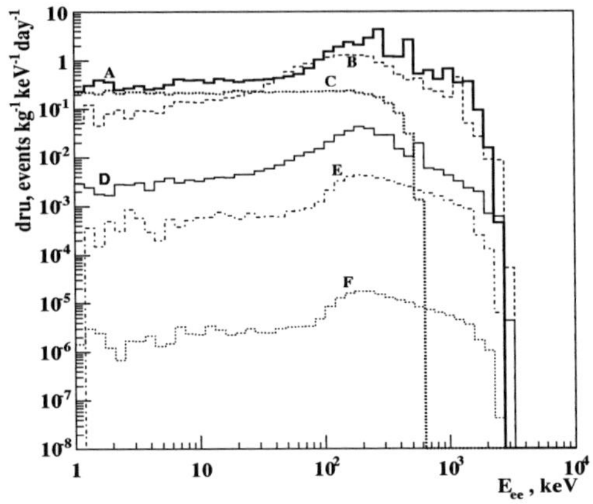

Figure 6: Energy deposition spectra in liquid xenon due to gammas from ^{222}Rn (10 ppb in 10 m^3 of air inside the lead shielding around the detector) (A), ultra-low-background PMTs (B), ^{85}Kr (5 ppb of Kr, Coulomb corrected) (C) and copper vessel (D). Also shown are the energy depositions due to gammas from rock activity after 10 cm Pb + 40 g cm^{-2} CH_2 (E) and 20 cm Pb + 40 g cm^{-2} CH_2 (F). The graph is taken from Ref. [34].

The β-decays of some isotopes can give a significant contribution to the background. The case for ^{85}Kr in liquid xenon has been studied in Ref. [34]. It was found that with 5 ppb contamination of Kr the background from betas is comparable or even higher than that from PMTs probably requiring additional purification of commercially available Xe.

Alphas from U/Th decay chains including radon and its daughters can also contribute to background events in high-sensitivity detectors, especially in WIMP dark matter detectors. Although they typically deposit a few MeV in the target, converted into an MeV signal after correction for quenching, some of them can be emitted close to the surface of the target and so lose energy in a keV range. Originating in a sensitive volume, it is possible for only a small fraction of their energy to be deposited inside the target, the remaining energy in the vessel walls, which is not detected. Nuclear recoil from the alpha decay may add to the total energy deposition. In a similar way, alphas emitted in the vessel or PMTs close to the surface can deposit a large fraction of their energy there, enter the target and deposit a remaining small fraction in the fiducial volume, which will be detected. In both cases the events can be misinterpreted as low energy nuclear recoil events mimicking the WIMP signal, unless a topological study reveals that the events occured close to the surface of the target and/or the discrimination between nuclear recoils and alphas allows complete rejection of alpha events.

BACKGROUND FROM RADON

^{222}Rn arises in bulk materials from the decay chain of ^{238}U and emanates slowly from surfaces. Radon gas is highly diffusive and mixes rapidly through large volumes of air. Although no neutrons are produced by spontaneous fission by its progeny, its decay causes several concerns for high-sensitivity experiments. A potential radon-induced background comes from the decay in air surrounding the detector, producing alpha, beta and gamma radiation. The detector vessels will shield against the alpha and beta particles produced outside, but high-energy gammas can give a signal in the detector (see Figure 6 and Ref. [34] for discussion). Another concern is the deposition and nuclear-recoil implantation of daughter radioisotopes in detector elements prior to assembly. The radon daughters attach to surfaces of the detector components where they subsequently decay. This problem can be avoided by careful surface treatment of detector parts after manufacture, followed by storage in a radon-free environment before assembly.

5-10 Bq/m^3 is a typical radon concentration in a well ventilated underground laboratory without special radon protection. Rn will also emanate from the materials surrounding the detector adding to this rate. If the shielded volume around the detector is sufficiently ventilated, the latter contribution is not significant compared to the general levels in air.

Alphas from radon decay can also add to the neutron background in the vicinity of the detector, if radon penetrates through the shielding via the air gaps (large thickness of lead and hydrocarbon makes the diffusion of radon through the shielding unlikely). Materials surrounding a detector consist mainly of isotopes with a high threshold for (α,n) reactions, such as copper, iron, nitrogen and oxygen in air and carbon in hydrocarbon shielding. As a consequence of this, the neutron background from U/Th in these materials is dominated by spontaneous fission of ^{238}U, with (α,n) reactions contributing about 20%. Radon and its daughters are free from spontaneous fission and only (α,n) reactions are the source of neutrons from radon decay. The main threat from radon is that if a significant fraction of rock produced radon is allowed to penetrate the shields, the alpha emitting progeny can be deposited close to the detector.

Gammas and neutrons from radon decay can give a background rate comparable to that from other sources, such as PMTs or ^{85}Kr beta decay in a xenon-based dark matter detector [4, 34]. To suppress this rate, an appropriate protection against radon should be used, such as gas-tight sealing or radon should be completely removed from the laboratory air via radon-trap system, as done in the Modane Underground Laboratory – LSM.

ACTIVE VETO SYSTEMS

A design of an active veto system depends on the aims, type and configuration of the main (target) detector. For most detectors an active veto system is designed to reject muon-induced events and is usually placed outside the passive shielding. There is, however, another way of designing an active veto system which is used in ZEPLIN I/II liquid and two-phase xenon detectors [37] of WIMP dark matter. The performance of such a veto has been recently investigated in connection with the large-scale xenon detector [34, 38]. The concept is to surround the target vessel with an active veto system based on liquid scintillator probably with Gd loading (ZEPLIN I/II veto does not have Gd in it). Such a configuration can help in rejecting neutrons and gammas from detector components if they interact firstly in the target and then in the veto. It was also shown [4, 26] that this veto (if it surrounds the detector from all sides) will detect all muon events capable of generating a signal in the target.

In an ideal case the 30-40 cm thick veto system described above with 0.2-0.5% of Gd by weight (to detect gammas from neutron capture) can have up to 90% efficiency for neutron rejection [34, 38]. In reality it is difficult to build a big veto system as a single module. Gaps between modules, finite time gate for coincident pulses (trigger and neutron capture), capture of neutrons on other materials and other effects reduce the rejection power down to about 70% [34].

For gammas from detector components, such as PMTs or detector vessel, the veto efficiency is about 35-40% (with 100 keV veto energy threshold). This relatively low value is attributed to the fact that gammas can be absorbed within the detector without any energy deposition in the veto. Significant levels of beta-emitters (for example, ^{85}Kr) in the target would reduce the total efficiency to reject gamma-like (or electron-like) events, as β's will deposit energy only in the target and not in the veto.

SUMMARY

The background radiations for particle astrophysics experiments have been discussed. Neutrons produced in rock and detector components via spontaneous fission and (α,n) reactions have been considered in detail. The modified code SOURCES can be used to simulate neutron production in these processes although some questions about its reliability still remain. The library of (α,n) cross-sections and transition probabilities to the excited states should be updated according to the most recent experimental data and calculations. GEANT3, GEANT4 (with corrected inelastic cross-section on Chlorine) and MCNP have been shown to give similar results for low-energy neutron propagation in the rock and shielding. Such simulations allow the calculations of the required type and thickness of the shielding for any particular detector.

GEANT4 and FLUKA have been found to be in reasonable agreement for producing, propagating and detecting muon-induced neutrons. The biggest difference in the neutron yield occurs for heavy materials, for which both codes may also underestimate the yield compared to some experimental data. More experiments with accurate simulations of the set-ups could clarify the situation. Neutron fluxes and energy spectra above 1 MeV before and after shielding, generated by GEANT4 and FLUKA, agree well, also giving similar nuclear recoil rates in dark matter detectors. Full three-dimensional Monte Carlo of all particles produced in muon-initiated cascades from their production to the detection, is absolutely crucial for evaluating correct background rates.

Other backgrounds include gammas, beta and alphas, all can be simulated with GEANT4 and some other codes. Detailed study of these has been started for WIMP dark matter detectors.

Backgrounds associated with radon represent a special case because of the ability of radon to penetrate (diffuse) through the materials and generate gammas, neutrons (via (α,n) reactions) and alphas close to, or in the fiducial volume of the detector.

An active veto system can be reliably simulated with GEANT4 for any particular detector. A case for a WIMP detector has revealed rejection factors of about 70% and less than 40% for neutrons and gammas, respectively.

ACKNOWLEDGMENTS

Part of this work has been undertaken within the framework of the UK Dark Matter Collaboration (University of Edinburgh, Imperial College London, Rutherford Appleton Laboratory and University of Sheffield), is funded by PPARC and contributes to the collaboration-wide simulation efforts for WIMP dark matter detectors. The wide exchange of information and results between different research groups across Europe, joint work and mutual help with the Monte Carlo codes became possible thanks to the financial support from EU FP6 project ILIAS. The author is grateful to all members of the working group on background studies within the JRA1 and N3 activities of ILIAS for their contributions to this work without which this review would not be possible. The author wishes to thank the organisers for hospitality and very interesting workshop.

REFERENCES

1. J. A. Formaggio, and C. J. Martoff, *Annual Rev. Nucl. Part. Sci.*, **54**, 361–412 (2004).
2. W. B. Wilson et al., SOURCES-4A, Technical Report LA-13639-MS, Los Alamos (1999).
3. B. E. Watt, *Phys. Rev.*, **87**, 1037 (1952).
4. M. J. Carson et al., *Astroparticle Phys.*, **21**, 667–687 (2004).
5. http://www-nds.iaea.org/
6. M. J. Carson et al., *Nucl. Instrum. & Meth. in Phys. Res. A*, in press (2005).
7. V. Chazal et al. *Astroparticle Phys.*, **9**, 163–172 (1998).
8. G. Chardin, and G. Gerbier. *Proc. 4th Intern. Workshop on the Identification of Dark Matter* (York, UK, 2-6 September 2002), ed. N. J. C. Spooner and Vitaly Kudryavtsev (World Scientific, Singapore, 2003), pp. 470–476.
9. S. Agostinelli et al., *Nucl. Instrum. & Meth. in Phys. Res. A*, **506**, 250 (2003); see also the GEANT4 web-page at CERN: http://geant4.web.cern.ch/geant4.
10. H. J. Kim et al., *Astroparticle Phys.*, **20**, 549–557 (2004).
11. P. Belli et al., *Nuovo Cimento A*, **101**, 959 (1989).
12. F. Arneodo et al., *Nuovo Cimento A*, **112**, 819–831 (1999).
13. H. Wulandari et al., *Astroparticle Phys.*, **22**, 313–322 (2004).
14. J. M. Carmona et al., *Astroparticle Phys.*, **21**, 523–533 (2004).
15. S. Scholl et al., To appear in: *Proc. 5th Intern. Workshop on the Identification of Dark Matter* (Edinburgh, UK, 6-10 September 2004), ed. N. J. C. Spooner and Vitaly Kudryavtsev (World Scientific, Singapore, 2005); see also http://www.shef.ac.uk/physics/idm2004.html for the collection of presentations.
16. MCNP-4B, Technical Report LA-12625-M (Ed. J. F. Briesmeister), Los Alamos, 1997.
17. R. Lemrani, ILIAS internal communication.
18. M. De Jesus, ILIAS internal communication.
19. H. M. Araújo, ILIAS internal communication.
20. V. A. Kudryavtsev, N. J. C. Spooner, and J. E. McMillan, *Nuclear Instrum. & Meth. in Phys. Res. A*, **505**, 688–698 (2003).
21. A. Fassò, A. Ferrari, and P. R. Sala, *Proceedings of the MonteCarlo 2000 Conference* (Lisbon, October 23-26, 2000), Ed. A.Kling, F.Barao, M.Nakagawa, L.Tavora, P.Vaz (Springer-Verlag, Berlin, 2001), p. 159; A. Fassò, A. Ferrari, J. Ranft and P. R. Sala, *ibid.* p. 995.
22. P. Antonioli et al., *Astroparticle Phys.*, **7**, 357–368 (1997).
23. V. A. Kudryavtsev et al., *Phys. Lett. B*, **471**, 251–256 (1999).
24. V. A. Kudryavtsev et al., *Phys. Lett. B*, **494**, 175–180 (2000).
25. H. Wulandari et al., To appear in: *Proc. 5th Intern. Workshop on the Identification of Dark Matter* (Edinburgh, UK, 6-10 September 2004), ed. N. J. C. Spooner and Vitaly Kudryavtsev (World Scientific, Singapore, 2005); see also http://www.shef.ac.uk/physics/idm2004.html for the collection of presentations.

26. H. M. Araújo et al., *Nuclear Instrum. & Meth. in Phys. Res. A*, in press (2005); hep-ex/0411026.
27. M. Bauer et al., To appear in: *Proc. 5th Intern. Workshop on the Identification of Dark Matter* (Edinburgh, UK, 6-10 September 2004), ed. N. J. C. Spooner and Vitaly Kudryavtsev (World Scientific, Singapore, 2005); see also http://www.shef.ac.uk/physics/idm2004.html for the collection of presentations.
28. Y.-F. Wang et al., *Phys. Rev. D*, **64**, 013012 (2001).
29. M. Aglietta et al., *Proc. 26th Intern. Cosmic Ray Conf.* (Salt Lake City), **2**, 44 (1999); hep-ex/9905047.
30. V. Chazal et al., *Nuclear Instrum. & Meth. in Phys. Res. A*, **490**, 334–343 (2002).
31. L. Bergamasco, S. Costa, and P. Picchi, *Il Nuovo Cimento A*, **13**, 403–412 (1973).
32. G. V. Gorshkov et al., *Sov. J. Nucl. Phys.*, **7**, 470–474 (1968); *ibid.*, **12**, 187–190 (1971); *ibid.*, **13**, 450–452 (1974); *ibid.*, **18**, 57 (1974).
33. R. Heaton et al. *Nucl. Geophys.*, **4**, 499 (1990).
34. M. J. Carson et al., Submitted to *Nuclear Instrum. & Meth. in Phys. Res. A*; M. J. Carson et al., To appear in: *Proc. 5th Intern. Workshop on the Identification of Dark Matter* (Edinburgh, UK, 6-10 September 2004), ed. N. J. C. Spooner and Vitaly Kudryavtsev (World Scientific, Singapore, 2005); see also http://www.shef.ac.uk/physics/idm2004.html for the collection of presentations.
35. P. F. Smith, and J. D. Lewin, *Phys. Reports*, **187**, 203 (1990).
36. UKDM website http://hepwww.rl.ac.uk/ukdmc/Radioactivity/index.html.
37. J. Alner et al., *New Astronomy Reviews*, in press (2005); N. J. T. Smith, talk at the 6th UCLA Symposium "Sources and Detection of Dark Matter and Dark Energy in the Universe" (Marina del Rey, USA, 18-20 February 2004), http://www.physics.ucla.edu/hep/dm04/dm04.htm; J. White, ibid.
38. C. Bungau et al., *Astroparticle Phys.*, **23**, 97–115 (2005).

^{11}C Background in Liquid Scintillator Detectors

Andrea Pocar [1]

Physics Department, Stanford University, Stanford, CA 94305 USA

Abstract. Cosmogenic ^{11}C produced in muon showers is one of the main backgrounds for the detection of *pep* and CNO solar neutrinos in underground organic liquid scintillator detectors. Experimental data available for the effective cross section for ^{11}C by muons indicate that ^{11}C is in fact the dominant background for the observation of such neutrinos. ^{11}C decays are expected to total a rate 2.5 (20) times higher than the combined rate of *pep* and CNO neutrinos in Borexino (KamLAND) in the energy window preferred for the *pep* measurement, between 0.8 and 1.3 MeV.

Background from ^{11}C in organic liquid scintillator detectors can be reduced if a neutron is emitted when the ^{11}C nuclide is created. ^{11}C decays can be tagged on a one-by-one basis using a three-fold coincidence with the parent muon track and the subsequent neutron capture on protons. The efficiency of such background reduction critically relies on the emission of a free neutron associated with ^{11}C production.

In order to verify the hypothesis, first suggested by Deutsch, that a neutron is in fact always emitted when ^{11}C is produced, we perform a detailed *ab initio* calculation of the production of cosmogenic ^{11}C, taking into consideration all relevant production channels. Results of the calculation are compared with the effective cross sections measured by target experiments in muon beams. "Bilnd" channels without a neutron in the final state account for only about 5% of ^{11}C production modes.

An estimation of the effectiveness of the one-by-one tagging of ^{11}C events, in light of such "blind" channels is performed for KamLAND, Borexino, and a possible scintillator experiment at SNOLab. Both KamLAND and Borexino can significantly improve their *pep* and CNO solar neutrino signal and could perform a 3% mesurement of the *pep* solar neutrino flux in five years. At SNOLab depths the muon flux is low enough that ^{11}C background would be much smaller than the *pep* and CNO neutrino signal and hence negligible.

Keywords: muon-induced nuclear reactions, organic scintillator, neutron moderation and diffusion, solar neutrinos, low background
PACS: 25.20.-x, 25.30.Mr, 26.65.+t, 28.20.Gd, 96.40.Tv, 96.60.-j

^{11}C BACKGROUND FOR PEP AND CNO SOLAR NEUTRINO DETECTION

Since the definitive discovery of neutrino oscillations in the solar sector [2, 3], the interest in solar neutrino physics has shifted towards precision and real-time measurements of the low energy solar neutrino flux and energy spectrum. Organic scintillator experiments such as Borexino [4] and KamLAND [5] will measure the presumably abundant ^{7}Be solar neutrino flux in the near future. A proposal for a liquid scintillator at SNOLab is also being discussed [6].

pep and CNO solar neutrinos. An interesting possibility for organic scintillator detectors is to detect *pep* and CNO solar neutrinos. As was recently pointed out [7], a measurement of the flux of *pep* neutrinos would yield essentially equivalent information on neutrino oscillation parameters and the other solar neutrino fluxes as a measurement of *pp* solar neutrinos at a comparable level of experimental uncertainty. *pep* solar neutrinos would probe the main fusion chain in the sun by which essentially all solar energy is generated. Moreover, given the low theoretical uncertainty on the *pep* neutrino flux, its measurement could allow investigating the matter-vacuum transition region for solar neutrino oscillations between 2 and 3 MeV [7]. The observation of such transition would provide a further test of the MSW-LMA solution. Due to this transition, the probability that electron neutrinos emitted by the sun have not oscillated into other neutrino species when they arrive on earth for *pep* neutrinos is about a factor of two larger than the one measured at higher energy for ^{8}B neutrinos (i.e. the survival probability is $\sim$60%). According to the BP04 Standard Solar Model [8] in the MSW-LMA oscillation scenario [7], 2.1 and 6.6 events are expected per day in 100 tons of liquid scintillator for *pep* and CNO neutrinos respectively (the CNO events go down to 3.5 if the recent results from LUNA are considered [9]). The *pep* neutrinos are emitted as a mono-energetic 1.44 MeV line; their elastic scattering on electrons gives a Compton electron spectrum with a 1.22 MeV endpoint. The expected spectrum of scattered electrons

[1] the work presented in this paper is taken from an original paper co-authored with Cristiano Galbiati, Davide Franco, Aldo Ianni, Laura Cadonati, and Stefan Schönert [1].

CP 785, *Topical Workshop on Low Radioactivity Techniques,* edited by B. Cleveland, R. Ford, and M. Chen

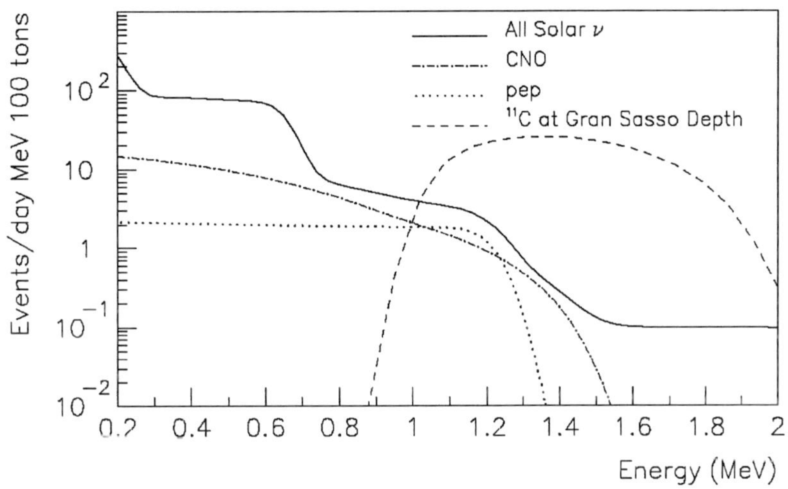

FIGURE 1. Recoil electron energy spectrum from *pep* (dotted line) and CNO neutrinos (dash-dotted line), for the MSW-LMA solution of the solar neutrino problem. The total spectrum for all solar neutrinos (including *pp*, ^{7}Be, and ^{8}B neutrinos) is also shown (continuous line). Neutrino fluxes are from the BP04 Standard Solar Model [7, 8]). The background signal expected from cosmogenic ^{11}C at Gran Sasso depth (from [10]) is superimposed (dashed line). The energy resolution of the detector is assumed to be 5% at 1 MeV with $1/\sqrt{E}$ energy dependence.

from different low energy solar neutrino sources is shown in fig. 1. The figure also shows the expected cosmogenic ^{11}C background at Gran Sasso depth. Information on depth, muon flux, neutron production rate and expected ^{11}C production rate for Kamioka, Gran Sasso, and SNOLab are summarized in table 1; an extensive list of references for the values reported in the table can be found in [1].

^{11}C background. ^{11}C is a β^+ emitter (0.96 MeV end point), therefore the visible energy in a scintillator detector lies between 1 and 2 MeV. This study focuses on the 0.8 to 1.3 MeV window for the recorded energy in the detector, where the expected *pep* + CNO rate in 100 tons is of 2 events per day (1.5 ev/day/100 tons with the LUNA results). About 35% of the ^{11}C decays fall in this energy range. The inclusive ^{11}C production rate by muons was measured on a beam experiment at CERN [10]. In the energy region of interest, about 5 ^{11}C events per day per 100 tons are expected at Gran Sasso depth, more than twice the *pep* + CNO signal. What makes ^{11}C a difficult background to deal with is its 30 minute mean life which makes it essentially impossible to eliminate by simply correlating it with incoming muons (this is possible for other cosmogenic isotopes such as ^{8}Li, ^{9}Li, ^{11}Be, ^{8}B, ^{12}B, and ^{9}C, all with mean lives below one minute). Cosmogenic ^{7}Be would also represent a similar problem (77 day mean life), but it was shown to have a small effective production cross section in organic liquid scintillators [10] (0.1 events/day/100 tons at Gran Sasso).

TABLE 1. Depth (D), residual muon flux(Φ_μ), average muon energy ($\langle E_\mu \rangle$), neutrons capture rate (N), and expected ^{11}C production rate (P) at KamLAND, Borexino, and SNOLab. The data on the muon flux, muon energy and neutron production rate reported in the table are from a variety of experimental measurements and phenomenological extrapolations, as explained in [1] and references therein. The ^{11}C production rates are obtained by extrapolating data from 100 and 190 GeV muon beams on an organic liquid scintillator target to higher energy [10].

	Depth [m.w.e.]	Φ_μ [μ/m^2 h]	$\langle E_\mu \rangle$ [GeV]	N	P [cts/d 100 tons]
KamLAND	2700	9.6	285	300	107
Borexino	3800	1.2	320	40	15
@ SNOLab	6000	0.012	350	0.43	0.15

In this study, other sources of background are neglected and the focus is exclusively on cosmogenic ^{11}C produced *in situ* by muon-triggered reactions. In practical terms, we are assuming a ^{238}U and ^{232}Th contamination in the scintillator of 10^{-17} g/g, and a ^{nat}K contamination of 10^{-15} g/g which combined would yield less than one count per day in 100 tons in the observation energy window. We are also neglecting contamination from radon daughters out of equilibrium with ^{238}U, as well as the experiment-dependent γ-ray background from the building materials of the detector, assuming it be significantly smaller than the *pep* neutrino rate.

^{11}C PRODUCTION IN LIQUID SCINTILLATOR

When ^{11}C production is accompanied by the emission of a free neutron, it is possible to tag the ^{11}C decays on a one-by-one basis and eliminate them from the data set. One needs to identify the position and time of the emitted free neutron captured on protons in the scintillator. Neutron capture on protons results in the emission of an easily detectable 2.2 MeV γ ray. In hydrocarbons, neutrons can also be captured on ^{12}C resulting in γ rays of combined energy 4.9 MeV [11]. The cross section for the capture on ^{12}C is $\sim 1\%$ of the cross section for capture on protons [12]. Once the neutron capture time and position are known, one can apply a cut in space and time around every capture point. Events within a time t from the double muon+neutron coincidence and inside a sphere of radius r from the neutron capture point are rejected. The rejection efficiency depends on the choices of t and r. This technique, originally suggested by Deutsch [13], has recently been applied in the 4-ton prototype Counting Test Facility of Borexino [14].

The effectiveness of this background reduction technique relies on having a free neutron emitted in ^{11}C production from muon showers in organic liquid scintillator. It's effectiveness can be quantified once the percentage of ^{11}C nuclei produced *without* a neutron being emitted is known. In order to estimate it, we have performed a detailed *ab initio* calculation of the production of cosmogenic ^{11}C, taking into consideration all relevant production channels. The cross section for these reactions were collected, when available, from existing literature. We then performed a full simulation of muon-induced showers with the particle transport code FLUKA [15, 16]. The FLUKA code has been used in the past to calculate the production rate of neutrons by muons in liquid scintillator at several depths, and has been found to reproduce experimental results very well [17]. Recently, FLUKA has also been used to calculate the distance between the parent muon track and the capture point of neutrons produced in scintillator by muon-induced cascades at Gran Sasso depth on protons [18], obtaining good agreement with the experimental data from the LVD experiment [19].

Muons of typical energy at Kamioka, Gran Sasso, and SNOLab were generated in liquid scintillator (the results presented here are for trimethylbenzene (C_9H_{12}, 0.88 g/cm^3)). Results should not vary greatly with other organic solvents, given that typical values of density and mass ratio between carbon and hydrogen are similar. The secondary particles produced in the shower were tracked and their cumulative range as a function of energy in the scintillator recorded. Their range was then convoluted with the cross section for the various channels and a ^{11}C production rate for each channel extracted. Details of the simulation can be found in [1]. This estimation is compared to the inclusive ^{11}C production rate at lower energies [10] in order to check the overall validity of the method. It finally yields a ^{11}C production rate for various reactions, including those in which a free neutron in the final state is not present.

^{11}C production channels. The main reactions that can produce ^{11}C are listed below. A detailed bibliography for the cross sections of the reactions can be found in [1] and is omitted here.

a) ^{12}C(γ,n)^{11}C
b) ^{12}C(n,2n)^{11}C
c) ^{12}C(p,p+n)^{11}C
d) ^{12}C(p,d)^{11}C
e) ^{12}C(π^-,π^-+n)^{11}C
f) ^{12}C(π^+,π+N)^{11}C:
g) ^{12}C(e,e+n)^{11}C
h) ^{12}C(μ,μ+n)^{11}C

For all the reactions with a neutron in the final state, the energy threshold is $\sim$20 MeV (i.e. the neutron binding energy in ^{12}C). The N in the final state of reaction (f) indicates a nucleon and can be either a proton or a neutron. Due to the possibile charge exhcange occurring in the strong meson-nucleon interaction, the fragments in the final state can be either (π^++n) or (π^0+p). No data are available on the relative composition of the final state, but it is expected from

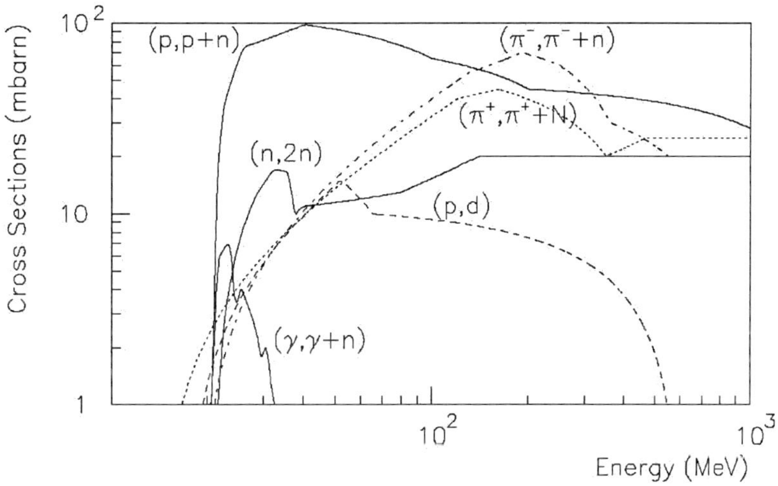

FIGURE 2. Cross sections of the main channels for ^{11}C production from ^{12}C as a function of energy.

theoretical considerations that the frequency of the invisible channel, with a proton in the final state, should account for 2/3 of all the π^+-induced reactions. The cross sections for channels (a) - (f) are are plotted as a function of energy in figure 2. The cumulative ranges as a function of energy for secondary particles produced in muon showers are shown in figure 3.

For the direct interaction of electrons, positrons, and muons with a nucleus through a virtual photon (reactions (g) and (h), also known as spallation), the cross section can be calculated using the von Weizsäcker approximation [20]. In case of high energy, ultra-relativistic charged particles inducing nuclear reactions with low momentum transfer, the cross section can be written as [21]:

$$\sigma_{e,\mu} = (\alpha/\pi)\,[2\ln(E/m) - 1]\,\sigma_{-1} \tag{1}$$

where $\sigma_{-1} = \int d\nu\sigma_\gamma(\nu)/\nu$ is the inverse energy weighted moment of the photodisintegration cross section σ_γ (reaction (a)).

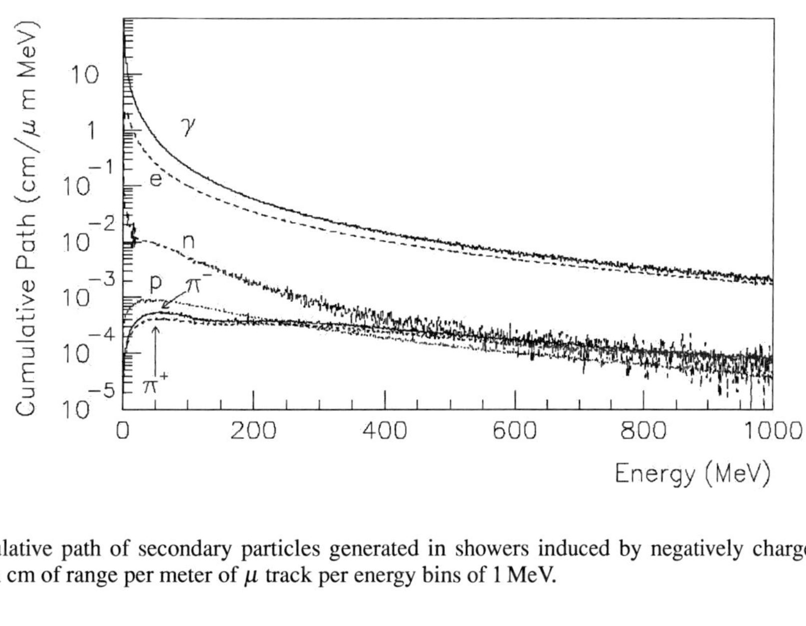

FIGURE 3. Cumulative path of secondary particles generated in showers induced by negatively charged muons at 320 GeV. Results are quoted in cm of range per meter of μ track per energy bins of 1 MeV.

TABLE 2. Production rates for ^{11}C in muon induced showers in trimethylbenzene. The calculated total production rates are compared with the experimental values available at 100 and 190 GeV from Ref. [10], and with the extrapolated values at the mean muon energy for KamLAND and Borexino (also from Ref. [10]) and at SNOLab. The procedures used to determine the expected rate from the invisible channels and the systematic error affecting our calculation are outlined in the text.

E_μ [GeV]	100	190	285	320	350
Process	Rate $[10^{-4}/\mu\,\mathrm{m}]$				
^{12}C$(p,p{+}n)^{11}$C	1.8	3.2	4.9	5.5	5.6
^{12}C$(p,d)^{11}$C	0.2	0.4	0.5	0.6	0.6
^{12}C$(\gamma,n)^{11}$C	19.3	26.3	33.3	35.6	37.4
^{12}C$(n,2n)^{11}$C	2.6	4.7	7.0	8.0	8.2
^{12}C$(\pi^+,\pi{+}N)^{11}$C	1.0	1.8	2.8	3.2	3.3
^{12}C$(\pi^-,\pi^-{+}n)^{11}$C	1.3	2.3	3.6	4.1	4.2
^{12}C$(e,e{+}n)^{11}$C	0.2	0.3	0.4	0.4	0.4
^{12}C$(\mu,\mu{+}n)^{11}$C	2.0	2.3	2.4	2.4	2.4
Invisible channels	0.9	1.6	2.4	2.7	2.8
Total	28.3	41.3	54.8	59.9	62.2
1σ systematic	1.9	3.1	4.4	5.0	5.2
Measured	22.9	36.0			
1σ experimental	1.8	2.3			
Extrapolated			48.4	52.6	56.2

Results. The ^{11}C production rates for the reactions listed above for the Kamioka, Gran Sasso, and SNOLab depths are reported in table 2, where a comparison is given with the results for inclusive production at a muon beam [10]. These results were obtained with 320 GeV negative muons.The production rate in the invisible channels (no neutron in the final state) has been calculated by adding the rate from the ^{12}C$(p,d)^{11}$C reaction to 2/3 of the rate from the ^{12}C$(\pi^+,\pi{+}N)^{11}$C reaction.

The error quoted in table 2 accounts for the systematic error, coming from two main sources. The most important source is the uncertainty in the knowledge of the cross section for the ^{12}C$(n,2n)^{11}$C reaction. Our best estimate for the systematic error is 50% of the production rate in such channel, which is between 5% and 7% of the total production rate depending on the energy. We also make the conservative assumption that the systematic error associated with the cross sections used by FLUKA to calculate the range of the secondaries accounts for 5% of the total production rate. We combine the two as independent sources of error. The statistical error associated with our Monte Carlo calculation is 0.6% and is negligible with respect to the systematic error.

The total calculated rate is systematically 20% higher than the measured rates on a beam at 100 and 190 GeV, even though the values are still within twice the combined experimental and systematic uncertainties. Possibly, the slight discrepancy is due to the difference in the geometry between the experiment and the simulation: in the beam experiment, the muon shower is built up in 240 cm of concrete and 200 cm of water placed in front of the scintillator target [10], while for our calculation we simulated the muon-induced interactions in a bulk scintillator volume. We also calculated the production rates for positive muons (320 GeV). The difference in the production of ^{11}C, for all of the channels considered, by μ^+ and μ^- at 320 GeV is within the statistical error of our Monte Carlo calculation and can be neglected in this study. We also took into account the possibility of producing ^{11}C nuclides from ^{13}C, which has a natural isotopic abundance of $\sim$1% [22]. ^{11}C production from this source was found to be negligible [1].

The particle content of muon-induced showers is dominated by γ rays and electrons (see fig. 3). Around 25 MeV, at the giant dipole resonance for ^{12}C where the cross section for γ rays is quite large, the cumulative range of γ rays is two to three orders of magnitude larger than the corresponding values for neutrons, and four to five orders of magnitude larger than for other hadrons. As a consequence, the dominant process for the production of ^{11}C nuclides is the (γ,n) exchange reaction. It accounts for $\sim$60% of the total production rate, even if the hadronic cross sections for ^{11}C production are up to a factor 10 larger than that at the peak value for γ rays. Electrons are not as effective as γ rays because their cross section is suppressed by a factor $\alpha \sim 1/137$. Other hadronic channels with a neutron in the final state and the muon-induced photodisintegration, also carrying a neutron, account for an additional 35% of the total production rate.

The ^{11}C production rate in the two invisible channels (the (p,d) and the (π^+,π^0+p) exhange reactions), accounts for only about 5% of the total production rate. The ^{11}C reduction technique outlined above is thus little affected by "blind" channels. A more precise evaluation of the potential for detecting *pep* and CNO solar neutrinos by KamLAND, Borexino, and a possible experiment at SNOLab is given in the next section.

ESTIMATION OF ^{11}C BACKGROUND REDUCTION

As shown above, at least one neutron is produced in association with 95% of the ^{11}C nuclides. This fraction of the ^{11}C background can be eliminated with the muon+neutron tagging. The remaining 5% cannot be reduced with this technique.

Events occurring a time t after the neutron capture and within a radius r from the capture point must be rejected; t should be equal to a few times the mean life of ^{11}C. The information carried by the neutron capture does not tell us about the position of the ^{11}C birthplace. Therefore, r should be set to a few times the average neutron range. To estimate the range of neutrons associated with ^{11}C production, we simulated the energy distribution of the neutrons produced in association with ^{11}C nuclides for Borexino, KamLAND, and at SNOLab. We used the resulting distribution to source neutrons into FLUKA, calculating the distribution of the range of neutrons produced in association with ^{11}C. The average energy of neutrons produced in association with ^{11}C is much lower than the average energy of all neutrons produced in muon-induced showers: this is due to the dominant production mechanism being the photoneutron reaction at the giant dipole resonance of ^{12}C at 23 MeV. Correspondingly, the average range of neutrons associated with ^{11}C production is also much lower than the average range of all the neutrons produced in the shower. In our estimation the assumption is made that the displacement of the ^{11}C isotope within the detector before it decays and the spatial resolution of the detector are both much smaller than the neutron range of interest. Due to the $(n,2n)$ exchange reaction that yields two neutrons in the final state, an average of 1.14 neutrons are created in interactions producing ^{11}C nuclides. Those neutrons, if sufficiently energetic, can also trigger nuclear reactions knocking off other neutrons: our calculation indicates that an average number of about 1.2 neutrons are captured in the scintillator for each neutron produced in a ^{11}C-forming reaction. Further details of the calculation are found in [1].

The efficiency ε of rejecting ^{11}C events tagged with the muon+neutron coincidence is equal to the combined efficiencies for the cut in space, $\zeta(r)$, and in time, $\eta(t)$:

$$\varepsilon = \zeta(r)\,\eta(t) \tag{2}$$

where $\eta(t)$ is the efficiency for the rejection of ^{11}C events when a cut of time t is applied around the neutron capture point. Because of the exponential time decay distribution of ^{11}C, we obtain:

$$\eta(t) = 1 - e^{-t/\tau} \tag{3}$$

where τ is the mean life of ^{11}C, 30 minutes. In equation 2, $\zeta(r)$ is the rejection efficiency for ^{11}C when a spherical cut of radius r around the neutron capture point is applied, and is given by:

$$\zeta(r) = \int_0^r n(s)ds \bigg/ \int_0^\infty n(s)ds \tag{4}$$

where $n(s)$ is the range distribution of neutrons.

In order to quantify the effectiveness of the background reduction technique we introduce a figure of merit R, independent of the experiments and their locations. R is defined as the ratio of the *pep* + CNO neutrino signal rate (S) and the residual background rate (B) from ^{11}C after subtraction of all events identified through the muon+neutron coincidence (both S and B are computed in the *pep* energy window [0.8-1.3 MeV]). Note that the figure of merit R accounts only for the remaining background from ^{11}C, but other possible and independent sources of background are neglected in its definition. The general expression for R is:

$$R = S/B = \frac{S/B_0}{F + (1-F)(1-\zeta\eta)} \tag{5}$$

TABLE 3. Optimal values of the dead mass-time fraction (D) for KamLAND, Borexino and at SNOLab as a function of the figure of merit R (defined in the text) after the reduction with the technique outlined in the text. In Borexino the ratio between signal rate and raw background before any ^{11}C rejection (B_0) is 0.4; in KamLAND, S/B_0 is 0.05; for a detector at SNOLab, S/B_0 would be 36.

S/B_0	0.05	0.4	36
$R = S/B$	KamLAND D [%]	Borexino D [%]	@ SNOLab D [%]
0.1	0.4		
0.2	11.6		
0.3	50.6		
0.4	87.4	<0.1	
0.5	98.8	<0.1	
0.8	>99.9	0.1	
1		0.3	
2		6.7	
3		27.8	
4		58.3	
5		85.3	
8		>99.9	
100			<0.1
500			2.6

where $F \simeq 0.05$ is the fraction of ^{11}C production rate in invisible channels, S the signal rate (discussed in the opening section), and B_0 is the raw ^{11}C background rate, dependent on the location of the experiment. Formulæ 2 and 5 show that there is a one-to-one correspondance between the figure of merit R and the combined rejection efficiency ε.

Using equation 5 we estimate that the optimal signal to background ratio achievable in Borexino (KamLAND) is 8 (1) in case all of the ^{11}C associated with a neutron are successfully tagged. However, he optimal limit cannot be reached due to dead mass-time limitations. We define the dead mass-time fraction D as the fraction of (mass$\times$time) data taking lost to the space and time cuts around a neutron capture event. Treating η and ζ as independent variables, t and r can be derived by inverting equations 3 and 4 and become functions of η and ζ respectively. We can then compute the dead mass-time fraction D corresponding to the chosen values of the cuts ζ and η as:

$$D = 1 - e^{-\frac{4}{3}\pi\rho r^3 tN} \tag{6}$$

where ρ is the scintillator density and N is the neutron capture rate per unit mass in the detector. The expected neutron capture rates for the experiments taken into consideration are summarized in table 1.

In table 3 we show the optimal values of the dead mass-time fraction D for the three experiments for fixed and given values of the ratio R between signal and reduced ^{11}C background. The values are calculated by minimizing the dead mass-time fraction D while keeping the product of the two efficiencies ζ, η constrained to the value of ε corresponding to the chosen value of R.

Improvements to the figures in table 3 could be obtained by optimizing the geometrical cut around the neutron capture point, as suggested, for example, in [14]. Once the total number of ^{11}C decays inside the detector is known, its residual number (after one-by-one tagging) can also be statistically subtracted from the data set.

CONCLUDING REMARKS

The data in table 3 show that both KamLAND and Borexino could improve their raw signal (pep + CNO neutrinos) to background (^{11}C) ratio by a significant factor by subracting ^{11}C events one by one from the data set. Borexino can improve its signal to background ratio from 0.4 to 2 (4) while losing 7% (58%) of the data to dead mass-time. KamLAND can improve its signal to background ratio from 0.05 to 0.2 (0.3) while losing 12% (51%) of data to dead mass-time. Both Borexino and KamLAND could measure the combined rate from *pep* and CNO neutrinos in the [0.8-1.3 MeV] window with a 3% statistical accuracy in five years, provided that internal and external background rates are kept within figures negligible compared to the signal rate. A possible experiment located at SNOLab would experience a very low muon flux and hence ^{11}C background would be negligible with respect to the expected rate from *pep* and CNO neutrinos.

Finally, it is worth noting how the technique used in this work to evaluate cosmogenic production of ^{11}C in liquid scintillator detectors starting from the cross sections of individual production channels and simulating the cumulative range of secondary particles associated with muon showers deep underground can be applied to other materials of interest to the low background physics community. It has indeed already been applied to estimate backgrounds in enriched ^{136}Xe for neutrinoless $\beta\beta$ searches [23].

REFERENCES

1. C. Galbiati et al., *Phys. Rev. C* (2005), in print (arXiv:hep-ph/0411002).
2. SNO Collaboration, Q. R. Ahmad et. al.l, *Phys. Rev. Lett.*, **89**, 011302 (2002).
3. KamLAND Collaboration, K. Eguchi et. al.l, *Phys. Rev. Lett.*, **90**, 021802 (2003).
4. Borexino Collaboration, G. Alimonti et al., *Astropart. Phys*, **16**, 205–234 (2002).
5. Y. Kishimoto, "Liquid Scintillator Purification," in *Low Radioactivity Techniques 2004*, Sudbury, ON, Canada, 2004, these proceedings.
6. M. Chen, "SNO Liquid Scintillator Project," in *NOW 2004*, Conca Specchiulla, Bari, Italy, 2004, proceedings to be published in Nuclear Physics B - Proceeding Supplements.
7. J. N. Bahcall and C. Peña-Garay, *J. High Energy Phys.*, **11**, 4 (2003).
8. J. N. Bahcall and C. Peña-Garay, *Phys. Rev. Lett.*, **92**, 121301 (2004).
9. A. Formicola et al., *Phys. Lett. B*, **591**, 61 (2004).
10. T. Hagner et al., *Astropart. Phys.*, **14**, 33 (2000).
11. R. C. Reedy and S. C. Frankle, *At. Data Nucl. Data Tables*, **80**, 1 (2002).
12. S. F. Mughabghab, M. Divadeenam, and N. E. Holden, *Neutron Cross Sections, Part A, Z = 1-60*, Academic Press, 1981.
13. M. Deutsch, "Proposal to NSF for a Borexino Muon Veto," Massachusetts Institute of Technology, 1996.
14. D. Franco for the Borexino Collaboration, "^{11}C measurement and CNO flux at Borexino," in *NOW 2004*, Conca Specchiulla, Bari, Italy, 2005, proceedings to be published on Nuclear Physics B - Proceeding Supplements.
15. A. Fassò et al., "Electron-photon transport in FLUKA: Status," in *Monte Carlo 2000 Conference*, October 2000, Lisbon, Portugal, A. Kling, F. Barao, M. Nakagawa, L. Tavora and P. Vaz eds., Springer-Verlag Berlin, 2001.
16. A. Fassò et al., "FLUKA: Status and Prospective for Hadronic Applications," in *Monte Carlo 2000 Conference*, October 2000, Lisbon, Portugal, A. Kling, F. Barao, M. Nakagawa, L. Tavora and P. Vaz eds., Springer-Verlag Berlin, 2001.
17. Y-F. Wang et al., *Phys. Rev. D*, **64**, 013012 (2001).
18. V. A. Kudryavtsev, N. J. C. Spooner, and J. E. McMillan, *Nucl. Inst. Meth. A*, **505**, 688 (2003).
19. LVD Collaboration, M. Aglietta et al., "Measurement of the neutron flux produced by cosmic-ray muons with LVD at Gran Sasso," in *26th International Cosmic Ray Conference*, Salt Lake City, UT, USA, Vol. 2, p. 44 (arXiv:hep-ex/9905047), 1999.
20. C. F. von Weizsäcker, *Z. Phys.*, **88**, 612 (1934).
21. J. Delorme, *Phys. Rev. C*, **52**, 2222 (1995).
22. C. M Lederer and V. S. Shirley editors, *Table of Isotopes*, 7th edition, Wiley-Interscience Publisher, New York, 1978.
23. T. Bender, "Cosmogenic Nuclide production in Liquid Xenon," Junior Paper, Physics Departemnt, Princeton University, January 2005, 2005.

POSTERS

Study of the muon-induced neutron background with the LVD detector

H. Menghetti* and M. Selvi*

*Bologna University and INFN
on behalf of the LVD collaboration

Abstract. High energy neutrons, generated as a product of cosmic muon interaction in the rock or in the detector passive material, represent the most dangerous background for a large list of topics like reactor neutrino studies, the search for SN relic neutrinos, solar antineutrinos, etc. [1].

Up to now there are few measurements of the muon-produced neutron flux at large depth underground. Moreover it is difficult to reproduce the measured data with Monte Carlo simulation because of the large uncertainties in the neutron production and propagation models.

We present here the results of such a measurement with the LVD detector, which is well suited for the detection of neutrons produced by cosmic-ray muons, reporting the neutron flux at various distances from the muon track, for different neutron energies ($E > 20$ MeV) and as a function of the muon track length in scintillator.

Keywords: Neutron, Cosmic rays, Underground detectors
PACS: 28.20.Gd, 13.85.Tp

DETECTOR DESCRIPTION

LVD [2] is a neutrino telescope located in the Hall A of the Gran Sasso INFN laboratory, in data acquisition since 1992, whose main goal is the observation of neutrinos from a Supernova core collapse. It is composed of three independent towers, each of them made of 280 scintillator counters, 1.5 m^3 each, for a total scintillator (C_nH_{2n}, $<n>= 9.6$) mass of about 1000 tons. The detector is also equipped with a tracking system made of limited streamer tubes, interleaved to the scintillator counters, which allows to reconstruct muons; the depth of the LVD site corresponds to a mean muon energy of about 270 GeV.

The observation of neutrinos is made mainly through the inverse beta decay reaction of electron anti neutrinos on protons:

$$\overline{\nu_e} + p \rightarrow n + e^+ \tag{1}$$

$$n + p \rightarrow d + \gamma \tag{2}$$

The prompt signal from the positron and the delayed signal from the 2.2 MeV gamma from the neutron capture (mean lifetime $\tau \sim 185$ μs) constitute the double signature for this reaction. The positron signal is detected with a high energy threshold level (HET), set at 7 MeV for the external counters and 4 MeV for the "core" counters, while the 2.2 MeV gamma is detected with a 1 MeV low energy threshold level (LET), activated by the high energy signal for a time duration of about 1 ms.

NEUTRON SIGNATURE

The LVD apparatus can detect neutrons with the same signature as the inverse beta decay reaction. High energy neutrons could cause a liquid scintillator proton to recoil (prompt HET signal), and are then thermalized and finally captured by the liquid scintillator protons with the emission of the 2.2 MeV gamma (delayed LET signal). Taking into account the energy transfer in the interaction between neutron and proton, the proton quenching and the value of the high energy threshold in the core of the detector, we estimate that the neutrons detected in this way have energies greater than about 20 MeV.

The background to the neutron detection is due to the accidental coincidences between the high energy signals and the low energy ones. This background, however, has a flat distribution of the delay between the two signals and can

CP 785, *Topical Workshop on Low Radioactivity Techniques,* edited by B. Cleveland, R. Ford, and M. Chen

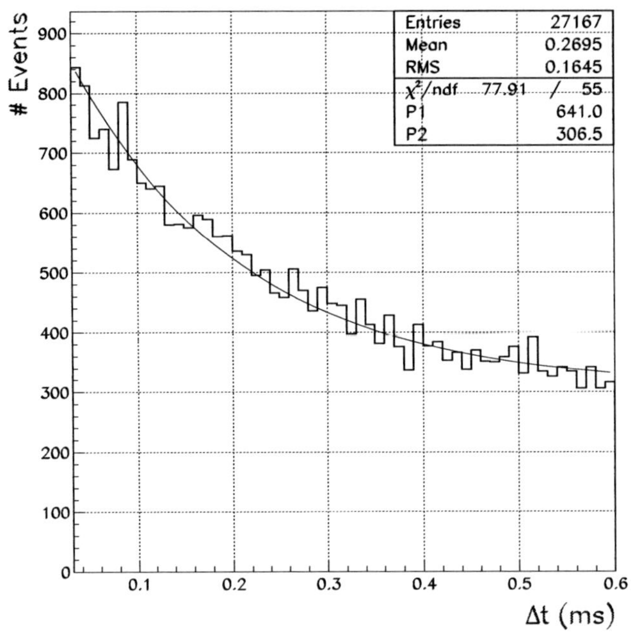

FIGURE 1. Time delay distribution between the high energy signals and the following low energy one detected in the same counter of the detector. From the distribution we can separate the neutron interaction from the accidental coincidences (see text).

be estimated by fitting the time delay distribution. An example of this distribution is shown in figure 1; we can fit the data with the curve

$$\frac{dN}{dt} = P1 \cdot e^{-t/\tau} + P2$$

where $\tau = 185\ \mu$s. From the first parameter we obtain the number of neutron interactions, while the second takes into account the number of accidental coincidences.

ANALYSIS AND RESULTS

We have analyzed the neutron production in association to single muon events, that is events with only one reconstructed track, from 1994 to 2002, for a total sample of more than 7 million single muon events.

First we have evaluated the production of neutrons per counter per event at various distances from the muon track; the distance is defined as the distance between the reconstructed muon track and the center of the counter where the neutron is detected. Notice that in the counters traversed by the muon track we require, in addition to the high energy signal associated to the muon a second one associated to the recoiling proton. Neutron candidates are selected with the procedure described in the previous section; at each distance the background contribution has been evaluated by fitting the time delay distribution between the HET signal and the LET ones. The result obtained is shown in figure 2; we were able to evaluate the neutron production up to 22 m from the muon track. Due to the non homogeneous distribution of the scintillator in the LVD detector the behavior observed in figure 2 has to be studied with a detailed Monte Carlo simulation which is under development.

To estimate the neutron energy spectrum, we studied the number of neutrons detected as a function of the energy released in the scintillator from the recoiling proton. The result is shown in figure 3; the data are well fitted by the power law spectrum:

$$\frac{dN}{dE} = A \times E^{-\alpha}$$

where $A = (1.6 \pm 0.1) \times 10^{-5}$ and $\alpha = (1.19 \pm 0.02)$; the errors are statistical only.

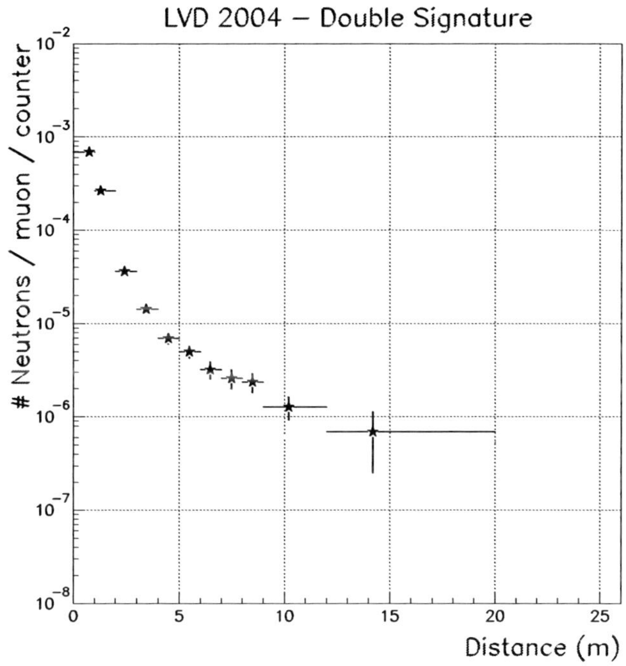

FIGURE 2. Number of neutrons detected per muon per counter as a function of the distance from the muon track.

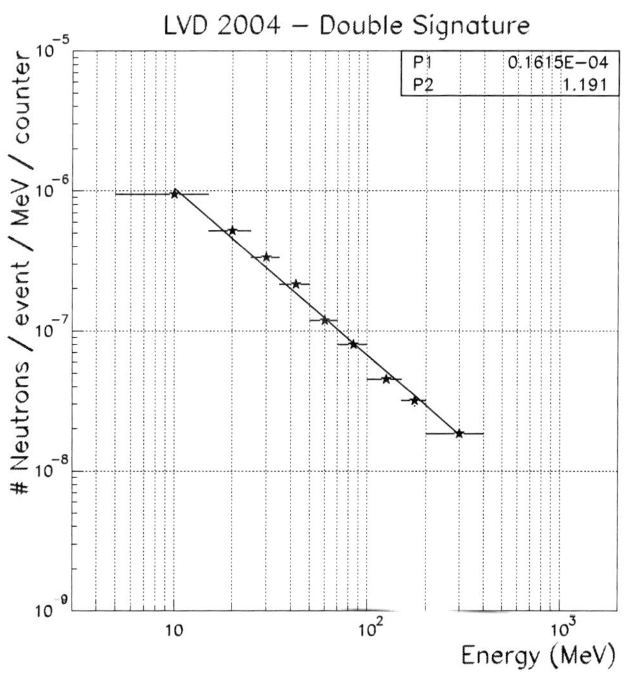

FIGURE 3. Number of neutrons detected per muon per counter as a function of the proton recoil energy.

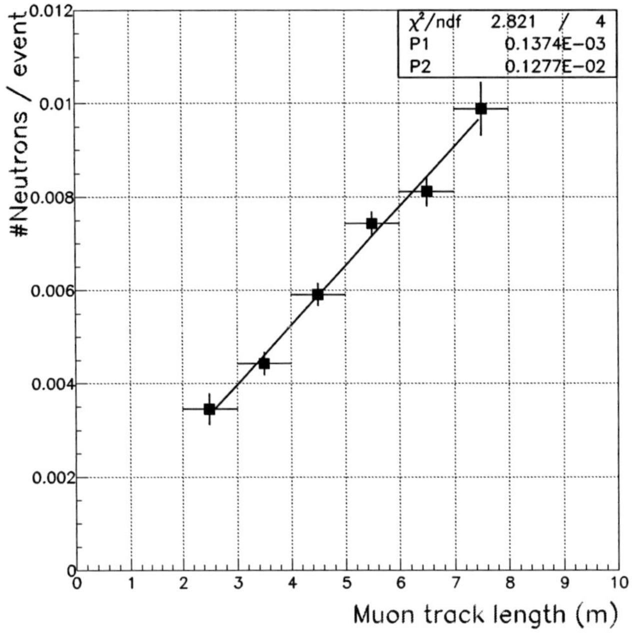

FIGURE 4. Number of neutrons per event detected as a function of the muon track length in scintillator; the main production of neutron is due to the muon interaction with the detector nuclei.

Finally we evaluate the neutron production as a function of the muon track length in scintillator. The preliminary result is shown in figure 4.

The data are well fitted by:

$$y = 0.14 \times 10^{-3} + 0.13 \times 10^{-2} \cdot L$$

where the first parameter takes into account the neutron production in the rock, as it is independent from the muon track length inside the liquid scintillator, while the second parameter takes into account the increase in the neutron production with the muon track length in scintillator. Comparing the two values we can conclude that the neutron production in the core of the experiment is mostly due to the interaction of muons with the detector nuclei (Fe,C).

CONCLUSION

We report the neutron production from muon interaction in the LVD detector for single muon events and selecting neutrons with a double signature, that is neutrons of energies larger than 20 MeV. For this sample of events we were able to measure the neutron flux up to 22 meter from the muon track and the neutron energy spectrum. We also find that the neutrons selected are produced in the muon interaction with the detector material.

REFERENCES

1. Khalchukov, *Il Nuovo Cimento C*, **6**, 320 (1983).
2. Aglietta, *Il Nuovo Cimento A*, **105**, 1793 (1992).

A low background facility at the Gran Sasso underground laboratory: the prototype of Borexino

A. Ianni

Laboratori Nazionali del Gran Sasso, I-67010 Assergi, Italy

Abstract.
The prototype of the Borexino solar neutrino experiment is a low counting rate detector located at the underground Gran Sasso Laboratory. The detector was designed and built to achieve ultra-purity in an organic liquid scintillator on a massive scale. A number of ancillary plants and screening facilities are needed to reach low background in the sub-MeV region. The detector and its screening facilities are briefly described in this paper.

Keywords: massive low counting detector, high-purity scintillator
PACS: 26.65.+t;29.40.Mc;85.60.Ha;43.75.Yy

THE COUNTING TEST FACILITY

The Counting Test Facility (CTF) (see [1] for further details) is a 4-ton unsegmented liquid scintillator detector located at the Gran Sasso underground Laboratory. CTF is equipped with 100 photomultipliers (PMTs) 8" in diameter and an active muon veto system. At present the CTF is filled with PC (C_9H_{12}) with 1.5 g/l of PPO. The scintillator is inside a 500 μm thick nylon vessel 1 m in radius. A second nylon vessel is present outside the inner one to decrease the diffusion of Rn toward the core of the detector. A volume of 1000 tons of high purity water produced by a dedicated plant [2] shields the scintillator against neutrons from the surrounding rock and high energy gamma's mainly from the PMTs. This water volume is contained inside a Water Tank made from carbon steel coated with an epoxy resin on the inside with a diameter of 11 m and an height of 10 m. The CTF was designed and built with the following goals: achieve a radio-purity in U and Th at the level of 10^{-16} g/g, perform spectroscopy below 1 MeV in a multi-ton detector, and address technical solutions to measure ^{7}Be solar neutrinos.

The CTF has been in operation in its present configuration since the beginning of 2002. Data collected with the detector allowed to identify besides U and Th other important sources of background, namely, ^{85}Kr (Q_β=0.687 MeV, $T_{1/2}$=10.756 y), ^{39}Ar (Q_β=0.565 MeV, $T_{1/2}$=263 y) and ^{210}Pb. In particular, ^{210}Pb comes from ^{222}Rn which may diffuse from low-purity to high-purity volumes within the experimental apparatus. ^{210}Pb daughters, ^{210}Po and ^{210}Bi, can be a dominant background below 1 MeV.

The CTF has shown that any detector which aims to perform spectroscopy with a ton-scale target mass needs a number of ancillary plants to properly handle the mentioned sources of background. Moreover, the CTF has shown how effective can be the self-shielding design for a detector which aims to search for low energy solar neutrinos. As a matter of fact, an important source of background comes from the water, the PMTs and other materials inside the water tank. High-purity water is at the level of 10^{-14}-10^{-15} g(U,Th)/g with an amount of Rn which depends on the apparatus. In the CTF the Rn in water is at the level of a few mBq/m^3 thanks to a dedicated purification plant [2]. Surrounding material within the water tank are in the range of 10^{-10}-10^{-12} g(U,Th)/g. The goal of low energy solar neutrino detectors is to have a high-purity target mass at the level of 10^{-16} g(U,Th)/g or better and to shield the external background with a proper design. In Figure 1 is shown the prompt signal spectrum from the CTF with an exposure of 204 days. The energy scale is determined using the yield of 275 p.e./MeV which is the measured one with 75 working PMTs. From this Figure the effect of the screening due to the detector design can be seen. Figure 1 shows that with proper shielding the background above 3 MeV is about zero. This background is low because of: the purity of water, the distant location of the PMTs, the self-shielding of the scintillator, and, in spite of the relatively small size, the depth of the Gran Sasso Laboratory. This latter point is of particular interest. As a matter of fact in [4] it is shown how cosmic muons can produce radioactive isotopes on a carbon-based target. This cosmogenic background can be dominant above 1 MeV and limit pep and CNO detectability as shown in [5]. Above 3 MeV the background is measured to be $(1.2\pm0.2)\times10^{-5}$ events/day kg with a radial cut at 60 cm. Below 1 MeV the signal is dominated by

CP 785, *Topical Workshop on Low Radioactivity Techniques,* edited by B. Cleveland, R. Ford, and M. Chen

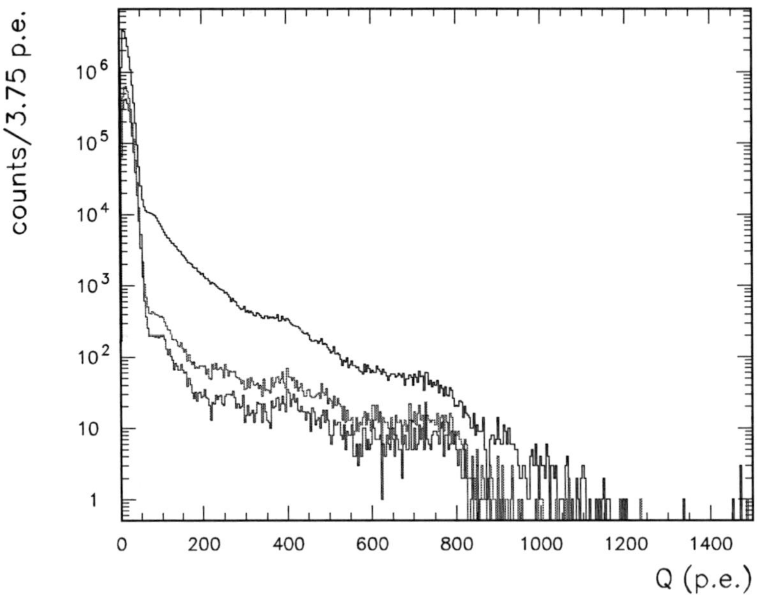

FIGURE 1. Prompt signal spectrum in the CTF for 204 days of exposure. The upper line is for the whole scintillator volume (3.7 tons). The middle line is with a radial cut at 70 cm. The lower line is with a radial cut at 60 cm. The energy scale is such that 1 MeV corresponds to 275 p.e.

the low energy sources of background reported above. In particular, in the region [0.25,0.8] MeV the background is measured to be $(21.4\pm0.3)\times10^{-3}$ events/day kg with a radial cut at 60 cm and is mainly due to Kr and Pb daughters.

To properly handle the background sources in a detector such as the CTF a number of ancillary plants and screening facilities are needed. In the following we report about some of the screening facilities and refer the reader to more specific papers on the purification plants to deal with internal backgrounds [3, 2].

SCREENING FACILITIES FOR THE CTF

In this Section a brief review of the screening facilities used for the CTF is presented.

A blank nitrogen layer is present inside the CTF water tank above the water shielding volume. The nitrogen flows continuously and a constant overpressure of about 10 mbar is set between the tank and the outside experimental hall. This nitrogen layer limits the Rn diffusion into the CTF apparatus. In the experimental hall the air has a concentration of Rn at the level of 50-100Bq/m^3, which depends on the ventilation inside the experimental hall, on the activities in the same hall, and ultimately on the season. With the aim to monitor the Rn in the nitrogen, a dedicated set-up has been designed and installed by the water tank [6]. The set-up consists of an electrostatic chamber (30kVolts) of 0.42 m^3 with an alpha detector. Nitrogen coming from the water tank is de-humidified and flushed into the chamber. Here the alpha detector collects the alpha's from the ^{222}Rn chain (^{218}Po and 214Po) to determine the Rn activity. The sensitivity of the instrument is at the level of 0.5-1 mBq/m^3. The Rn in the CTF blank is measured at the level of 20 mBq/m^3. In Figure 2 is shown an example of a spectrum from the Rn chamber. From left to right the peaks corresponds to signals due to ^{210}Po, ^{218}Po and ^{214}Po. In particular, the peak due to ^{210}Po comes from the collected contamination of Pb in the alpha counter after more than 3 years of operation. This system is easy to couple with other apparatus and has been

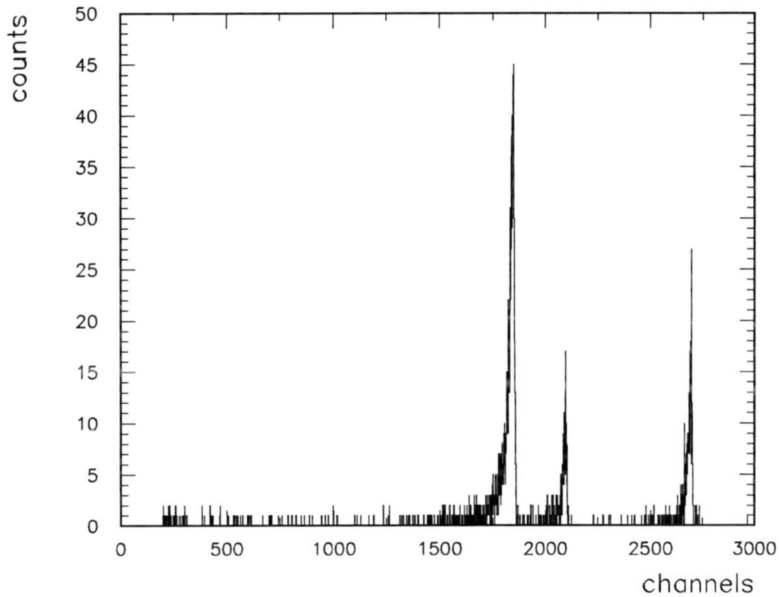

FIGURE 2. Sample spectrum from the Rn chamber. From left to right the peaks are due to alpha's from ^{210}Po, ^{218}Po, and ^{214}Po.

used to monitor Rn during many operations including the filling of the Borexino vessels.

Another screening facility used by the CTF is the low background set-up of the Gran Sasso laboratory. This consists of 10 HpGe detectors located each inside an acrylic vessel. The vessel is flushed with nitrogen to reduce Rn contamination. Moreover, detectors are equipped with copper and lead shielding. The lead in particular is a low activity one with less than 20 Bq/kg in ^{210}Pb. These detectors have been used to screen materials used in the CTF apparatus and have a sensitivity of the order of 10^{-10} g/g in U and Th. A compilation of selected results obtained with HpGe detectors and details of the set-up can be found in [7].

One of the most dangerous backgrounds in low counting experiments is Rn. One way to remove Rn and other noble gases from the scintillator is by stripping using high-purity nitrogen. This technique is described in [2]. The Rn stripping plant is another screening facility coupled to the CTF detector.

Rn emanation from the inner surface of vessels used to store and handle the scintillator is a source of background which requires a dedicated screening set-up. In the CTF vessels were tested against Rn emanation using collection on charcoal and internal proportional counting [7]. A sensitivity of the order of 50μBq has been reached.

It is easy to understand that Rn is often not in equilibrium with U because it can diffuse. A second important isotope can be also found not in equilibrium, ^{226}Ra. The presence of Ra in water can be a serious background problem because, in spite of a low level of U, a high level of Ra can produce Rn and ultimately Pb. For this reason in the CTF the amount of Ra in water was measured with a dedicated screening facility [7]. This consists of a 480 l electropolished stainless steel tank equipped with a sparging system. It operates by sparging the Rn from the water using Rn-free nitrogen and then concentrating the Rn in charcoal. The Rn in the charcoal will be measured with a low level proportional counter. A sensitivity of the order of 1mBq/m^3 of Ra has been achieved.

A detector such as the CTF with a target mass of 4 tons requires a fluid handling system with vessels, pumps, filter housings and many meters of pipes. The system is used to move the scintillator, to purify it, and to fill the detector. The cleanliness of such a system must be under control in order to avoid particulates with U, Th, and Rn daughters to go into the detector while filling. Because of possible contamination during the installation and welding of parts a precision cleaning is necessary to reduce the particulates and ensure a high corrosion resistance. A dedicated system has been designed and constructed for the CTF and Borexino, the so-called Cleaning Module (CM). The CM consists of a pump, a 300 l vessel, filters, and a heater. Cleaning solutions can be poured into the vessel and circulated through the equipment to be treated. A method used for precision cleaning and particulate counting has been studied following standard recommendations [8].

REFERENCES

1. Borexino coll., Astrop. Phys., 8, 1998, pp. 141. Borexino coll., NIM A406, 1998, pp. 411.
2. Borexino coll., Astrop. Phys., 16, 2002, pp. 205-234.
3. J. Benziger, et al., NIM A417, 1998, pp. 278.
4. T. Hagner, et al., Astrop. Phys., 14, 2000, pp. 33.
5. C. Galbiati, et al., Rev. Phys. C, 2005, 005800.
6. J. Kiko, NIM A460, 2001, pp. 272-277.
7. Borexino coll., Astrop. Phys., 18, 2002, 1.
8. *Standard Specification for Chemical Passivation Treatments for Stainless Steel Parts*, ASTM, A 967 - 99. *Standard Practice for Cleaning, Descaling, and Passivation of Stainless Steel Parts, Equipment, and Systems*, ASTM, A 380 - 99.

Cosmogenic activation of materials

J. Amaré*, B. Beltrán*, S. Capelli†, F. Capozzi†, J. M. Carmona*, S. Cebrián*, O. Cremonesi†, E. García*, I. G. Irastorza*, H. Gómez*, G. Luzón*, M. Martínez*, J. Morales*, A. Ortiz de Solórzano*, M. Pavan†, C. Pobes*, J. Puimedón*, A. Rodríguez*, J. Ruz*, M. L. Sarsa*, L. Torres* and J. A. Villar*

*Laboratorio de Física Nuclear y Altas Energías, Universidad de Zaragoza, E-50009 Zaragoza, Spain
†Dipartimento di Fisica dell'Università di Milano-Bicocca e Sezione di Milano INFN, I-20126 Milano, Italy

Abstract. The problem of cosmogenic activation produced at sea level in materials typically used in underground experiments looking for rare events is being studied. Several nuclear data libraries have been screened looking for relevant isotope production cross-sections and different codes which can be applied to activation studies have been reviewed. The excitation functions for some problems of interest like production of ^{60}Co and ^{68}Ge in germanium and production of ^{60}Co in tellurium have been obtained taking into account both measurements and calculations and a preliminary estimate of the corresponding rates of production at sea level has been performed.

Keywords: double beta decay, cosmogenic activation, germanium, tellurium
PACS: 23.40.-s, 24.10.Lx, 25.40.-h

INTRODUCTION

Searches for rare events like the Double Beta Decay (DBD) require ultra-low background conditions. Working in deep underground locations and using active and passive shields reduce very efficiently the background levels, and consequently, intrinsic radioactive impurities in the materials of the set-up, like those induced cosmogenically at sea level, may become problematic (see for instance proposals of several projects of DBD experiments [1, 2, 3, 4]). To quantify the cosmogenic activity in a material there are three basic ingredients: the history of the material (time of exposure at sea level, if it has flown or been stored underground, ...), the flux and spectrum of cosmic rays and the cross-sections of isotope production. Cosmic neutron spallation is the most significant process in cosmogenics. One of the main difficulties to obtain activation yields relevant for DBD experiments is the lack of accurate information on the isotope production cross-sections in targets of interest.

In this work, several nuclear data libraries have been screened looking for production cross-sections relevant for materials typically used in DBD experiments (like Ge and Te) as a first step. Different codes which can be applied to activation studies have been also reviewed. Then, the excitation functions for some particular problems of interest in DBD searches like production of ^{60}Co and ^{68}Ge in germanium and production of ^{60}Co in tellurium have been obtained, taking into account different sources based on both measurements and calculations. A preliminary estimate of the corresponding rates of production due to cosmic neutrons at sea level has been made, allowing an evaluation of activity yields for different conditions of exposure to cosmic rays. The considered nuclides are the most significant cosmogenic isotopes in germanium and tellurium contributing to the background for DBD searches with ^{76}Ge (Q=2039 keV) and ^{130}Te (Q=2529 keV).

AVAILABLE DATA ON PRODUCTION CROSS SECTIONS

Available information on cross sections for isotope production induced by bombarding targets of interest in DBD experiments with neutrons and protons has been collected from different nuclear data libraries and the literature.

CP 785, *Topical Workshop on Low Radioactivity Techniques,* edited by B. Cleveland, R. Ford, and M. Chen

Results from measurements as well as from calculations have been considered.

- The EXFOR/CSISRS database[1] contains useful results from measurements, mainly with proton beams, for natural germanium and tellurium as well as for other materials like copper or lead. The limited range of products and specially of energies for which cross section data are available make difficult to carry out a complete activation study based only on experimental data but this information is essential to check the reliability of calculations.
- MENDL-2 and MENDL-2P [5] libraries cover a very wide range of target and product nuclides, either for neutrons or protons as projectiles, but only for energies up to 100 MeV for neutrons (MENDL-2) and 200 MeV for protons (MENDL-2P). In this libraries, excitation functions have been calculated using codes ALICE-92 and ALICE-IPPE and checked against some experimental data.
- The library from Refs. [6, 7] gives excitation functions including available experimental data and calculated results (with HMS-ALICE, CEM95 and LAHET) for selected targets (including germanium isotopes) and products. Energies up to 1.7 GeV have been considered for both neutrons and protons.

COMPUTATIONAL CODES FOR ACTIVATION

Computational codes which can be used to estimate activation yields have been analyzed. Some codes implement the semiempirical formulas of Silberberg & Tsao [8, 9, 10] for the isotope production cross sections, while other ones simulate by Monte Carlo the interaction between nucleons and nuclei to derive the induced nuclides.

Codes based on semiempirical formulas (like COSMO [11], Σ [12] or YIELDX [10]) offer a wide coverage of targets and products in a short calculation time. But the formulas have been deduced from proton-induced reactions and are applicable only above $\sim$ 100 MeV.

Monte Carlo codes (either specifically developed for hadronic interactions or general-purpose) allow neutron activation studies but may be very time-consuming (specially for some products). There are many families of codes applicable to activation studies like HETC, CEM, LAHET, ISABEL, INUCL, CASCADE, ALICE, SHIELD, MARS, GEM, LAQGSM, ... The predictive power of some of them when calculating activation yields in the energy range up to 5 GeV was estimated, evaluating the quality of calculated activation yields for many target elements by comparison with experimental data, in a study promoted by the Nuclear Energy Agency (NEA) [13]. Packages like FLUKA, GEANT4 and MCNP can be also used to estimate activation.

Detailed reports on available production cross section data and computational codes for activation have been written within the framework of the double beta decay activity (IDEA, Integrated Double-beta decay European Activities) of the ILIAS (Integrated Large Infrastructures for Astroparticle Science) project promoted by the European Union[2].

ACTIVATION IN GERMANIUM

Different excitation functions for ^{60}Co and ^{68}Ge in natural Ge have been compared in Figs. 1 and 2, using data from several sources:

- The only cross-sections measured with proton beams correspond to 660 MeV for both ^{60}Co and ^{68}Ge (Aleksandrov'91 [14]), 800 MeV for ^{60}Co (Norman'05 [15]) and from 25 to 64 MeV for ^{68}Ge (Horiguchi'83 [16]).
- Calculations at energies above 100 MeV have been made for both isotopes using YIELDX and results using CEM95 are also available from [6].
- Calculations at low energies (up to 100-200 MeV) from MENDL-2 and MENDL-2P and from the libraries in Refs. [6] and [17] using the HMS-ALICE code are taken into consideration for protons and neutrons independently.
- The excitation function calculated for ^{68}Ge using LAHET (ISABEL) by the Majorana Collaboration is also included [2].

---

[1] `http://www-nds.iaea.or.at/exfor/`

[2] Available at the IDEA website: `http://idea.dipscfm.uninsubria.it/`.

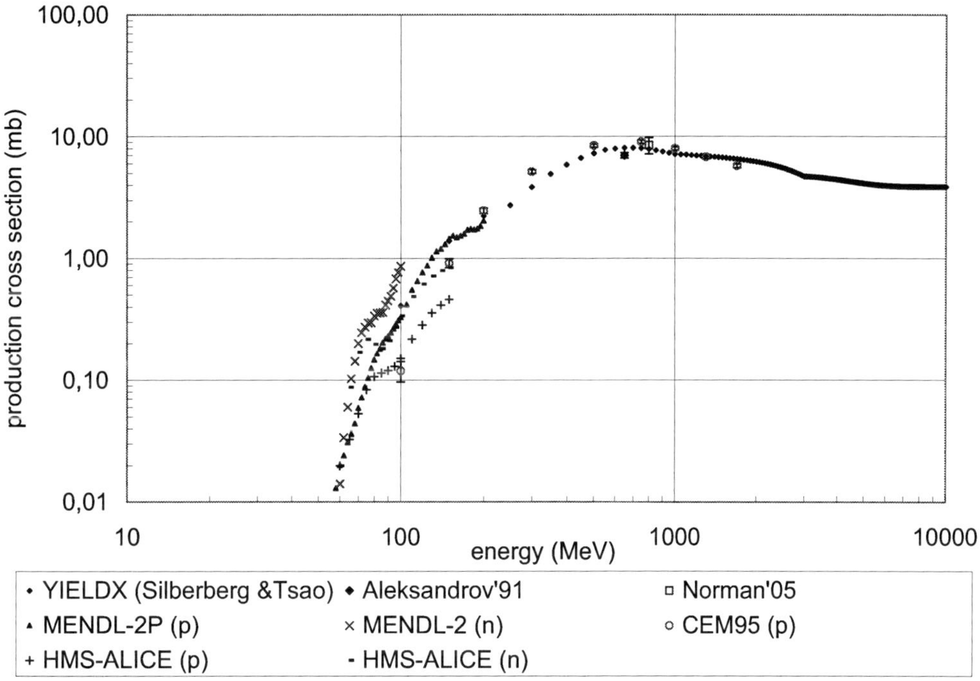

FIGURE 1. Comparison of excitation functions for ^{60}Co in natural germanium by nucleons.

TABLE 1. Rates of production (expressed in $kg^{-1}d^{-1}$) of ^{60}Co and ^{68}Ge in natural germanium by cosmic neutrons at sea level obtained in this work compared with previous estimates.

	^{60}Co	^{68}Ge
HMS-ALICE (<150 MeV), YIELDX ($\geq$150 MeV)	4.8	89
Refs. [20, 21]	4.8	$\sim$ 25
Ref. [19]	6.4	58

For ^{60}Co, the measurements at 660 and 800 MeV agree very well with YIELDX (within $\sim$10%). The deviation factor between results from MENDL libraries and those from [6, 17] at low energies is below $\sim$3.

For ^{68}Ge, the best agreement with measurements is given by YIELDX at high energy and by results from [6, 17] at low energy. Unfortunately, for this nuclide there are significant discrepancies between different estimates specially at low energies.

A tentative estimate of the rate of production of ^{60}Co and ^{68}Ge in natural germanium has been made considering the parametrization of the cosmic neutron spectrum at sea level from Ref. [18] and the excitation functions given by the estimates which produce the best agreement with the scarce experimental data: below 150 MeV, HMS-ALICE from [6, 17] and above 150 MeV, YIELDX. The obtained rates of production are shown in Table 1 together with previous estimates taken from the literature: using the Σ code [19] and a MC code [20, 21]. For ^{60}Co, the three estimates agree reasonably despite using different approaches. However, for ^{68}Ge the rate of production calculated here is significantly higher than the others; this is due to the excitation function considered at low energies. Due to the shapes of the neutron spectrum and the excitation function, the low energy region gives the largest contribution to the activation yield of ^{68}Ge. The excitation function for neutrons using HMS-ALICE from [6, 17] has been chosen because the calculations in this library for protons give a good agreement with measurements from Horiguchi'83 [16] (see Fig. 2).

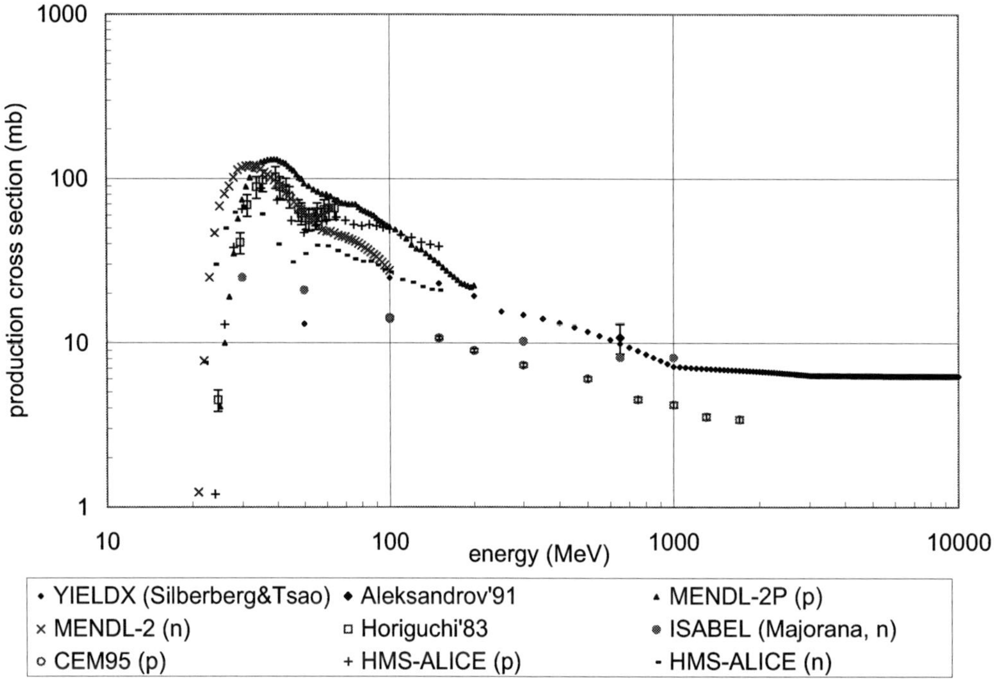

FIGURE 2. Comparison of excitation functions for ^{68}Ge in natural germanium by nucleons.

ACTIVATION IN TELLURIUM

The production of ^{60}Co in natural tellurium has been studied. Due to the high difference in the mass number of the target isotopes and the product, no information is available in libraries like MENDL-2. Excitation functions using a modified version of COSMO and YIELDX have been evaluated and compared with measurements made with protons at 0.8 and 1.85 GeV (Norman'05 [15, 22]), as shown in Fig. 3. The value at 800 MeV from [15] is just an upper limit for the cross-section. Since YIELDX is in good agreement with the experimental data, this estimate, together with the cosmic neutron spectrum from [18], has been used to obtain a rate of production of 0.06 $kg^{-1}d^{-1}$. This activation yield is almost one order of magnitude smaller than using the modified COSMO. Consequently, a longer exposure to cosmic rays at sea level could be allowed in the plan of construction and transport of TeO_2 crystals for the CUORE experiment [3].

CONCLUSIONS

A study of cosmogenic activation in materials for DBD experiments is underway. Relevant isotope production cross-sections have been identified in several nuclear data libraries and different computational codes which can be used to evaluate activation yields have been examined. Particular problems like production of ^{60}Co and ^{68}Ge in germanium and production of ^{60}Co in tellurium have been studied using all the available information for cross-sections, obtaining some rates of production which are significantly different than previous estimates.

ACKNOWLEDGMENTS

The ILIAS activity is funded by the European Union. We deeply acknowledge C. H. Tsao for providing the YIELDX routine and S. G. Mashnik for sharing with us the results of his calculations for germanium.

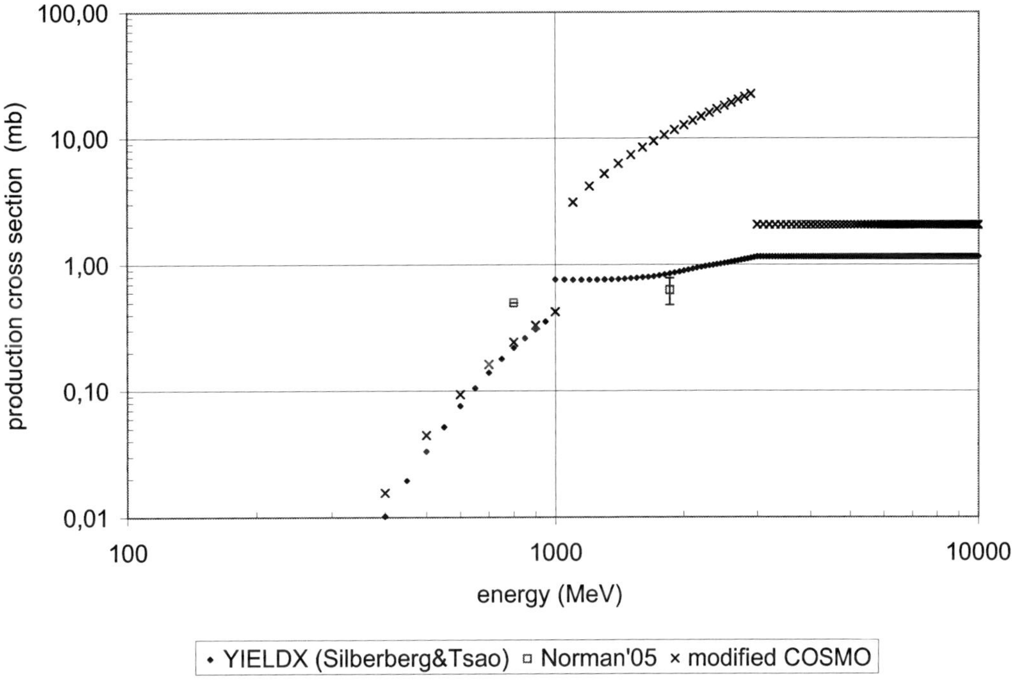

FIGURE 3. Comparison of excitation functions for ^{60}Co in natural tellurium by nucleons. The value at 800 MeV from Norman'05 is an upper limit for the cross-section.

REFERENCES

1. H. V. Klapdor-Kleingrothaus et al., "GENIUS: a SuperSensitive Germanium Detector System for Rare Events", [arXiv:hep-ph/9910205].
2. The Majorana Collaboration, "White Paper on the Majorana Zero-Neutrino Double-Beta Decay Experiment", [arXiv:nucl-ex/0311013].
3. R. Ardito et al., " CUORE: A Cryogenic Underground Observatory for Rare Events", [arXiv:hep-ex/0501010].
4. I. Abt et al., "A new ^{76}Ge Double Beta Decay Experiment at LNGS", [arXiv:hep-ex/0404039].
5. Yu.N. Shubin et al., "Cross-section data library MENDL-2 to study activation as transmutation of materials irradiated by nucleons of intermediate energies", report INDC(CCP)-385 (International Atomic Energy Agency, May 1995).
6. S. G. Mashnik et al., Proc. Fourth Workshop on Simulating Accelerator Radiation Environments (SARE4), Knoxville, Tennessee, September 14-16, 1998, LA-UR-98-6000, [arXiv:nucl-th/9812071].
7. K. A. Van Riper, S. G. Mashnik, and W. B. Wilson, *Nucl. Instrum. Methods*, **A463**, 576 (2001), LA-UR-99-360, [arXiv:nucl-th/9901073].
8. R. Silberberg, and C. H. Tsao, *Astrophys. J. Suppl. Ser.*, **25**, 315 (1973), ibid p. 335.
9. R. Silberberg, and C. H. Tsao, *Astrophys. J. Suppl. Ser.*, **35**, 129 (1977), *Astrophys. J. Suppl. Ser.*, **58**, 873(1985), *Phys. Rep.*, **191**, 351 (1990).
10. R. Silberberg, and C. H. Tsao, *Astrophys. J.*, **501**, 911 (1998).
11. J. Martoff, and P. D. Lewin, *Comput. Phys. Commun.* **72**, 96 (1992).
12. J. Bockholt, PhD thesis, University of Heidelberg, 1995.
13. "International Codes and Model Intercomparison for Intermediate Energy Activation Yields". NSC/DOC(97)-1, January 1997.
14. Yu. V. Aleksandrov et al., *Bull. Russian Acad. Sci. - Phys. Ser.*, **59** 895 (1995).
15. E. B Norman et al., *Nucl. Phys. B (Proc. Suppl.)*, **143**, 508 (2005).
16. T. Horiguchi et al., *Int. J. Appl. Rad. Isot.*, **34** 1531 (1983).
17. A.J. Koning et al., "Neutron and Proton Transmutation-Activation Cross Section Libraries to 150 MeV for Application in Accelerator-Driven Systems and Radioactive Ion Beam Target-Design Studies", ECN Report ECN-R-98-012, Petten, The Netherlands (1998). M.B. Chadwick et al., "LA150 Documentation of Cross Sections, Heating, and Damage", LA-UR-99-1222.
18. J. F. Ziegler, *IBM J. R&D* **42**, 1 (1998).
19. H. V. Klapdor-Kleingrothaus et al., *Nucl. Instrum. Methods*, **A481** 149 (2002).
20. H. Miley et al., *Nucl. Phys. B (Proc. Suppl.)*, **28A**, 212 (1992).
21. F. T. Avignone et al., *Nucl. Phys. B (Proc. Suppl.)*, **28A**, 280 (1992).
22. D. W. Bardayan et al., *Phys. Rev.*, **C55**, 820 (1997).

MIT-EAPS Neutron Activation Analysis and Radiometric Laboratory Contribution to Geosciences: Past, Present, and Future

Ila Pillalamarri

Department of Earth Atmospheric and Planetary Sciences,
Massachusetts Institute of Technology, Cambridge, MA 02139
http://web.mit.edu/pila/www/INAAHOMEPAGE.HTM
pila@mit.edu

Abstract. The Instrumental Neutron Activation Analysis (INAA) and Radiometric Laboratory's current system is described. This laboratory has been in continuous operation for the past thirty years. A review is provided about the laboratory's analytical participation in trace element geochemical studies of the earth's upper mantle, trace impurity studies of high purity materials, the provenance study of archaeological glass beads, trace multi-element analyses of standard reference materials, the preparation of synthetic analytical standards for Neutron Activation Analysis, and providing a training course in nuclear analytical techniques for environmental samples. The multi-element analysis by INAA consists of determining elements like the rare earths La, Ce, Nd, Sm Eu, Tb, Dy, Ho, Yb, Lu, and also As, Ba, Cl, Co, Cr, Cs, Dy, Fe, Hf, Hg, K, Mn, Na, Ta, Th, U. The projected future of the laboratory is explained in terms of its resources, expertise in high precision analysis of trace impurities for the material selection that is to be used in rare event physics experiments. For example, this 'surface' laboratory can be efficiently interfaced/integrated with a deep underground low background counting facility, especially in the initial stages.

Keywords: neutron activation analysis, trace impurity, standard reference materials, trace element geochemistry, gamma spectroscopy

INTRODUCTION

The Neutron Activation Analysis (NAA) Laboratory of the department of Earth, Atmospheric and Planetary Sciences, called the MIT-EAPS NAA laboratory, has been in existence since the 1970s. In 1980, the gamma spectroscopy system of the NAA Laboratory was expanded and computerized by a major equipment grant from the National Science Foundation obtained by Prof. Fred Frey. As a result, Dr. Ila Pillalamarri set up a Canberra Series 80 system with PDP11 Computer for routine geological analyses [Ref.1]. The Instrumental Neutron Activation Analysis (INAA) Laboratory has been maintained as a state of the art system by Prof. Fred Frey and Dr. Ila Pillalamarri, with upgrades in 1985 to Canberra's Series 85 system; in 1990 to Canberra's Series 90 system coupled with the DEC Micro Vax 3600 computer; and in 2000 to a Genie 2000 ESP system coupled with the Alpha Work Station. The laboratory has been equipped with new hyperpure Ge detectors as well for the gamma spectroscopy system. The Laboratory has been a center for trace element analyses for geochemists and other applied scientists not only from the New England area but also from other parts of the USA and other countries such as Australia, Belgium, China, and Egypt. I present this paper in two parts. In the first part, I describe the various past and present activities of the NAA laboratory. In the second part I present the vision of the laboratory for the future.

PART I: MIT-EAPS NEUTRON ACTIVATION ANALYSIS LABORATORY FOR TRACE ELEMENT STUDIES IN GEOSCIENCES PAST AND PRESENT (1970-2005)

The NAA Laboratory is a dedicated facility for trace element analysis of a variety of geological samples since the 1970s. Since then the facility is in continuous operation for various research activities such as

1) Trace element geochemistry of the earth's upper mantle,
2) Trace element analysis of standard reference materials,
3) Trace element analysis of impurities in high purity materials,

CP 785, *Topical Workshop on Low Radioactivity Techniques,* edited by B. Cleveland, R. Ford, and M. Chen

4) Trace element analysis of archaeological and plant materials,
5) Trace element analysis of environmental materials: Training course

I provide below detailed description of each of the above mentioned five activities.

1. Trace Element Geochemistry of the Earth's Upper Mantle (1974-2005)

The earth can be visualized as consisting of three concentric layers, namely, crust, mantle, and core. The outer most layer is, the crust, 100 km thick. The next lower layer is, the mantle, 2900 km thick. The inner most layer is, the core, consisting of a liquid outer core, 2200 km thick, and a solid inner core 1250 km thick. A pictorial representation can be seen in the reference provided by: The Earth [This Dynamic Earth, United States Geological Survey] http:/ /pubs.usgs.gov publications/text/ inside.html.

The Earth's crust is broken into several plates each moving relative to each other. Since 2500 millions ago, there have been tectonic and magmatic processes created by the moving plates.

Currently there are seven major plates each about 80 km thick, namely, 1) Eurasian Plate, 2) Indo-Australian Plate, 3) Pacific Plate, 4) North American Plate, 5) South American Plate, 6) African Plate, and 7) Antarctic Plate; there are also three minor plates: 1) Cocos Plate, 2) Nazca Plate and 3) Arabian Plate and many more smaller Plates. Figure and Reference: http://vulcan.wr.usgs.gov/Glossary/PlateTectonics/Maps/map_plate_tectonics_world.html

The prominent ocean ridges are: Mid Atlantic Ridge, Juan de Fuca Ridge, Southwest Indian Ridge, Southeast Indian Ridge, Central Indian Ridge, Pacific Antarctic Ridge, Reference to figure and text:
http://pubs.usgs.gov/publications/text/baseball.html
http://pubs.usgs.gov/publications/text/developing.html#anchor10564457

The major geochemical research studies of the oceanic ridges and volcanic lavas were conducted under the Ocean Drilling Project, the Deep Sea Drilling Project, and the Hawaii Scientific Drilling Project. Most of the funding was provided by the National Science Foundation. Some of the major studies conducted under these projects are listed below:

Some major projects during 1970–2001:
- Mid Ocean Ridge
- Ninety East Indian Ridge
- Kerguelen Archipelago
- Emperor Seamounts

Some major projects during 1981-2002:
- Lavas from Haleakala
- Lavas from Kahaoolawe
- Lavas from Koolau
- Lavas from Mauna Kea
- Lavas from Molokai

Some major projects during 1970-1995:
- Studies of volcanic lavas in Chile and Iceland.

These studies involved, as quoted by Prof. Fred Frey, "volcanic rocks formed in intraplate, divergent plate and convergent plate tectonic settings. Research has been on Andean lavas in Chile and Argentina, and on the largest terrestrial volcanoes, (known as hotspot volcanoes) which form long, linear chains of volcanoes, such as the Hawaii-Emperor Ridges in the Pacific Ocean and the Ninetyeast Ridge in the Indian Ocean. Our laboratory is active in the Hawaii Scientific Drilling Project and in studies of lavas recovered from the Kerguelen Plateau, a large igneous province, sampled during Ocean Drilling Program Leg 183." The MIT-EAPS Neutron Activation Analysis Laboratory played a major role in all these studies by providing the trace abundances, especially of rare earth elements (REEs), transition elements and other elements, with high precision and accuracy consistently for the past thirty years. Some highlights of the studies are given below.

Mid Ocean Ridge Basalts (MORB):

Trace element compositions of basalts are studied extensively by many groups of scientists. These studies provided major compositional information about the processes of melt generation, transport and crystallization that create the oceanic crust.

The scientific studies provided information about

1) global variability of the chemical compositions of MORB,
2) compositional variability exhibited by basalts sampled from individual segments of the global ridge system,

3) links between compositional variability observed temporal and spatial relationships,
4) determination of chemical signatures of basalts that convey information about the nature of the mantle source,
5) characteristics of the melt transport processes,
6) development of new methods for determining the relative ages of very young basalts.
Reference: http://www.agu.org/revgeophys/kinzle01/node3.html

Some names of the research studies are:
"Atlantic Ocean Floor: Geochemistry and petrology of basalts from Legs 2 and 3 of the Deep Sea Drilling Project", 1974 [Ref. 2];
"Oldest Atlantic seafloor: Basalts from western North Atlantic margin and eastern North America", 1977 [Ref. 3];
"Tholeiitic and alkali basalts from the Mid-Atlantic Ridge at 43°N", 1979.[Ref. 4];
"Petrologic evolution of Pre-1 m.y. old western north Atlantic lithosphere, in: The Western North Atlantic Region", 1986, [Ref. 5]
"Geochemical characteristics of basaltic glasses from the AMAR and FAMOUS axial valleys Mid-Atlantic Ridge (36°-37°N)", 1993, [Ref. 6]

Ninety East Indian Ridge – Kerguelen Archepelago:
The location and the sites explored are given on the web site
http://www.ga.gov.au/odp/publications/183_SR/005/005_f1.htm
Some names of the research studies are:
"Rare earth distributions in lherzolite and garnet pyroxenite and the constitution of the upper mantle", 1971 [Ref. 7];
"Petrology and geochemistry of basalts and related rocks from DSDP Leg 22 sites 214 and 216, Ninety-east ridge,
"Petrology and geochemistry of basalts and related rocks from sites 214, 215, 216, DSDP Leg 22, Indian Ocean" [Ref. 9];
"Geochemical results for basalts from sites 253 and 254", 1974, [Ref. 10];
"The origin of lavas from the 90°E Ridge, Eastern Indian Ocean, An evaluation of fractional crystallization models", 1980, [Ref.11];
"Geochemical and petrologic data for basalts from sites 756, 757 and 758: Implications for the origin and evolution of the Ninetyeast Ridge, Proceedings of the Ocean Drilling Program Leg 121," 1991 [Ref. 12]
"Petrogenesis of the flood basalts forming the northern Kerguelen Archipelago: Implications for the Kerguelen plume, Journal of Petrology", 1998 [Ref. 13]
"Origin and Evolution of a Submarine Large Igneous Province: the Kerguelen Plateau and Broken Ridge, southern Indian Ocean", [Ref. 14]

Hawaii Scientific Drilling Project information and maps are available on the web site:
http://expet.gps.caltech.edu/press.html
Some names of the research studies are:
"Evolution of Mauna Kea volcano: Inferences from Lava Compositions Recovered in the Hawaiian Scientific Drilling Project", [Ref. 15]

Lavas forming Haleakala, Koolau, Kahoolawe, Maunakea Volcanoes, Hawaii
Some names of the research studies are:
"Origin of Hawaiian tholeiite and alkalic basalt "1983 [Ref. 16];
"Trace element and isotope geochemistry of lavas from Haleakala Volcano, East Maui:
"Implications for the origin of Hawaiian basalts", 1985 [Ref. 17];
"Geochemical evolution of Kohala volcano, Hawaii", 1987 [Ref. 18];
"Petrogenesis, and rare-earth element enrichment of tholeiitic and alkalic basalts from Kahoolawe", 1992 [Ref. 19];
"Geochemical characteristics of Koolau volcano: Implications of intershield differences among Hawaiian volcanoes", 1994 [Ref. 20];

Lavas forming Andean Volcanoes in Chile and Iceland
Some names of the research studies are:
"Andesites from Central-South Chile: Trace Element Abundances and Petrogenesis", 1976 [Ref. 21];
"Petrology and geochemistry of the Antuco volcanic lavas, A basaltic volcano of the southern Andes (37°25'S)", 1981 [Ref. 21];
"Geochemical variations in volcanic rocks from central-south Chile (33°-42°S): Implications for their petrogenesis", 1984 [Ref. 22];

"Multiple sources for basaltic arc rocks from the southern volcanic zone of the Andes: Trace element and isotopic evidence for contributions from subducted oceanic crust, mantle and continental crust", 1986, [Ref. 23];
"Recent volcanism in the Puyehue-Cordon Caulle Region, southern Andes, Chile (40.5°S)", 1988, [Ref. 24];
"Geologic history of the active Azufre-Planchon-Peteroa volcanic center (35°15'S) with implications for the origin of compositional gaps", 1989 [Ref. 25];
"Evolution of Icelandic Central Volcanoes: Evidence from the Austurhorn Intrusion, Southeastern Iceland", 1992 [Ref. 26].

2. Trace Element Analysis of Standard Reference Materials

MIT-EAPS INAA Laboratory participated consistently in the trace element analysis of Standard Reference Materials such the second gene ration USGS standards AGV2, BCR2, BHVO2, DTS2 and GSP2 by instrumental neutron activation analysis [Ref. 27]. We prepared synthetic "In-house" standards using high purity silica powder and carefully prepared stock solutions with known amounts of concentrations simulating the matrices of typical basalt and granite.

TABLE 1. Standard Reference Materials analyzed by INAA

Year	Reference Standards
1988	In-House Reference Standards
1988	Miscellaneous Standards
1989	In-House Reference Standards
1990	Miscellaneous Standards
1992	ARHCO Standard
1994	In-House Reference Standards
1996	USGS2-1 AGV2, BCR2, DTS2
1997	USGS2-2 BHVO2, GSP2
1998	READING Standards
1998	PRIS S2

3. Trace Element Analysis of Impurities in High Purity Materials

Instrumental neutron activation analysis has been successfully used to determine the trace element concentrations of impurities in high purity materials.

1) Trace element analysis of spectroscopically pure silicon dioxide by instrumental neutron activation analysis, 1989: Large amounts of silicon dioxide are added to geological samples with low trace elements, such as quartz feldspar and olivine, the trace element contents of the silicon dioxide are a significant contaminant. Rare earth element impurities at the level of 1 to 100 ppb were determined with about 10% to 30% error. [Ref. 28].
2) Trace element analysis of high purity sorbent materials for the removal of trace heavy metals in drinking water by inductively coupled plasma mass spectrometry (ICPMS) and instrumental neutron activation analysis (INAA), an EPA project, 1991.
3) Trace element analysis of high purity graphite, 1994:
   Determination of Chlorine at ppm level and Co, Cr, Cs and Sc at ppb level in reactor grade graphites: Special procedures were set up and performed (Batelle Project 1994).

4. Trace Element Analysis of Archeological and Plant Materials

- Multielement analysis of food spices by instrumental neutron activation analysis was done to examine the elemental composition of some typical spices, (A pilot project) 1980, [Ref. 29]
- Lichens are sensitive to environmental pollution. A pilot project was done to study the feasibility of determining multi elements in lichens by instrumental neutron activation analysis, 1988, [Ref. 30]

- Application of instrumental neutron activation analysis in archaeology:
  Indium and Tin are potentially useful elements in archaeological provenience studies of ancient tin and bronze artifacts. In (10-15000 ppm) and Sn (20% to 90%) are quantitatively determined in Mexican cassiterite samples, 1994, [Ref. 31].

5. Trace Element Analysis of Environmental Materials: Training Course

The NAA facility routinely participates in the MIT 12.119 credit course teaching analytical techniques for the analyses of environmental samples. MIT conducts what is called an Independent Activities Period (IAP); during that time demonstrations and lectures about neutron activation analysis would be given. Recently the (credit) course was 12.091 taught [Ref. 32] which is published on the web site of MIT Open Course Ware. Trace element analysis done on samples of some recent 12.119 course projects:

- Soil samples near Army Lab, Natick, MA, 1996, 1997.
- Sediment samples from Aberjona River and Mystic Lakes Winchester & Woburn, MA 2000.
- Sediment samples form Paint Shop Pond, Wellesley, MA, 2001.
- Sediment samples from Pettaquamscutt Estuary, RI, 2002.
- Sediment samples Lower Mystic Lake, Winchester, MA, 2003.
- Sands from beach, Gloucester, MA, 2004.

PART II: MIT-EAPS NEUTRON ACTIVATION ANALYSIS & RADIOMETRIC LABORATORY FUTURE

MIT-EAPS INAA Laboratory has been very active in geosciences for the past 30 years, amassing a wealth of experience and expertise in sub ppm level trace element analysis of a wide variety of samples.
It is important to note that:

- three orders of magnitude reduction in background can be achieved readily in the shielded gamma-ray detector spectra,
- this reduction translates to a one sigma factor in excess of 30 (which is SQRT1000),
- therefore smaller samples can be counted underground which is an application for nanotechnology materials.
- Similarly, if the total activity of a neutron activated sample is well below the detector background, like in high purity organic matrices, LBC INAA is the future way to go.

MIT-EAPS NAA & Radiometric Laboratory

- is currently set up for ***future*** participation in materials testing for K, U and Th in the Low Background Counting (LBC) Facility set up for the site specific deep underground science and engineering laboratory (DUSEL) initiative,
- is well equipped and maintained with state of the art equipment for testing materials which will be used in the fabrication of ***future*** detectors for rare event particle physics experiments,
- is useful in ***future*** materials research involving nanotechnology,
- is useful in establishing the ***future*** second generation standard reference materials useful for analytical techniques to be developed deep underground.

MIT-EAPS INAA & Radiometric Laboratory's future is with Low Radioactivity Techniques.

REFERENCES

1. P. Ila and F. A. Frey, Atomkernenergie Kern-technik supplement **44** , 710-716 (1984).
2. F.A., Frey, W.B. Bryan, and G. Thompson, J. Geophys. Res. **79**, 5507-5528 (1974).
3. W. B. Bryan, F.A. Frey, and G. Thompson, Contrib. Min. Petrol. **64**, 223-242 (1977).
4. T. Shibata, G. Thompson, and F.A. Frey, Contrib. Mineral. Petrol. **70**, 127-141 (1979).

5. W. B. Bryan and F. A. Frey, "Petrologic evolution of Pre-1 m.y. old Western North Atlantic lithosphere in the Western North Atlantic Region", in The Geology of North America, Geol. Soc. Amer., Boulder, CO, edited by B.E. Tucholke and P.R. Vogt, 1986, pp. 271-296.
6. F. A. Frey, F.A., N. Walker, D. Stakes, S. R. Hart, and R. Nielsen, Earth Planet. Sci. Letts. **115**, 117-136 (1993).
7. J. B. Reid, J.B., and F.A. Frey, J. Geophys. Res. **76,** (5), 1184-1196 (1971).
8. G. Thompson, G., W. B. Bryan, and F. A. Frey, EOS **54** (11), 1019-1021 (1973).
9. G. Thompson, G., W.B. Bryan, F.A. Frey, and C.M. Sung, Petrology and geochemistry of basalts and related rocks from sites 214, 215, 216, DSDP Leg 22, Indian Ocean, in von der Borch, C.C., Sclater, J.G., et al., Initial Reports of the Deep Sea Drilling Project, Vol. 22, Washington, D.C.
10. F. A. Frey and C.M. Sung, Geochemical results for basalts from sites 253 and 254 in Davies, T.A., Luyendyk, B.T., et al., Initial Reports of the Deep Sea Drilling Project, Vol. 26, Washington, D.C. (U.S. Government Printing Office), 1974, pp. 567-572.
11. J. N. Ludden, G. Thompson, W.B. Bryan, and F.A. Frey, J. Geophys. Res. **85**, 4405-4420 (1980).
12. F. A. Frey, W.B., H. Davies, H. and D. Weis, Geochemical and petrologic data for basalts from sites 756, 757 and 758: Implications for the origin and evolution of the Ninetyeast Ridge, Proceedings of the Ocean Drilling Program Leg 121, (eds., Weissel, J., Peirce, J., and Alt, J. et al.), 1991, pp. 611-659.
13. H. –J. Yang, H.-J., F. A. Frey, D. Weis, A. Giret, D. Pyle, and G. Michon, Journ. of Petr. **39**, 711-748 (1998).
14. F. A. Frey, M. F. Coffin, P. J. Wallace, Earth and Planet. Sci. Letts. **176,** 73-89 (2000).
15. H.-J. Yang, F. A. Frey, F.A., J. M. Rhodes and M. O. Garcia, J. Geophys. Res. **101**, 11,747-11,767, (1996).
16. C. -Y. Chen and F. A. Frey, Nature **302**, 785-789 (1983).
17. C. -Y. Chen and F.A. Frey, J. Geophys. Res. **90, B10**, 8743-8768 (1985).
18. M. Lanphere, M. and F.A. Frey Frey, Contrib. Mineral. Petrol**.** **95**, 100-113 (1987).
19. R. V. Fodor, F. A. Frey, G. R. Bauer, G.R. and D. A. Clague, Contrib. Mineral. Petrol. **110**, 442-462 (1992).
20. F. A. Frey, M. O. Garcia, and M. F. Roden, Geochim. Cosmochim. Acta, **58**, 1441-1462 (1994).
21. L. Lopez-Escobar, L., F.A. Frey, and M. Vergara, "Andesites from Central-South Chile: Trace Element Abundances and Petrogenesis" in Proc. of the Symposium on Andean and Antarctic Volcanology Problems (ed. O. Gonzales-Ferran), IAVCEI Spec. Issue, (1976) pp. 725-761.
22. R. L. Hickey, D.C. Gerlach, and F.A. Frey, "Geochemical variations in volcanic rocks from central-south Chile (33°-42°S): Implications for their petrogenesis", in Andean Magmatism: Chemical and Isotopic Constraints (edited by R. Harmon and B. Barreiro) Shiva Pub., Cheshire, England, 1984, pp. 72-95
23. R. L. Hickey, F. A. Frey, D. C. Gerlach, and L. Lopez-Escobar, J. Geophys. Res. **91**, 5963-5983 (1986).
24. D. C. Gerlach, F. A. Frey, and H. Moreno,, J. Petrol. **29,** 332-382 (1988).
25. D. Tormey, F. A. Frey, L. Lopez-Escobar, Assoc. Geol. Argentina Rev., 1989, XLIV (1-4), 420-430 (1989).
26. T. Furman, P. S. Meyer, P.S. and F. A. Frey, Bull. Volcano., **55**, 45-62 (1992).
27. P. Ila and F. A, Frey, Modern Trends in Activation Analysis, MTAA10, Bethesda, MD, U. S. A., April 19-23, 1999, Jour. Radioanal. and Nucl. Chem. **244**, 599-602 (2000).
28. P. Ila and F. A. Frey, Jour. Radioanal. and Nucl. Chem. **131**, 37-42 (1989).
29. P. Ila and P. Jagam, EUROANALYSIS III,Dublin, Ireland, 1978, p347, Jour. Radioanal. and Nucl. Chem. **57**, 205-210 (1980).
30. P. Ila, Jour. Radioanal. and Nucl. Chem. **120,** 247-252 (1988).
31. P. Ila and A. M. Macfarlane, Jour. Radioanal. and Nucl. Chem. **182**, 427-435 (1994).
32. I. Pillalamarri, MIT IAP 12.091 credit course, "Trace Element Analysis of Geological, Biological & Environmental Materials by Neutron Activation Analysis: An Exposure", January 3–19, 2005.
http://ocw.mit.edu/OcwWeb/Earth--Atmospheric--and-Planetary-Sciences/12-091January--IAP--2005/CourseHome/index.htm

LRT2004 List of Participants

Last Name	First Name	Affiliation
Aalseth	Craig	Pacific Northwest National Laboratory
Abt	Iris	MPI fuer Physik, Heidelberg
Aharmim	Bassam	Laurentian University
Barnabé Heider	Marik	University of Montreal
Bartoszek	Larry	Bartoszek Engineering
Bauer	Dan	Fermilab
Bredeweg	Todd	Los Alamos National Laboratory
Chen	Mark	Queen's University
Clark	Ken	Queen's University
Cleveland	Bruce	SNOLAB
Cushman	Priscilla	University of Minnesota
Dai	Xiongxin	Carleton University
Danevich	Fedor	Institute for Nuclear Research, Kiev, Ukraine
Doane	Patrick	Université de Montréal
Duncan	Fraser	Queen's University
Farine	Jacques	Laurentian University
Fleurot	Fabrice	Laurentian University
Ford	Richard	SNOLAB
Formaggio	Joseph	University of Washington
Gaitskell	Richard	Brown University
Galindo-Uribarri	Alfredo	Oak Ridge National Laboratory
Gornea	Razvan	University of Montreal
Gove	Harry	University of Rochester
Graham	Kevin	Queen’s University
Guénette	Roxanne	University of Montreal
Hahn	Richard	Brookhaven National Laboratory
Hallin	Aksel	Queen's University/SNO
Hallman	Doug	Laurentian University
Heusser	Gerd	Max-Planck-Institut fuer Kernphysik, Heidelberg
Hime	Andrew	Los Alamos National Laboratory
Inoue	Kunio	RCNS, Tohoku University
Inzhechik	Lev	Kurchatov Institute, Moscow
Irastorza	Igor	CEA / Saclay
Kim	Yeongduk	Physics Department, Sejong University
Kishimoto	Tadafumi	Department of Physics, Osaka University

Kishimoto	Yasuhiro	Research Center for Neutrino Science, Tohoku University
Knoepfle	Karl-Tasso	MPI Kernphysik, Heidelberg
Koga	Masayuki	Tohoku University
Kormos	Laura	Queen's University
Krauss	Carsten	Queen's University
Kudryavtsev	Vitaly	University of Sheffield
Lan	Chunlin	Queen's University
Lange	Richard	Brookhaven National Laboratory
Leung	Michael	Princeton University
Litherland	Albert	IsoTrace Laboratory, University of Toronto
Loaiza	Pia	Laboratoire Souterrain de Modane
Moriyama	Shigetaka	Institute of Cosmic Ray Research
Martin	Ryan	Queen's University
McDonald	Art	Queen's University
McFarlane	Ken	SNOLAB
Menghetti	Helenia	Bologna University and Bologna INFN, Italy
Mifflin	Cathy	Carleton University
Miley	Harry	Pacific Northwest National Laboratory
Miramonti	Lino	Milan University & INFN
Mitsui	Tadao	Research Center for Neutrino Science, Tohoku University
Nakahata	Masayuki	Kamioka observatory, ICRR, Univ. of Tokyo
Nir-El	Yoram	Queen's University
Noble	Tony	Queen's University
Novikov	Vladimir	Queen's University
Orzel	Chad	Union College
Pillalamarri	Ila	Massachusetts Institute of Technology
Peltoniemi	Juha	University of Oulu
Pirro	Stefano	INFN-Milano
Pocar	Andrea	Stanford University
Rau	Wolfgang	TU Munich
Schoenert	Stefan	MPIK Heidelberg
Selvi	Marco	Bologna University & INFN
Shutt	Tom	Princeton University
Simgen	Hardy	Max Planck Institute for Nuclear Physics, Heidelberg
Sinclair	David	Carleton University
Skensved	Peter	Queen’s University
Storey	Connie	Queen's University
Suvorov	Yury	INFN-Gran Sasso Laboratories

Takeuchi	Yasuo	Kamioka Observatory, ICRR, Univ. of Tokyo
Viktor	Zacek	Universite de Montreal
Virtue	Clarence	Laurentian University
Wichoski	Ubi	University of Montreal
Wright	Alexander	Queen's University
Yeh	Minfang	Brookhaven National Laboratory
Zuzel	Grzegorz	Max Planck Institute for Nuclear Physics

T

U

V

Y

Z